Proceedings of the Twelfth Asia-Pacific International Conference on

Gravitation, Astrophysics, and Cosmology

Dedicated to the Centenary of Einstein's General Relativity

Proceedings of the Twelfth Asia-Pacific International Conference on

Gravitation, Astrophysics, and Cosmology

Dedicated to the Centenary of Einstein's General Relativity

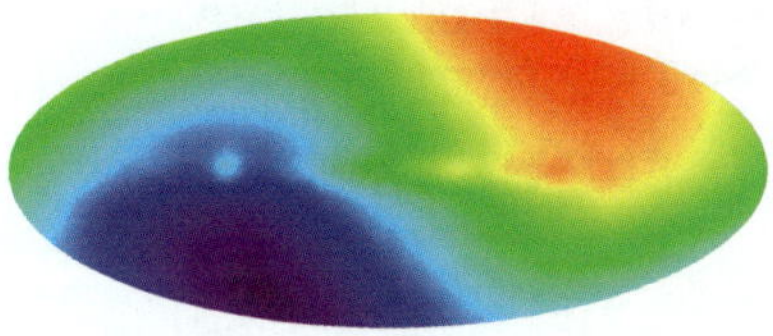

Editors

Vitaly Melnikov

Peoples' Friendship University of Russia (PFUR), Moscow, Russia

Jong-Ping Hsu

University of Massachusetts Dartmouth, USA

World Scientific

NEW JERSEY • LONDON • SINGAPORE • BEIJING • SHANGHAI • HONG KONG • TAIPEI • CHENNAI • TOKYO

Published by

World Scientific Publishing Co. Pte. Ltd.

5 Toh Tuck Link, Singapore 596224

USA office: 27 Warren Street, Suite 401-402, Hackensack, NJ 07601

UK office: 57 Shelton Street, Covent Garden, London WC2H 9HE

British Library Cataloguing-in-Publication Data
A catalogue record for this book is available from the British Library.

GRAVITATION, ASTROPHYSICS, AND COSMOLOGY
Proceedings of the Twelfth Asia-Pacific International Conference

ISBN 978-981-4759-80-9

In-house Editor: Christopher Teo

Typeset by Stallion Press
Email: enquiries@stallionpress.com

Printed in Singapore

The conference is dedicated to
the centenary of Einstein's
General Relativity theory

Preface

The year 2015 enjoyed a world-wide celebration for the centennial of Einstein's development of his momentous theory of gravity. We should mention also Hilbert's discovery of the general-coordinate-invariant action for gravity at the same time. In his paper on "The Foundations of Physics," Hilbert began with admirable and piercing statements: "The powerful posing of problems by Einstein, as well as his penetrating method devised for their solution, and the deep thoughts and original conceptualizations by means of which Mie builds his electrodynamics have opened new paths for the investigation of the foundations of physics." What Hilbert referred to included Einstein and Grossmann's original 1913 work: "Outline of a Generalized Theory of Relativity and of a Theory of Gravitation, I. Physical Parts, by A. Einstein, II. Mathematical Part, by M. Grossmann." These works are milestones in our present understanding of the universe. It is interesting to note that in 1915, mathematician Hilbert had already seriously aimed at a unified theory of physics, including gravity.

From June 28 to July 4, 2015, at the Peoples' Friendship University of Russia (PFUR) in Moscow, more than 130 leading and young Russian participants were joined by 45 distinguished and young guests from Australia, Brazil, Canada, China, Germany, India, Italy, BLTP JINR, Japan, Korea, Mexico, NIS, Taiwan, USA and others to participate in the International Conference on Gravitation, Astrophysics and Cosmology (ICGAC-12).

The Russian Gravitational Society (RGS) organized the ICGAC-12 as an open conference for physicists, graduate and post-graduate students. It aimed to serve as a common platform around the world for exchange and communication of ideas and works in these research fields. This conference was most successful in attracting 160 speakers—far more than any previous ICGAC conference. There were 40 plenary speakers and 120 sectional speakers. Part of these presentations are collected and published in this book. These papers reflect the current status and activities in gravitation, astrophysics and cosmology around the globe. The main topics were classical gravity, GR extensions , classical and quantum cosmology, dark matter and dark energy, black holes, wormholes, strings, branes, higher spin fields and quantum gravity, extra dimensions and variation of constants, experimental studies of gravity, GW and fundamental physics space projects, relativistic astrophysics, white dwarfs, neutron stars and gamma ray bursts.

The series of ICGAC conferences were initiated by Yong Min Cho in Korea in 1993. They served a significant need and played an increasing role as the information hub for the exchange of ideas for all physicists and researchers.

We believe that the ICGAC-12 was a delightful and successful conference. Its content was in line with contemporary level of world research in gravitation, astrophysics and cosmology.

We would like to express our gratitude to RGS, the Institute of Gravitation and Cosmology, PFUR, Center for Gravitation and Fundamental Metrology, the Russian Research Institute for Metrological Service (VNIIMS) in Moscow, and the Bogoliubov Laboratory of Theoretical Physics, JINR, Dubna, for the organization of the conference. We are grateful also for the support from the Russian Foundation for Basic Research, PFUR, VNIIMS, the University of Massachusetts Dartmouth Physics Department and the Professor Leung Memorial Fund, as well as APCTP within the APCTP-BLTP Joint Program.

We also thank all colleagues, graduate, post-graduate students and the staff of PFUR participating in this conference for their help and efforts to make ICGAC-12 successful, which were reflected in the conference presentations and discussions. The delightful banquet had bountiful food and Russian wines, as well as tolerable laughter and exciting noises. Dmitry Galtsov even spontaneously went to a piano and played Chopin's music to entertain colleagues. What a night! Most fun of all, the spectacular views of the Moscow River Cruise were incomparable. The only regret might be that some Russian main organizers were very busy in handling with participants' financial support from APCTP and others, and could not enjoy the cruise.

At the meeting of the Organizing Committee members, it was discussed and suggested that the future of the ICGAC conference should aim at more participants from different parts of the world, not only from the Asia-Pacific region. We hope that the proceedings in this volume will be a valuable document and will serve as a step towards fulfilling our goal and expectation.

J.P. Hsu and V.N. Melnikov

A. Einstein

M. Grossmann

D. Hilbert

CONTENTS

Section 1. Classical Gravity and GR Extensions 183

Plenary Sessions

Gravitation, Astrophysics and Cosmology

A test of an alternative origin of the cosmic microwave background anisotropy

Jong-Ping Hsu

Physics Department, University of Massachusetts Dartmouth
North Dartmouth, MA 02747, USA
E-mail: jhsu@umassd.edu

Leonardo Hsu

Department of Postsecondary Teaching and Learning
University of Minnesota, Minneapolis, MN 55414, USA
E-mail: lhsu@umn.edu

The dipole anisotropy of the cosmic microwave background (CMB) is usually attributed to the motion of the solar system relative to the CMB. However, such an interpretation requires that the Planck distribution $P(\omega, T)$ be non-Lorentz invariant. We discuss an alternative interpretation called the 'Big Jets' model, based on a Lorentz-invariant blackbody distribution $B(k_\lambda U^\lambda/T_{inv})$, which reduces to the Planck distribution in the non-relativistic limit. This model, which implies that the CMB anisotropy must originate from a real, physical anisotropy of the universe, can be tested by examining the shape of the invariant blackbody distribution $B(k_\lambda U^\lambda/T_{inv})$ and the dipole spectrum as measured by the COBE, WMAP, and Planck satellite experiments. The Big Jets model (a) provides a way to restore matter-antimatter symmetry in the universe by proposing that instead of a big bang, the universe began with two big jets, one with more matter and one with more anti-matter, and (b) suggests that it is the remnant of the anti-matter jet, detectable only as an extremely distant blackbody, that produces the observed CMB dipole anisotropy.

Keywords: Anti-matter galaxies; blackbody; dipole anisotropy of CMB.

A self-consistent understanding of the anisotropy of the cosmic microwave background (CMB) is important because its interpretation is potentially intertwined with the relativistic transformation properties of temperature and the Planck law, as well as the question of the missing anti-matter half of the universe. The CMB has been investigated by satellite experiments such as the Cosmic Background Explorer (COBE), the Wilkinson

Microwave Anisotropy Probe (WMAP) and the European Space Agency's Planck satellite.[1–5] In the conventional interpretation, one assumes that the observed temperature dipole anisotropy (at the level of 10^{-3}) is due entirely to Doppler boosting of the CMB monopole and thus deduces that motion of the solar system relative to the CMB is characterized by a velocity V towards the galactic coordinates (ℓ, b) with[2]

$$V = V_{CMB} = 369.0 \pm 0.9\ km/s, \qquad (\ell, b) = (263.99^o, 48.26^o), \qquad (1)$$

which is roughly in the direction of the constellation Leo. However, this interpretation requires two unusual properties. The first is that the Planck distribution is non-Lorentz-invariant. When interpreting CMB satellite data, the Planck distributions are assumed to have the forms ($c = \hbar = 1$),

$$P(\omega, T) = \frac{1}{[exp(\omega/k_B T) - 1]} \quad and \quad P(\omega', T') = \frac{1}{[exp(\omega'/k_B T') - 1]},$$

for F and F' frames respectively. Superficially, this may appear to be Lorentz invariant. However, we note that only ratios of corresponding components of two four-vectors such as $dp_x dp_y dp_z/E$ are Lorentz invariant. The fractions ω/T and ω'/T' are not such ratios and thus, are not Lorentz invariant.[6]

A second unusual property is that the temperature $T' = T'(\theta)$ in F' is assumed to be angle-dependent, $T'(\theta) = T\sqrt{1-V^2}/(1 + V cos\theta')$, where θ' is the angle between V and the line of sight.[3,5] This is necessary to force there to be a Planck spectrum in F' frame. However, such a directional temperature is absent from virtually all discussions of relativistic temperature transformations based purely on thermodynamic considerations.[7,8]

To obviate the need for such unusual properties, let us consider a Lorentz-invariant law for blackbody radiation, expressed in terms of the differential of the covariant energy-momentum tensor $dT_{\mu\nu}$ of photons.* We assume that it takes the form

$$dT_{\mu\nu} = 4(2\pi)^{-3} k_\mu k_\nu B(k_\lambda U^\lambda / T_{inv}) \delta(k_\alpha k^\alpha) d^4k, \qquad (c = \hbar = 1), \qquad (2)$$

$$B(k_\mu U^\mu / T_{inv}) = \frac{1}{[exp(k_\mu U^\mu / k_B T_{inv}) - 1]}, \quad k_\mu U^\mu = \eta_{\mu\nu} k^\mu U^\nu, \qquad (3)$$

*The component dT_{00} corresponds to the energy density inside the cavity of a blackbody. The scalar $B(k_\lambda U^\lambda / T_{inv})$ is a generalized Planck distribution.

where $\eta_{\mu\nu} = (1, -1, -1, -1)$ and $U^\mu = dx^\mu/ds$ is the 4-velocity of the blackbody, as measured in an inertial frame F.[8] The temperature is assumed to be a scalar, as it is in most discussions of relativistic thermodynamics.[8,9] When the blackbody is at rest in F, $U^\mu = (1, 0, 0, 0)$ and $k_\mu U^\mu = k_0 = \omega$, and the invariant distribution (3) reduces to the usual form, $B(k_\mu U^\mu/T) \rightarrow P(k_0, T)$. This particular frame may be identified with the CMB frame, which is just an ordinary inertial frame.

In contrast to the Planck distribution $P(\omega, T)$, the invariant blackbody distribution $B(k_\mu U^\mu/T_{inv})$ in (3) is explicitly invariant in all inertial frames. If it is isotropic in one frame, then it is isotropic in all inertial frames. Therefore, the observed CMB dipole anisotropy cannot be due to the motion of the solar system, in contrast with the usual interpretation based on a non-Lorentz invariant Planck distribution.[3,5]

In view of the fact that Lorentz-invariance seems to underpin almost all of physics, we explore and discuss a new 'Big Jets' model[10,11] of the universe, which has the potential to resolve the incompatibility between the observed CMB dipole anisotropy and a Lorentz-invariant CMB radiation spectrum, as well as provide an answer to the question surrounding the missing anti-matter of the universe, while remaining consistent with satellite data.

High energy particle experiments frequently observe two-jet particle production events. This has stimulated us to explore the implications of a universe that originated from two Big Jets rather than a single Big Bang because it naturally allows the preservation of cosmic matter-antimatter symmetry. These two Big Jets may be pictured as two gigantic fire balls moving rapidly away from each other. From the vantage point of an observer in one of the fire balls, the evolution of the universe resembles that of a single Big Bang. While there is both matter and anti-matter within each fire ball, the process of matter/anti-matter annihilations could lead to one fireball dominated by matter while the other is dominated by anti-matter.

Now, billions of years later, the two fire balls have expanded to very large size, one composed principally of galaxies and the other composed principally of anti-galaxies (which may be called Yang and Yin respectively). Presumably, the distance between Yang and Yin is so large that radiation from individual stars or galaxies in Yin are too faint to be detected by us. Instead, we can only detect the microwave radiation emitted by the entire anti-galaxy cluster Yin, acting like a single large blackbody. In this model, we suggest the CMB dipole anisotropy is caused by the microwaves emitted from Yin blackbody, which have propagated in the direction from (l, b) in

(1) toward us.

To make this model more quantitative, let us approximate Yin and Yang as two spherical blackbodies, each with radius $\approx R_Y$, separated by a distance $\approx D_{YY}$. Using the distance to the most remote astronomical object yet observed, we estimate, $R_Y \geq 13\ Gly$. The temperatures of the spheres are $T_{Yin} = T_{Yang} = T_{CMB} = 2.726K$. A microwave receiver in space, such as might be found on the COBE, WMAP, or Planck satellites, is modeled as a blackbody with a temperature T_{rec}. From the conservation of energy (i.e., the power absorbed and the power emitted by the receiver blackbody are equal) and the Stefan-Boltzmann law, $W = \sigma T^4$,

$$\Omega_{rec} T_{rec}^4 = \Omega_{CMB} T_{CMB}^4 + \frac{\pi R_Y^2}{D_{YY}^2} T_{Yin}^4, \tag{4}$$

when the receiver is pointed in the direction of Yin, where $\Omega_{rec}, \Omega_{CMB}$ and $\pi R_Y^2/(D_{YY}^2)$ are solid angles. The left-hand side of (4) is the radiation energy emitted by the receiver blackbody and the two terms on the right-hand side of (4) denote the radiation energies absorbed by the receiver from the Yang and Yin blackbodies, respectively. Thus, T_{Yin} in (4) is the temperature of the blackbody Yin at the time when it emitted the radiation energy, roughly 'D_{YY}/c' years ago. As the receiver is turned away from Yin, the magnitude of the second terms decreases, producing the observed dipole anisotropy. In order to get a rough estimate of the extremely large distance D_{YY} in (4), we approximate $T_{Yin} \approx T_{CMB}$ (in the absence of other data) and use the relations $\Omega_{rec} = 4\pi$ and $\Omega_{CMB} = 4\pi$, where we consider T_{rec} to be the effective temperature observed by the receiver. It follows from (4) that the maximum observed temperature is

$$T_{rec} \approx T_{CMB} \left(1 + \frac{R_Y^2}{16 D_{YY}^2}\right). \tag{5}$$

Since the dipole anisotropy observed from satellite experiments is $(3.355 \pm 0.008) \times 10^{-3}K$,[5] and $T_{CMB} = 2.73K$,

$$T_{rec} = T_{CMB} + 3.355 \times 10^{-3}K \approx T_{CMB}\left(1 + 1.24 \times 10^{-3}\right), \tag{6}$$

in the direction specified by the galactic coordinate (1) to the solar system. Combining (5) and (6), we obtain (as an order-of-magnitude estimate)

$$D_{YY} \approx 7R_Y. \tag{7}$$

Based on the Big Jets model, one may picture the universe as a 'dumbbell' with one end dominated by a gigantic cluster of baryonic galaxies (or matter) and the other dominated by a gigantic cluster of anti-baryonic galaxies (or anti-matter). Given our limited knowledge about the universe, it seems premature to conclude that our inability to observe as much anti-matter as matter implies the non-existence of the anti-matter,[11,12] as opposed to saying that we have observed only one end of the dumbbell universe.

One way the Big Jets model might be tested experimentally is by examining the dipole anisotropy of the CMB radiation more closely. In the Big Bang model paired with the conventional interpretation using a non-invariant $P(\omega, T)$, the distribution of the dipole radiation is approximated by the Taylor series expansion[4,13]

$$P(\omega, T') \approx P(\omega, T_{CMB}) + P_{BigBang}, \tag{8}$$

where

$$P_{BigBang} \approx V cos\theta' \; T_{CMB} \left(\frac{\partial P(\omega, T')}{\partial T'} \right)_{T'=T_{CMB}}, \tag{9}$$

and $T' \approx T_{CMB}(1 + V cos\theta')$.

In the Big Jets model, suppose Yang is modeled as a spherical blackbody of radius R_Y. Electromagnetic radiation emitted from the Yin blackbody in the direction of Yang is completely absorbed by Yang, so that Yin cannot be resolved as an ‘object. This radiation also causes the Yang hemisphere facing Yin to be slightly warmer than the opposite hemisphere, giving Yang a dipole distribution roughly given as

$$T^{ob} \approx T_{CMB} + T_{Yin} \left(\frac{R_Y}{D_{YY}} \right)^2 cos\theta, \tag{10}$$

where θ may be defined to be zero in direction connecting the center of the two blackbodies and T_{Yin} is the temperature of the Yin blackbody at the time when its radiation was emitted.

In the presence of the Yin blackbody in the Big Jets model, the invariant distribution $B(k_\lambda U^\lambda / T^{ob})$ can be approximated by the Taylor series expansion, $f(x, y) = f(\xi, \eta) + (x - \xi)\partial f(\xi, \eta)/\partial x + ...$ Thus, we have

$$B(k_\lambda U^\lambda / T^{ob}) \approx B(k_\lambda U^\lambda / T_o) + B_{BigJets}, \qquad T_o \approx T_{CMB}, \tag{11}$$

where the second term determines the CMB dipole spectrum due to the Yin blackbody, and is given by

$$B_{BigJets} \approx T_{Yin} \left(\frac{R_Y}{D_{YY}} \right)^2 cos\theta \left(\frac{\partial B(k_\lambda U^\lambda / T^{ob})}{\partial T^{ob}} \right)_{T^{ob}=T_o}. \tag{12}$$

This anisotropy of the invariant blackbody distribution $B(k_\lambda U^\lambda/T^{ob})$ is a fixed pattern determined by the position of the Yin blackbody and is independent of the velocity of an observer. In view of the approximate relation (7), the angles θ in (12) and θ' in (9) are roughly the same if the observer is not too far away (say, several billion light-years) from the center of the Yang blackbody, which is assumed to have an extremely large but finite volume.

Measurements of the CMB dipole spectrum by satellites over a range of frequencies spanning the peak of the blackbody spectrum are consistent with the prediction of the differential of the non-invariant Planck distribution with respect to temperature and evaluated at the CMB temperature as given by (8) and (9).[4] Such measurements could also be compared with the prediction of the Big Jets model (12), which is similar to (9) except that the non-invariant Planck distribution $P(\omega, T)$ is replaced by the invariant blackbody distribution (3). For comparison purposes, one may take the non-relativistic limit for the 4-velocity, $U^\lambda \approx (1, 0, 0, 0)$, in (11) and (12). It is possible for currently available satellite measurements to be consistent with both (9) and (12).

The crux of the matter in this discussion is the incompatibility between the Lorentz invariant blackbody distribution (3) and the non-Lorentz invariant Planck distribution used in the conventional interpretation of CMB data, which includes a velocity-dependent dipole anisotropy and spectrum. If satellite data is shown to be incompatible with (12), then one must conclude the Lorentz-Poincaré invariance, at least as currently conceived, is not a universal principle at all scales. Either special relativity can be violated at the cosmic scale or our idea of what Lorentz-Poincaré invariance implies must be modified.

Naturally, many more details remain to be worked out and verified for the physical implications of the Big Jets model. However, an advantage of such a model is that it preserves the fundamental conservation laws of particle physics concerning the relative abundances of matter and antimatter while also explaining the dipole anisotropy of CMB without the need for the law describing CMB radiation to violate the principle of relativity.[†] It would be very interesting if an extremely distant anti-galaxies blackbody Yin turned out to be hidden in the CMB dipole anisotropy.

[†]Based on a broader theory of relativity, one can re-define the evolution variable within the framework of 4-dimensional symmetry, in which the distribution of CMB radiation is invariant.[14–16]

This paper is respectfully dedicated to the memory of Prof. Susumu Okubo, whose lasting influence JP acknowledges with gratitude. We would like to thank colleagues for discussions. The work is partially supported by the Jing Shin Research Fund of the UMassD Foundation.

References

1. See, for example, Particle Data Group, Chin. Phys. C. **38**, 090001 (2014) (http://pdg.lbl.gov).
2. G. Hinshaw et al., Astrophys. J. Supp. **180**, 225 (2009).
3. See, for example, S. Weinberg, *Gravitation and Cosmology, Principles and Applications of the General Theory of Relativity* (John Wiley and Sons, 1972) pp. 520-522.
4. D. J. Fixsen et al., Astrophys. J. **420**, 445 (1994).
5. D. Scott and G. F. Smoot, '27. COSMIC MICROWAVE BACKGROUND', Sec. 27.22, for The Review of Particle Physics, K. A. Olive et al (Particle Data Group) in ref. 1.
6. L. Landau and E. Lifshitz, *The Classical Theory of Fields*, (Addison-Wesley, 1951), p. 31.
7. See, for example, A. Einstein Jahrb. Rad. u. Elektr, 1907 **4**: 411
8. P.T. Landsberg, Nature 212 (1966) 571.
9. N.G. van Kampen, Phys. Rev. **173** 295 (1968).
10. J. P. Hsu and L. Hsu, *Space-Time Symmetry and Quantum Yang-Mills Gravity* (World Scientific, 2013) p. 225.
11. J. P. Hsu and K.O. Cottrell, Chinese Phys. C (to be published, 2015).
12. For a very interesting discussion on cosmology, see T. Rothman and E. C. G. Sudarshan, *Doubt and Certainty* (With ill. by Shannon K. Comins, Helix Book, Reading, Mass. 1998) pp. 214-264.
13. D. J. Fixson, Astrophys. J. **707**, 916 (2009). arXDiv:0911.1955.
14. J. P. Hsu, Nuovo Cimento, **B**, **75**, 185 (1983).
15. J. P. Hsu and L. Hsu, Phys. Lett. A, **196**, 1, 1994.
16. J. P. Hsu and T. N. Sherry, Foundations of Phys. **10**, 57 (1980).

Multidimensional integrable models of gravitation and cosmology

V. D. Ivashchuk* and V. N. Melnikov**

Center for Gravitation and Fundamental Metrology, VNIIMS, 46 Ozyornaya ul. Moscow 119361, Russia
**ivas@vniims.ru*

Institute of Gravitation and Cosmology, Peoples' Friendship University of Russia, 6 Miklukho-Maklaya ul., Moscow, 117198, Russia
*** melnikov@phys.msu.ru*

Review of the motivation and main results in multidimentional gravitation and cosmology is presented. Special attention is devoted to results within the model with scalar fields and fields of forms in the billiard approach for obtaining cosmological solutions with branes and integrable configurations with flux- and black branes. In case of the quantum billiard with branes it is shown that the basis solutions for wave functions vanish in the limit of the formation of billiard walls (i.e., at the singularity) for the $D = 11$ model which mimics the $D = 11$ supergravitational cosmology. Another fruitful approach - to multidimensional gravity with higher derivatives is mentioned, which leads to a unified description of inflation and the present accelerated expansion of the Universe. Some of these models explain possible spatial and temporal variations of the fine structure and the gravitational constants.

Keywords: Comological billiards; fluxbranes; black branes.

1. Introduction

In the developement of relativistic gravitation and dynamical cosmology we may notice three distinct stages. First, investigation of models with matter sources in the form of a perfect fluid, as was originally done by Einstein and Friedmann. Second, studies of models with sources as different physical fields, starting from electromagnetic and scalar ones, both in classical and quantum cases, see Refs. 1, 2. And third, which is really topical now, application of ideas and results of unified models for treating fundamental problems of cosmology, black hole and wormhole physics, especially in high energy regimes and for explanation of the present acceleration of the Universe, dark matter and dark energy problems. Multidimensional gravitational models play an essential role in the latter approach.

The necessity of studying multidimensional models of gravitation and cosmology appeared from the idea of unification of all known fundamental physical interactions, which is now applied for solving the other unresolved problems of modern physics, such as dark energy, and also for describing possible temporal variations of fundamental physical constants.[3–6]

Having no definite successful theory of unification now, in our approach[5–9] we adopted models based on multidimensional Einstein equations with different sources: cosmological constant, perfect and viscous fluids, scalar and electromagnetic fields and their interactions, dilaton and moduli fields, fields of forms (related to branes) etc. Our program's main objective was and is to obtain self-consistent integrable models in cosmological and spherically symmetric cases. Then we tried to single out models which do not contradict the available experimental or observational cosmological data and bounds on variations of the gravitational and fine structure constants. In our models[5,6] we use a product of n Einstein spaces. Then we perform reduction to a sigma-model and/or Toda-like system, further to Liouville, Abel, generalized Emden-Fowler equations, etc. and try to find exact solutions. We suppose that extra spaces are stable or dynamically either compactified like a torus or are large but with barriers, walls etc. So, we realized the program in arbitrary dimensions (since 1988)[5–9] in cosmology obtaining exact solutions to multidimensional Einstein equations with sources: a cosmological constant Λ, Λ plus a scalar field (we singled out nonsingular, dynamically compactified, inflationary solutions, 1994); a perfect fluid (PF), a PF plus a scalar field (e.g., nonsingular, inflationary solutions); a viscous fluid (e.g., nonsingular solutions with generation of mass and entropy, quintessence and coincidence in a 2-component model); oscillating behavior near the singularity, billiards in Lobachevsky space; in all cases mentioned above, solutions with Ricci-flat factor spaces were obtained for any n, also with curvature in one factor space; with curvatures in 2 factor spaces only for the total dimensions $D = 10, 11$; with fields: scalar, forms of arbitrary rank (1998);[10–12] inflationary solutions generated by forms (p-branes);[13] first billiards for models with branes (1999); quantum generalization (solutions to WDW-equation)[14] for certain cases where classical solutions were obtained; solutions for models with dilatonic fields with potentials,[15,16] billiard behavior for them also and recently quantum billiards for D-dimensional cosmological models of generalized Bianchi-I type.

Some of these integrable models were also used for calculation of temporal variations of the effective gravitational constant. Comparison with the present experimental bounds allowed us to choose particular models or to single out some classes of solutions.

A class of solutions depending on the radial variable in any dimensions was found as well: generalized Schwarzchild, Tangherlini solutions where black holes (BHs) were singled out, also with a minimal scalar field φ; generalized Reissner-Nordstrom solutions (BHs were also singled out), the same was done with φ (no BHs); multi-temporal solutions; for dilaton-like interaction of φ and electromagnetic fields (BHs exist only for a special case); stability studies (stable solutions only for the above BH case[17]); for obtained dilaton-forms (p-brane) interaction, the solutions' stability was found only in some cases, e.g. for a one-form, in particular. PPN parameters for most models were calculated.

Here we consider a model governed by the action

$$S = \int d^D x \sqrt{|g|} \Big\{ R[g] - h_{\alpha\beta} g^{MN} \partial_M \varphi^\alpha \partial_N \varphi^\beta - \sum_{a \in \triangle} \frac{\theta_a}{n_a!} \exp[2\lambda_a(\varphi)](F^a)^2 \Big\}, \tag{1}$$

where $g = g_{MN}(x) dx^M \otimes dx^N$ is a metric, $\varphi = (\varphi^\alpha) \in \mathbb{R}^l$ is a vector of scalar fields, $(h_{\alpha\beta})$ is a constant symmetric non-degenerate $l \times l$ matrix, $\theta_a = \pm 1$, $F^a = dA^a$ is a n_a-form ($n_a \geq 1$), λ_a is a 1-form on $\mathbb{R}^l$: $\lambda_a(\varphi) = \lambda_{a\alpha}\varphi^\alpha$, $a \in \triangle$, $\alpha = 1, \dots, l$. Here $\triangle$ is some finite set.

In Section 2 we present a brief overview of main recent results on the (classical and quantum) billiard approach for cosmological models with branes and their relation to the singularity problem. We remind the reader that a brane is a physical object that generalizes the notion of a point particle to higher dimensions. An important motivation of considering p-brane solutions comes from the unification theory of physical interactions.[18]

In Section 3 we overview two subclasses of the so-called brane solutions: fluxbrane and black brane ones, defined on product manifolds containing several Ricci-flat factor spaces.

2. Classical and Quantum Billiard Approach with Branes

Now we present the main results of the billiard approach for D-dimensional cosmological-type models, where $D = 4$.

The billiard approach in classical gravity originally appeared in the dissertation of Chitre[19] for the explanation the Belinskii-Lifshitz-Khalatnikov-oscillations[20] in the Bianchi-IX model.[21] In this approach a simple triangle billiard in the Lobachevsky space H^2 was used. In Ref. 22 the billiard approach for $D = 4$ was extended to the quantum case. Namely, the solutions to Wheeler-DeWitt (WDW) equation $H\Psi = 0$ were reduced to the problem of finding the spectrum of Laplace-Beltrami operator on Chitre's triangle billiard. Such approach was also used in Ref. 23 in a context of studying the large-scale inhomogeneities of metric near the singularity.

A straightforward generalization of the Chitre's billiard to the multidimensional case was performed in Refs. 24–26 where multidimensional cosmological model with multicomponent "perfect" fluid and n Einstein factor spaces was studied. In Ref. 26, a search for oscillating behaviour near the singularity was reduced to the problem of proving the finiteness of the billiard volume. In Ref. 26 the inequalities on Kasner parameters were found and the "quantum billiard" approach was considered. The classical billiard approach for multidimensional models with fields of forms and scalar fields was suggested in Ref. 27. These inequalities have played a key role in the proof of the never-ending oscillating behaviour near the singularity which takes place in effective gravitational models with forms and scalar fields induced by superstrings.[28,29] For reviews on the billiard approach, see Refs. 29, 30.

In Ref. 31 the quantum billiard approach for the multidimensional gravitational model with several forms was considered. The main motivation for this approach is coming from the quantum gravity paradigm. It should be noted that the asymptotic solutions to WDW equation presented in Ref. 31 are equivalent to the solutions obtained earlier in Ref. 26. In Refs. 26, 31 a "semi-quantum" approach was used: the gravity (of a toy model) was quantized but the matter sources (e.g. fluids, forms) were considered at the classical level. Such semi-quantum form of the WDW equation for the model with fields of forms and a scalar field was suggested in Ref. 32.

In our paper, Ref. 33, we have used another form of the WDW equation with enlarged minisuperspace which includes the form potentials[48]. We have suggested another version of the quantum billiard approach by deducing the asymptotic solutions to WDW equation for the model with fields of forms when non-composite electric brane ansatz has been adopted.

In Ref. 33 we have considered an example of 9-dimensional quantum billiard for $D = 11$ model with 120 four-forms which mimic space-like $M2$-brane solutions ($SM2$-branes) in $D = 11$ supergravity. It was shown that the wave function vanishes at the singularity.

In Ref. 34 the approach of Ref. 33 was substantially generalized to the case when scalar fields and dilatonic couplings were added, and new examples of quantum billiards with electric and magnetic S-branes in $D = 11$ and

$D = 10$ models (analogues of supergravitational models) were presented. In both examples of billiards magnetic branes do not participate in the formation of the billiard walls and do not change the classical asymptotic oscillating behaviour of scale factors and scalar field (for $D = 10$). But in quantum case the adding of magnetic branes changes the asymptotic behaviour of the wave function.

3. Fluxbrane and Black Brane Solutions

The first subclass considered contains multidimensional fluxbrane solutions, see Refs. 35–45. These solutions are generalizations of the well-known Melvin solution.[46] The fluxbrane solutions from Ref. 45 are governed by functions $\mathcal{H}_s(z) > 0$ defined on $(0, +\infty)$ and obeying the set of differential equations

$$\frac{d}{dz}\left(\frac{z}{\mathcal{H}_s}\frac{d}{dz}\mathcal{H}_s\right) = \mathcal{P}_s \prod_{l=1}^{m} \mathcal{H}_l^{-A_{sl}}, \tag{1}$$

with the boundary conditions $\mathcal{H}_s(+0) = 1$, $s = 1, ..., m$. Here $\mathcal{P}_s > 0$ for all s and (A_{sl}) is a nondegenerate matrix which governs the intersection rules of branes. In Ref. 45 the following hypothesis was suggested: solutions to Eqs. (1) with $\mathcal{H}_s(+0) = 1$ are polynomials if (A_{sl}) is the Cartan matrix for some finite-dimensional semisimple Lie algebra $\mathcal{G}$ of rank m. In Ref. 47, generalized "flux-S-brane" solutions were obtained. In this subclass certain solutions describing the present Universe acceleration were singled out.

Another class of solutions contains black branes which are generalizations of black hole solutions. Solutions of such type appear in models with antisymmetric forms and scalar fields, see Refs. 10–12, 48–54. For black brane solutions with 1-dimensional factor spaces (of Euclidean signatures), see Refs. 55–57.

The black brane solutions in the model with scalar field and fields of forms are governed by a set of functions H_s, $s = 1, ..., m$, obeying non-linear differential equations with certain boundary conditions. These equations depend on the non-degenerate $m \times m$ matrix A. It was conjectured in Ref. 50 that the moduli functions H_s should be polynomials if A is a Cartan matrix for some semisimple finite-dimensional Lie algebra $\mathcal{G}$ of rank m. In this case we deal with special solutions to open Toda chain equations related to the Lie algebra $\mathcal{G}$[58,59] which are integrable in quadratures. The conjecture from Ref. 50 was verified for the Lie algebras A_m, C_{m+1}, $m \geq 1$ in Refs. 51, 52 by using the solutions to Toda chain equations corresponding to the Lie algebras A_m from Ref. 60.

It was shown in Refs. 61,62 that the black brane solutions under consideration may be also found by using fluxbrane polynomials[45] which may be calculated for any (semi)simple finite-dimensional Lie algebra by MAPLE.[63] A similar approach appeared in Ref. 64 for special Toda charged black hole solutions corresponding to Lie algebras A_m. In Ref. 62 certain examples of electric black brane solutions related to Lie algebras A_2, C_2 and G_2 were considered.

We note that recently in Ref. 66 dilatonic black hole dyon solutions with an arbitrary dilatonic coupling constant and any sign of the scalar field kinetic term were considered. Two group of bounds on the gravitational mass and scalar charge (for fixed and arbitrary extremality parameter) were found by using a certain conjecture on the parameters of solutions.

4. Conclusion

We have presented a short review of our motivation and main results in multidimentional gravitation and cosmology and gave an overview (based on Refs. 33, 34) of the quantum billiard approach for D-dimensional cosmological-type models of (generalized) Bianchi-I type. Apart from formulation of the quantum billiard approach with branes, the main result of Ref. 34 is as follows: the basis solutions for wave functions vanish in the limit of the formation of billiard walls (i.e., at the singularity) for the $D = 11$ model which mimics the $D = 11$ supergravitational cosmology. The magnetic branes could not be neglected in calculations of quantum asymptotic solutions while in the classical case they did not change the asymptotic relations.[34]

We have also presented an overview of two families of composite brane solutions: fluxbrane and black brane ones, which correspond to semisimple Lie algebras in models with a scalar field and fields of forms. Intersection rules for branes are given by Cartan matrices for these Lie algebras. The metric of any solution contains $(n-1)$ Ricci-flat "internal" metrics, and certain restrictions on brane intersections are imposed.

The moduli functions of both type of solutions are given by fluxbrane polynomials which define special solutions to open Toda chain equations corresponding to semisimple Lie algebras. These polynomials may be calculated (in principle) for any simple or semisimple Lie algebra, e.g., by using MAPLE or MATHEMATICA. One can also use formal relations for A_m-polynomials which were obtained recently in Ref. 64.

It should be noted that "flux-S-brane" solutions governed by fluxbrane polynomials with negative P_s were considered earlier in cosmological (S-brane) solutions which describe an accelerated expansion of the 3-dimensional factor space.[67,68] Recently, another class of cosmological solutions which lead to exponential expansion of "our" 3-dimensional factor space and obey the observational constraints on the temporal variation of effective gravitational constant G was obtained in a D-dimensional gravitational model with the Gauss-Bonnet term.[69] Among them there are two exact solutions in dimensions $D = 22$ and 28 with constant G and also an infinite series of solutions for larger dimensions.

Another fruitful approach to multidimensional gravity should also be mentioned.[70,71] It deals with models of arbitrary dimensions with Lagrangians involving various curvature-nonlinear terms, such as arbitrary functions of the Ricci scalar, the Ricci tensor squared and the Kretschmann scalar. Under reasonable assumptions (above all, slow changes of basic variables as compared to the Planck scale), the original broad class of multidimensional theories with higher derivatives reduces to Einstein gravity with a set of effective scalar fields. Among other results, this approach was successfully applied for obtaining a unified description of early inflation and the present accelerated expansion of the Universe,[72] a possible explanation of the origin of the Higgs field,[73] and some models explaining spatial and temporal variations of the fine structure constant and the gravitational constant in connection with modern observations[74,75] (see also a contribution from K. Bronnikov and M. Skvortsova in the present volume).

References

1. K. P. Staniukovich and V. N. Melnikov, *Hydrodynamics, Fields and Constants in the Theory of Gravitation*, (Energoatomizdat, Moscow, 1983) 256 pp. [in Russian].
2. V. N. Melnikov, *Fields and Constants in the Theory of Gravitation*, CBPF MO-02/02, (CBPF, Rio de Janeiro, 2002) 145 pp.
3. V. N. Melnikov, *Int. J. Theor. Phys.* **33**, 1569 (1994).
4. V. N. Melnikov, Gravitation and cosmology as key problems of the millennium, in *Proc. Albert Einstein Century International Conference*, eds. J.-M. Alimi and A. Fuzfa, AIP Conference Proceedings, Vol. 861 (AIP, New York, 2006), p. 109.
5. V. N. Melnikov, Multidimensional classical and quantum cosmology and gravitation. Exact solutions and variations of constants, in *Cosmology and Gravitation*, ed. M. Novello (Editions Frontieres, Singapore,

1994), p. 147.
6. V. N. Melnikov, Multidimensional cosmology and gravitation, in *Cosmology and Gravitation. II*, ed. M. Novello (Editions Frontieres, Singapore, 1996), p. 465.
7. V. D. Ivashchuk and V. N. Melnikov, Multidimensional cosmological and spherically symmetric solutions with intersecting p-branes, in *Mathematical and Quantum Aspects of Relativity and Cosmology. Proc. of the Second Samos Meeting on Cosmology, Geometry and Relativity*, eds. S. Cotsakis, G.W. Gibbons., Lecture Notes in Physics, Vol. 537 (Springer-Verlag, New York, 2000), p. 214.
8. V. D. Ivashchuk and V. N. Melnikov, *Class. Quantum Grav.* **18**, R87 (2001).
9. V. N. Melnikov, Exact Solutions in Multidimensional Gravity and Cosmology III. (CBPF Publ., Rio de Janeiro, 2002).
10. K. A. Bronnikov, V. D. Ivashchuk and V. N. Melnikov, *Grav. Cosmol.* **3**, 203 (1997).
11. K. A. Bronnikov, *Grav. Cosmol.* **4**, 49 (1998).
12. K. A. Bronnikov, *J. Math. Phys.* **40**, 924 (1999).
13. M. A. Grebeniuk, V. D. Ivashchuk and V. N. Melnikov, *Grav. Cosmol.* **4**, 145 (1998).
14. U. Bleyer, V. D. Ivashchuk, V. N. Melnikov and A. I. Zhuk, *Nucl. Phys. B* **429**, 117 (1994).
15. V. D. Ivashchuk, V. N. Melnikov and A. B. Selivanov, *JHEP* **0309**, 059 (2003).
16. J.-M. Alimi, V. D. Ivashchuk and V. N. Melnikov, *Grav. Cosm.* **11**, 111 (2005).
17. K. A. Bronnikov and V. N. Melnikov, *Nucl. Phys. B* **584**, 434 (2000).
18. R. Blumenhagen, M. Cvetic, P. Langacker and G. Shiu, *Ann. Rev. Nucl. Part. Sci.* **55**, 71 (2005).
19. D. M. Chitre, Ph. D. Thesis (University of Maryland, 1972).
20. V. A. Belinskii, E. M. Lifshitz and I. M. Khalatnikov, *Adv. Phys.* **31**, 639 (1982).
21. C. W. Misner, *Phys. Rev.* **186**, 1319 (1969).
22. R. Graham and P. Szepfalusy, *Phys. Rev.* **D 42**, 2483 (1990).
23. A. A. Kirillov, *Int. J. Mod. Phys.* **D3**, 431 (1994).
24. V. D. Ivashchuk, A. A. Kirillov and V. N. Melnikov, *Russ. Phys. J.* **37**, 1102 (1994).
25. V. D. Ivashchuk, A. A. Kirillov and V. N. Melnikov, *JETP Lett.* **60**, 235 (1994).

26. V. D. Ivashchuk and V. N. Melnikov, *Class. Quantum Grav.* **12**, 809 (1995).
27. V. D. Ivashchuk and V. N. Melnikov, *J. Math. Phys.* **41** 6341 (2000).
28. T. Damour and M. Henneaux, *Phys. Rev. Lett.* **85**, 920 (2000).
29. T. Damour, M. Henneaux and H. Nicolai, *Class. Quantum Grav.* **20**, R145 (2003).
30. V. D. Ivashchuk and V. N. Melnikov, *Grav. Cosmol.* **15**, 49 (2009).
31. A. Kleinschmidt, M. Koehn and H. Nicolai, *Phys. Rev.* **D 80**, 061701 (2009).
32. H. Lü, J. Maharana, S. Mukherji and C. N. Pope, *Phys. Rev.* **D 57**, 2219 (1997).
33. V. D. Ivashchuk and V.N. Melnikov, *Grav. Cosmol.* **19**, 171 (2013).
34. V. D. Ivashchuk and V. N. Melnikov, *Eur. Phys. J. C* **74**: 2805 (2014).
35. G. W. Gibbons and D. L. Wiltshire, *Nucl. Phys. B* **287**, 717 (1987).
36. G. Gibbons and K. Maeda, *Nucl. Phys. B* **298**, 741 (1988).
37. F. Dowker, J. P. Gauntlett, G. W. Gibbons and G. T. Horowitz, *Phys. Rev. D* **53**, 7115 (1996).
38. D. V. Gal'tsov and O. A. Rytchkov, *Phys. Rev. D* **58**, 122001 (1998).
39. C.-M. Chen, D. V. Gal'tsov and S. A. Sharakin, *Grav. Cosmol.* **5**, 45 (1999).
40. M. S. Costa and M. Gutperle, *JHEP* **0103**, 027 (2001).
41. P. M. Saffin, *Phys. Rev. D* **64**, 024014 (2001).
42. M. Gutperle and A. Strominger, *JHEP* **0106**, 035 (2001).
43. C.-M. Chen, D. V. Gal'tsov and P. M. Saffin, *Phys. Rev. D* **65**, 084004 (2002).
44. R. Empharan and M. Gutperle, *JHEP* **0112**, 023 (2001).
45. V. D. Ivashchuk, *Class. Quantum Grav.* **19**, 3033 (2002).
46. M. A. Melvin, *Phys. Lett.* **8**, 65 (1964).
47. I. S. Goncharenko, V. D. Ivashchuk and V. N. Melnikov, *Grav. Cosmol.* **13**, 262 (2007).
48. V. D. Ivashchuk and V. N. Melnikov, *J. Math. Phys.* **39**, 2866 (1998).
49. S. Cotsakis, V. D. Ivashchuk and V. N. Melnikov, *Grav. Cosmol.* **5**, 52 (1999).
50. V. D. Ivashchuk and V. N. Melnikov, *Grav. Cosmol.* **5**, 313 (1999).
51. V. D. Ivashchuk and V. N. Melnikov, *Grav. Cosmol.* **6**, 27 (2000).
52. V. D. Ivashchuk and V. N. Melnikov, *Class. Quantum Grav.* **17**, 2073 (2000).
53. M. A. Grebeniuk, V. D. Ivashchuk and S.-W. Kim, *J. Math. Phys.* **43**, 6016 (2002).

54. M. A. Grebeniuk, V. D. Ivashchuk and V. N. Melnikov, *Phys. Lett. B* **543**, 98 (2002).
55. M. Cvetic and A. Tseytlin, *Nucl. Phys. B* **478**, 181 (1996).
56. I. Ya. Aref'eva, M. G. Ivanov and I. V. Volovich, *Phys. Lett. B* **406**, 44 (1997).
57. N. Ohta, *Phys. Lett. B* **403**, 218 (1997).
58. M. A. Olshanetsky and A. M. Perelomov, *Invent. Math.* **54**, 261 (1979).
59. B. Kostant, *Adv. in Math.* **34**, 195 (1979).
60. A. Anderson, *J. Math. Phys.* **37**, 1349 (1996).
61. V. D. Ivashchuk, *J. Geom. Physics* **86**, 101 (2014).
62. V. D. Ivashchuk and V. N. Melnikov, *Grav. Cosmol.* **20**, 182 (2014).
63. A. A. Golubtsova and V. D. Ivashchuk, *Grav. Cosmol.* **15**, 144 (2009).
64. H. Lü and W. Yang, *Class. Quantum Grav.* **30**, 235021 (2013).
65. J. Fuchs and C. Schweigert, *Symmetries, Lie algebras and Representations. A graduate course for physicists* (Cambridge University Press, Cambridge, 1997).
66. M. E. Abishev, K. A. Boshkayev, V. D. Dzhunushaliev and V. D. Ivashchuk, *Class. Quantum Grav.* **32**, 165010 (2015).
67. V. D. Ivashchuk, S. A. Kononogov and V. N. Melnikov, *Grav. Cosmol.* **14**, 235 (2008).
68. A. A. Golubtsova, *Grav. Cosmol.* **16**, 298 (2010).
69. V. D. Ivashchuk and A. A. Kobtsev, *Eur. Phys. J. C* **75**, 177 (2015).
70. K. A. Bronnikov and S. G. Rubin, *Phys. Rev. D* **73**, 124019 (2006).
71. K. A. Bronnikov and S. G. Rubin, *Black Holes, Cosmology, and Extra Dimensions* (World Scientific, 2012).
72. K. A. Bronnikov, S. G. Rubin and I. V. Svadkovsky, *Phys. Rev. D* **81**, 084010 (2010).
73. S. V. Bolokhov, K. A. Bronnikov and S. G. Rubin, *Phys. Rev. D* **84**, 044015 (2011).
74. K. A. Bronnikov, V. N. Melnikov, S. G. Rubin and I. V. Svadkovsky, *Gen. Rel. Grav.* **45**, 2509 (2013).
75. K. A. Bronnikov and M. V. Skvortsova, *Grav. Cosmol.* **19**, 114 (2013).

Flat potentials in models with chaotic inflation

D. S. Gorbunov

Institute for Nuclear Research of the Russian Academy of Sciences,
Moscow 117312, Russia
E-mail: gorby@ms2.inr.ac.ru

We discuss the naturality of the standard chaotic initial conditions in inflationary models with flat scalar potential, which are favorable by the present cosmological data.

Keywords: Chaotic inflation; Higgs; flat potential; Big Bang.

1. Inflationary Solution to the Hot Big Bang Problems

The Standard Model of particle physics (SM) and General Relativity (GR) fail to describe the cosmology properly. The major phenomenological problems are dark matter phenomena and the baryon (matter-antimatter) asymmetry of the Universe. There are several puzzling coincidences between cosmological parameters, among them the widely recognized is an order-of-magnitude equality of baryon, dark matter and dark energy contributions to the present energy density of the Universe. The vanilla ΛCDM model seems to exhibit lack of dwarf galaxies and other issues probably related to evolution of the matter perturbations at small scales.

There is a separate class of problems called *the problem of initial conditions for the Hot Big Bang Theory.* Since the homogeneous Universe may form spherical or hyperbolic 3-dimensional manifold, its flatness at present is mysterious. Both at radiation and at matter dominated stages of the Universe expansion a size of the causally connected region grows faster than a physical size does, so the Universe must be highly inhomogeneous at large scales, that is quite contrary to all observations. The Universe expansion is adiabatic, keeps the entropy conserved in a comoving volume, hence the entropy density in the very early Universe was enormously high. Both radiation and matter dominated stages imply initial singular energy density at the beginning of expansion. If SM is a remnant of some Grand Unified Theory, the corresponding phase transition(s) at the Hot stage generally populates the Universe with topologically defects; later the monopoles or domain walls become dominant and overclose the Universe.

The initial condition problems can be elegantly solved by introducing an (exponentially) fast expansion preceding the Hot stage. Then the curvature (if any) grows exponentially and hence the Universe is exponentially flat. An originally causally connected region whatever tiny becomes exponentially large and today we observe nothing but its minuscule part. Any heavy stable relics (like monopoles) are at exponentially large distances one from each other, so the late Hot Universe is safe. There are no initial singularities in the exponentially expanding Universe.

This stage, dubbed *inflation*, can be naturally organized by a slowly evolving homogeneous scalar field ϕ, *inflaton*. In the "old inflation"[4] the scalar field settles in a false vacuum (see left panel in Fig. 1) which provides

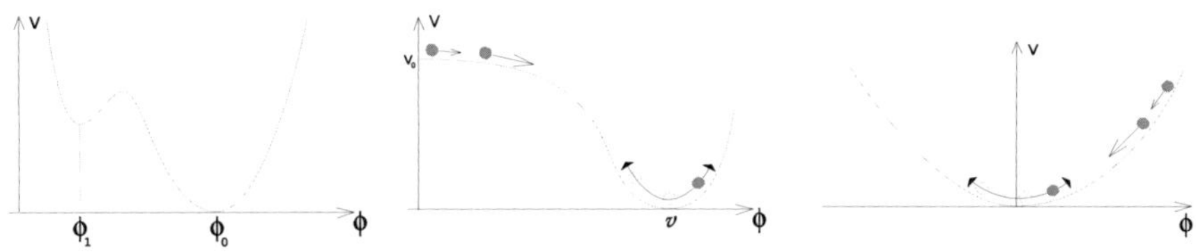

Fig. 1. Scalar potentials $V(\phi)$ of "old inflation" (left panel), "new inflation" (middle panel) and "chaotic inflation" (right panel).

with exactly exponential expansion terminated by tunneling to the true vacuum. The latter process is not effective and other variant had been put forward. In the "new inflation"[2,3] the scalar field slowly rolls down from the hilltop of the scalar potential (see middle panel in Fig. 1) that yields almost exponential expansion. In both models above the special starting positions of the scalar field may be attributed to a preceding Hot stage. Another idea—chaotic initial conditions—is utilized in models called *large field inflation*[1] (see right panel in Fig. 1).

2. Chaotic Inflation: Unlikeness of Models with Flat Potential

In the chaotic inflation field ϕ rolls down along the potential $V(\phi)$ starting from large (generally transplanckian) values. In the Robertson–Walker metric, $ds^2 = dt^2 - a^2(t)\mathbf{dx}^2$, with $a(t)$ being the scale factor, $\phi(t)$ solves equation

$$\ddot{\phi} - \frac{\Delta\phi}{a^2} + 3\mathcal{H}\dot{\phi} + V'(\phi) = 0\,. \tag{1}$$

The Hubble parameter $\mathcal{H} \equiv \dot{a}/a$ obeys the Friedmann equation

$$\mathcal{H}^2 = \frac{8\pi}{M_{\rm Pl}}\left(\frac{1}{2}\dot{\phi}^2 + \frac{1}{2a^2}(\nabla\phi)^2 + V(\phi)\right) \tag{2}$$

and plays a role of dynamical friction in Eq. (1). Actually, if the field evolves very slowly, its energy density is dominated by the scalar potential and the Hubble parameter remains almost constant, see Eq. (2), ensuring the almost exponential expansion. Then the gradient terms in Eqs. (1), (2), being suppressed by the scale factor, get nullified and the Hubble friction governs the field slow rolling towards the origin of scalar potential. The kinetic terms in Eqs. (1), (2) are negligible provided the so-called slow roll parameters are small,

$$\frac{M_{\rm Pl}^2}{16\pi}\left(\frac{V'}{V}\right)^2 \equiv \epsilon \ll 1\,, \quad \frac{M_{\rm Pl}^2}{8\pi}\frac{V''}{V} \equiv \eta \ll 1\,. \tag{3}$$

One can check that the latter conditions are satisfied for any power-law scalar potential if the scalar field takes a transplanckian value. Then the field goes down slowly and once $\epsilon, \eta \simeq 1$, the inflation terminates.

The chaotic initial conditions for the inflaton field imply[1,5] that the scalar field fluctuates in any Planck-size region, and all the contributions to the energy density–kinetic term, gradient term and potential term–are (on average) of the same order, which is Planckian, i.e.

$$\frac{1}{2}\dot{\phi}^2 \sim \frac{1}{2}(\partial_i\phi)^2 \sim V(\phi) \sim M_{\rm Pl}^4\,. \tag{4}$$

Then if in a particular Planck-size volume the potential term becomes somewhat dominant as result the fluctuations, the dynamics of the scalar field rapidly approaches the slow roll and the chaotic inflation begins in this region.

Inflation itself is almost untestable: the only exception is the prediction of exponentially flat Universe, that is in agreement with observations at present. However, it is a general prediction, inherent in the very paradigm of inflation. To recognize a particular inflationary model we need something else. Quite remarkably, the inflation provides with an appropriate feature. Namely, inflaton inhomogeneities (linear perturbations about the homogeneous inflationary solution) originated from the inflaton quantum fluctuations can seed the matter inhomogeneities in the late Universe. Their subsequent evolution initiates large scale structure formation and explain the anisotropy of cosmic microwave background radiation $\delta T/T \sim 10^{-4}$.

The inflaton perturbations solve Eq. (1) and behave in a very peculiar way: the amplitude drops inversely with the scale factor till the very moment when the physical 3-momentum of the mode $\propto 1/a$ matches the Hubble parameter, and then the mode amplitude remains constant, while the stretching proceeds. Hence the amplitude of the modes is almost momentum-independent, the small decrease in the Hubble parameter during inflation results in a small tilt of the otherwise flat scalar perturbation spectrum. Both the amplitude and the tilt depend on the inflaton model parameters, and analysis of the cosmological observables allows to discriminate between various inflationary models, for details see e.g. Ref. 6. The Figure 2 shows that the present data tend to favor the large field infla-

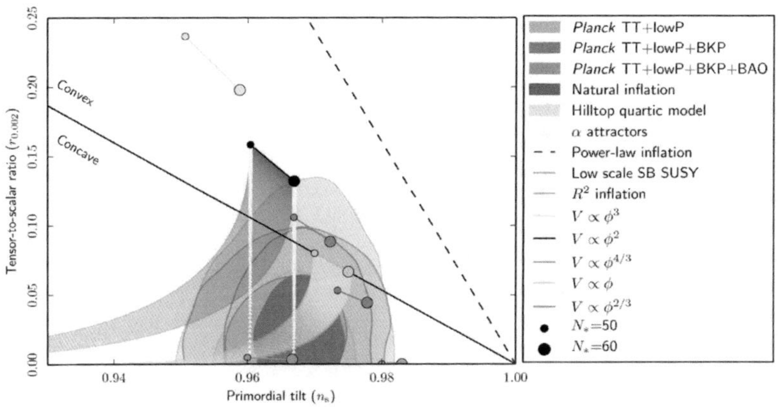

Fig. 2. Cosmological constraints[9] on inflationary models.

tion models with *concave* scalar potential. Among them are theoretically well motivated models with asymptotically flat potentials: R^2-inflation,[7] Higgs-inflation,[8] moduli inflation in supersymmetric extensions of the SM, etc. These models seem to be unlikely in the framework of the chaotic inflation[10]. Indeed, at the Planck epoch, when kinetic and gradient terms take the Planckian values according to conditions (4), the potential (being flat) is fixed by the cosmological observations at much smaller value,

$$\frac{1}{2}\dot{\phi}^2 \sim \frac{1}{2}(\partial_i \phi)^2 \sim M_{\rm Pl}^4 \gg V(\phi)\,, \tag{5}$$

so that slow roll never happens. In such a model the inflation requires a beginning with all terms of the same order that is much smaller than the Planckian value, which seems unnatural. Below we illustrate the problem with two prominent examples and resolve it by explaining the proper way to implement the chaotic initial conditions, for more details see Ref. 11.

3. An Example: R^2-Inflation

Let us consider the first well-developed inflationary model,[7] based on modified gravity with action

$$S^{JF} = -\frac{M_P^2}{2}\int\sqrt{-g}\,d^4x\left(R - \frac{R^2}{6\,\mu^2}\right),$$

which after the convenient Weyl transformation (from the Jordan frame to the Einstein frame),

$$g_{\mu\nu} \to \tilde{g}_{\mu\nu} = \chi\, g_{\mu\nu}\,, \qquad \chi = \exp\left(\sqrt{2/3}\,\phi/M_P\right),$$

casts into the standard Einstein–Hilbert term and terms describing the scalar degree of freedom, dubbed *scalaron*, emerging in the gravity sector,

$$S^{EF} = \int\sqrt{-\tilde{g}}\,d^4x\left[-\frac{M_P^2}{2}\,\tilde{R} + \frac{1}{2}\,\tilde{g}^{\mu\nu}\partial_\mu\phi\partial_\nu\phi - \frac{3\,\mu^2 M_P^2}{4}\left(1 - \frac{1}{\chi(\phi)}\right)^2\right],$$

The model has an exponentially flat potential, which approaches the value

$$V_0 \simeq 10^{-12}\,M_{\rm Pl}^4$$

fixed by matching the amplitude of scalar perturbations to $\sim 10^{-4}$. Naively, within the chaotic paradigm the model suffers from unlikeness problem (5).

This is not true in fact. Indeed, in the Jordan frame the chaotic initial conditions look like $|R^{JF}| \sim M_{\rm Pl}^2$, and matching to cosmological data infers

$$\chi^2 = 1 - \frac{R^{JF}}{3\,\mu^2} \sim 10^{12}\,.$$

The scalar curvature in the Jordan frame is related to that in the Einstein frame and scalaron as follows,

$$-R^{JF} = -\chi^2\,R^{EF} + \chi^2\,\frac{8\pi}{M_{\rm Pl}^2}\,g_{\mu\nu}^{EF}\partial^\mu\phi\partial^\nu\phi - \chi^2\,\frac{4\sqrt{3\pi}}{M_{\rm Pl}}g_{\mu\nu}^{EF}\,\partial^\mu\partial^\nu\phi\,.$$

The first and second terms in the r.h.s. above refer to the pure gravity and scalaron contributions to the energy density, respectively. Hence they must be of the same order within the paradigm of chaotic inflation. The third term happens to be of the same order provided the equations of motion.

Therefore one has

$$\left|R^{JF}\right|\chi^{-2} \sim \left|R^{EF}\right| \sim \dot{\phi}^2 M_{\rm Pl}^2 \sim (\partial_i\phi)^2 M_{\rm Pl}^2\,,$$

which means that all the terms in the Einstein frame are actually of the same order, consistent with the requirement of successful inflation,

$$\frac{1}{2}\dot{\phi}^2 \sim \frac{1}{2}(\partial_i\phi)^2 \sim V(\phi) \sim 10^{-12} \times M_{\rm Pl}^4\,.$$

There is no unlikeness problem in the R^2-inflation.

4. One More Example: Higgs Inflation

To cook the successful inflation with the SM Higgs doublet H one must introduce its non-minimal coupling to gravity,[8]

$$S = \int d^4x\sqrt{-g}\left(-\frac{M_P^2}{2}R - \xi H^\dagger H R + \mathcal{L}_{SM}\right)$$

or written in the unitary gauge $H^T = \left(0\,, (h+v)/\sqrt{2}\right)$ (neglecting $v = 246\,\mathrm{GeV}$)

$$S = \int d^4x\sqrt{-g}\left(-\frac{M_P^2 + \xi\,h^2}{2}R + \frac{(\partial_\mu h)^2}{2} - \frac{\lambda\,h^4}{4}\right)$$

Performing the Weyl transformation

$$g_{\mu\nu} = \Omega^{-2}\tilde{g}_{\mu\nu}\,, \qquad \Omega^2 = 1 + \frac{\xi\,h^2}{M_P^2}$$

one finds in the Einstein frame for the canonically normalized field χ

$$\frac{d\chi}{dh} = \frac{M_P\sqrt{M_P^2 + (6\xi+1)\xi\,h^2}}{M_P^2 + \xi\,h^2}$$

the scalar potential

$$U(\chi) = \frac{\lambda M_P^4\,h^4(\chi)}{4(M_P^2 + \xi\,h^2(\chi))^2} = \frac{\lambda M_P^4}{4\xi^2}\left(1 - \exp\left(-\frac{\sqrt{2}\,\chi}{\sqrt{3}M_P}\right)\right)^2$$

at relevant for inflation values $h \gg M_P/\sqrt{\xi}$. It looks exactly as the scalaron potential in the previous example of R^2-inflation, so one naively expects here the same unlikeness problem.

Actually, there is no unlikeness problem here as well. Similarly to the case of R^2-inflation, one starts with the chaotic initial conditions implemented in the Jordan frame,

$$\Omega^2 M_{\rm Pl}^2 R^{JF} \sim \xi\dot{h}^2 \sim \xi(\partial_i h)^2 \sim \lambda h^4 \sim M_{\rm Pl}^4 .$$

Here we account for the fact that the nonminimal gravity coupling to the Higgs field effectively changes the Higgs kinetic and gradient terms, and chose the dominant parts. Consequently,

$$R^{JF} \sim M_{\rm Pl}^2/\Omega^2$$

and cosmological observations fix $V_0^{EF} \sim 10^{-12} \times M_{\rm Pl}^4$, hence

$$\Omega^2 \sim 10^6 .$$

Then in the Einstein frame one arrives at the same picture: all the terms are of the same order needed for the successful inflation, namely

$$\frac{1}{2}\dot{\phi}^2 \sim \frac{1}{2}(\partial_i\phi)^2 \sim M_{\rm Pl}^4/\Omega^4 \sim 10^{-12} \times M_{\rm Pl}^4 .$$

There is no unlikeness problem in the Higgs-inflation either.

5. Conclusion: Initial Conditions in an Arbitrary Frame

We find that Chaotic paradigm works perfectly well for both R^2-inflation and Higgs-inflation: they do not suffer from the unlikeness problem. The presented analysis suggests a simple prescription for implementing the chaotic initial conditions in any model as follows. Start in any frame with all contributions to the energy density being of the same order (say, $\sim M_{\rm Pl}^4$ or any other value). Then in any other frame the corresponding contributions are always of the same (though another) order (say, $\sim 10^{-12} \times M_{\rm Pl}^4$ in the EF).

The work is supported by the RSCF grant 14-12-01430.

References

1. A. D. Linde, Chaotic Inflation, *Phys. Lett. B* **129**, 177 (1983).
2. A. D. Linde, A New Inflationary Universe Scenario: A Possible Solution of the Horizon, Flatness, Homogeneity, Isotropy and Primordial Monopole Problems, *Phys. Lett. B* **108**, 389 (1982).
3. A. Albrecht and P. J. Steinhardt, Cosmology for Grand Unified Theories with Radiatively Induced Symmetry Breaking, *Phys. Rev. Lett.* **48**, 1220 (1982).

4. A. H. Guth, The Inflationary Universe: A Possible Solution to the Horizon and Flatness Problems, *Phys. Rev. D* **23** 347 (1981).
5. A. D. Linde, The Inflationary Universe, *Rept. Prog. Phys.* **47**, 925 (1984).
6. D. S. Gorbunov and V. A. Rubakov, *Introduction to the theory of the early universe: Cosmological perturbations and inflationary theory* (Hackensack, USA: World Scientific, 2011)
7. A. A. Starobinsky, A New Type of Isotropic Cosmological Models Without Singularity, *Phys. Lett. B* **91**, 99 (1980).
8. F. L. Bezrukov and M. Shaposhnikov, The Standard Model Higgs boson as the inflaton, *Phys. Lett. B* **659**, 703 (2008).
9. Planck Collaboration (P. A. R. Ade *et al.*), Planck 2015 results. XX. Constraints on inflation, *e-Print:* arXiv:1502.02114 (2015).
10. A. Ijjas, P. J. Steinhardt and A. Loeb, Inflationary paradigm in trouble after Planck2013, *Phys. Lett. B* **723** 261 (2013).
11. D. S. Gorbunov and A. G. Panin, Are R^2- and Higgs-inflations really unlikely?, *Phys. Lett. B* **743**, 79 (2015).

Cosmology based on $f(R)$ gravity with $\mathcal{O}(1\,\text{eV})$ sterile neutrino

A. S. Chudaykin* and D. S. Gorbunov‡

Institute for Nuclear Research of the Russian Academy of Sciences,
Moscow, 60th October Anniversary prospect 7a, Russian Federation
Moscow Institute of Physics and Technology,
Dolgoprudny, Institutsky per. 9, Russian Federation
**E-mail: chudy@ms2.inr.ac.ru*
‡E-mail: gorby@ms2.inr.ac.ru

A. A. Starobinsky

L. D. Landau Institute for Theoretical Physics of the Russian Academy of Sciences,
Chernogolovka, Academica Semenova prospect, 1a, Russian Federation
E-mail: alstar@landau.ac.ru

R. A. Burenin

Space Research Institute of the Russian Academy of Sciences (IKI),
Moscow, ul. Profsoyuznaya, 84/32, Russian Federation
E-mail: rodion@hea.iki.rssi.ru

We refer [1] to the role of an additional $\mathcal{O}(1)$ eV sterile neutrino in modified gravity models. We found parameter constraints in particular $f(R)$ gravity model employing up-to-dated cosmological data: measurements of the cosmic microwave background (CMB) anisotropy, the CMB lensing potential, the baryon acoustic oscillations (BAO), the cluster mass function and the Hubble constant. We obtained for the sterile neutrino mass $0.47\text{ eV} < m_{\nu,\,\text{sterile}} < 1\text{ eV}$ (2σ) assuming that the sterile neutrino is thermalized in the early Universe and the active neutrinos are massless, not significantly larger than in the standard cosmology model within the same data set: $0.45\text{ eV} < m_{\nu,\,\text{sterile}} < 0.92\text{ eV}$ (2σ). But, if the mass of sterile neutrino is fixed and equals $\approx 1.5\,\text{eV}$ according to various anomalies in neutrino oscillation experiments, $f(R)$ gravity model is much more consistent with observation data than the ΛCDM one.

Keywords: Modified gravity; galaxy clusters; neutrino masses from cosmology.

1. Introduction

The Universe is known to be undergoing an accelerated expansion today. The present Dark Energy (DE) provides such Universe behaviour in ΛCDM model. Corresponding DE fraction is considered lately as evolving and unstable by analogy with the inflation stage in the early Universe. We work within $f(R)$ formalism[2] where the scalar curvature (Ricci curvature) R is replaced with a new phenomenological function $f(R)$ in the Einstein-Hilbert action. Cosmology based on modified gravity can explain cosmic acceleration today without introducing cosmological constant Λ, so $f(0) = 0$.

New theory must be viable and consistent with the past evolution of the Universe. Hence, it should satisfy following conditions in the relevant region of R where we want to use the theory[3]

$$f'(R) > 0, \qquad f''(R) > 0, \tag{1}$$

and for $R \gg R_0$

$$|f(R) - R| \ll R, \qquad |f'(R) - 1| \ll 1, \qquad f''(R)R \ll 1, \tag{2}$$

where prime denotes a derivative with respect to argument R and R_0 is the present Ricci curvature in the Universe. The last three conditions are necessary to provide the correct Newtonian limit for the matter- and radiation-dominated stages in the past and smallness of non-GR corrections to a space-time background metric of compact astrophysical objects at present epoch.

There is dramatic difference between standard cosmology and model based on $f(R)$ gravity: matter density perturbations in modified gravity grow faster on scales smaller than the Compton wavelength of the scalaron field that occurs at recent redshifts. Therefore if we add extra one sterile neutrino species in the model described by $f(R)$ gravity, the net result can be zero because modified gravity and rest mass of sterile neutrino play opposite roles in the evolution of matter density perturbations on small scales.[4] As we know small massive particles suppress structure formation below Jeans length through free-streaming effect.

If mixing angle is not extremely small, sterile component is produced and thermalized in the early Universe before the decoupling of active neutrinos. We add one sterile neutrino with mass O(1) eV to Standard Model by mixing with active ones. Importantly, one additional sterile neutrino species is not forbidden by the recent research of the primordial helium abundance[5]: an effective number of neutrinos is $N_{eff} = 3.58 \pm 0.40\,(2\sigma)$.

The light sterile neutrinos are required by several anomalies results in neutrino oscillation experiments.[6] For instance, the so-called gallium anomaly observed by GALLEX[7] and SAGE[8] experiments is explained very well if the electron neutrino oscillates into sterile neutrino of 1.5 eV mass.[9]

2. Background Universe

The $f(R)$ gravity is defined by the following action

$$S = \frac{1}{2\kappa^2} \int d^4x \sqrt{-g} f(R) + S_m \,, \tag{3}$$

where $\kappa^2/(8\pi) \equiv G$ is the Newton gravitational constant and S_m is the action of matter fields minimally coupled to gravity.

We consider the particular modified gravity model[3]

$$f(R) = R + \lambda R_s \left[\left(1 + \frac{R^2}{R_s^2} \right)^{-n} - 1 \right] , \tag{4}$$

where n, λ, R_s are model parameters of modified gravity. Discussion about different instabilities and how to cope with them you can find in Ref. 10.

From derived field equations by varying the action (3) with respect to space-time metric[a] $g_{\mu\nu}$ we find the effective equation-of-state parameter ω_{DE} for the DE fraction defined by

$$\omega_{DE} \equiv \frac{P_{DE}}{\rho_{DE}} = -1 + \frac{2\dot{H}(f'-1) - H\dot{f}' + \ddot{f}'}{-3H\dot{f}' + 3(H^2 + \dot{H})(f'-1) - (f-R)/2} , \tag{5}$$

where ρ_{DE} and P_{DE} are density and pressure for DE in the following equation $R_{\mu\nu} - \frac{1}{2} g_{\mu\nu} R = \kappa^2 \left(T_{\mu\nu}^{(M)} + T_{\mu\nu}^{(DE)} \right)$ written in Einsteinian form.

Figure 1 depicts the behaviour of ω_{DE} as a function of redshift z obtained in numerical analysis assuming $\Omega_m = 0.3$, $H_0 = 72$ km/s/Mpc. We observe generic feature for modified gravity models which obey $f''(R) > 0$ – the phantom boundary crossing (the equation-of-state intersects the constant value $\omega_{DE} = -1$) at small redshifts $z \leq 1$.

3. Matter Perturbation

Now we turn to evolution matter density perturbations in linear regime in modified gravity models using the quasi-static approximation and considering only sub-Hubble modes. From Ref. 11 we write

$$G_{eff}(t,k) \equiv \frac{G}{f'} \frac{1 + 4\frac{k^2}{a^2}\frac{f''}{f'}}{1 + 3\frac{k^2}{a^2}\frac{f''}{f'}} . \tag{6}$$

We introduce the effective scalaron mass[11] $M_s^2 \approx 1/3f''(R)$ in the quasi-GR regime ($f' \approx 1$). Hence, matter density fluctuations evolve by two

[a] We use the following metric signature $ds^2 = -dt^2 + a^2(t) d\vec{x}^2$, where $a(t)$ is the scale factor.

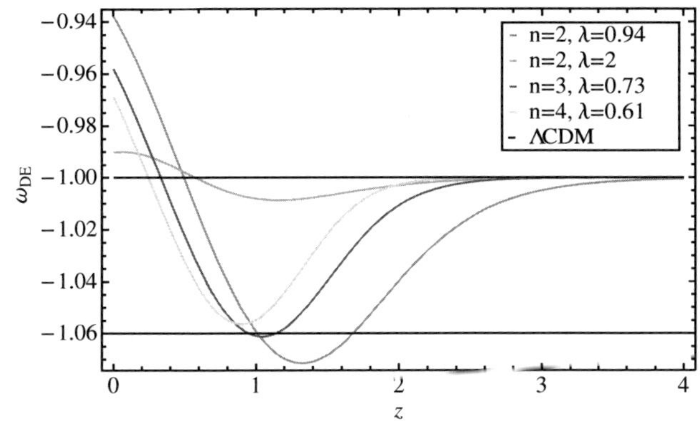

Fig. 1. Evolution of the equation-of-state parameter ω_{DE} for DE fraction in the $f(R)$ model (4) for different value of parameters.

different ways: $M_s \gg k/a$ and $M_s \ll k/a$. The first one corresponds to $G_{eff} \approx G$ (density evolution mimics that in GR), whereas the latter corresponds to $G_{eff} \approx 4G/3$ (enhancement of growth rate mentioned in Sec. 1). Consequently, the effective gravitational "constant" can be increased by 33 % irrespective of the function form of $f(R)$.

4. Parameter Constraints

We carry out the Markov Chain Monte Carlo (MCMC) analysis and compare the ΛCDM model with the $f(R)$ gravity described by eq. (4) with one additional massive sterile neutrino assuming active neutrinos are massless. Sterile species is taken to be thermalized and shares the same temperature as the active ones. We have modified the MGCAMB[12] plugged into CosmoMC package[13] that allows to describe $f(R)$ gravity by adopting (6). In the fit we use eight free parameters: $\Omega_b h^2$, $\Omega_c h^2$, $\theta_* \equiv 100 r_s/D_A(z_*)$, τ, n_s, $\ln(10^{10} A_s)$, $m_{\nu,\,\mathrm{sterile}}$ and λ. We also fixed $n = 2$ in gravity law (4).

In our simulations we use measurement of the CMB from the Planck satellite[14] supplemented with the low-ℓ polarization measurements[15] and extended with CMB measurements at high-ℓ by the Atacama Cosmology Telescope[16] and the South Pole Telescope.[17] We also consider different measurements of baryon acoustic oscillation: the LOWZ[18] and CMASS[19] samples of BOSS (SDSS DR11) in the redshift range $0.15 < z < 0.43$ and $0.43 < z < 0.7$, respectively, and also the 6dF Galaxy Survey[20] at $z = 0.106$. In addition, we use the Hubble constant measurement[21] and the full-sky lensing potential from maps obtained by Planck Collaboration.[22] Finally, we include observations of galaxy clusters[23] in resent redshifts. In the latter data we use a sample of 86 massive galaxy clusters in the ranges $z < 0.2$ and $z \approx 0.4 - 0.9$ with masses measured with about 10 % accuracy by the *Chandra* X-ray telescope (the subsample of distant massive clusters was taken from the *400d* X-ray galaxy cluster survey[24]).

On the left panel in Fig. 2 we see the modified gravity influence on the

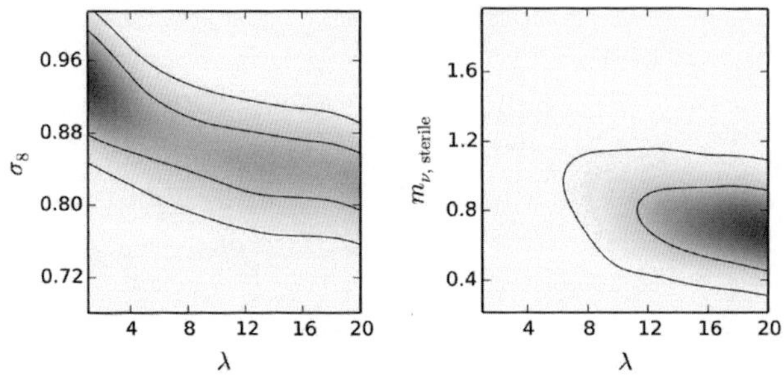

Fig. 2. The regions of parameter space consistent with full cosmological data without measurements of cluster mass function (left panel) and with that (right panel) at the 65 % and 95 % confidence levels for $f(R)$ gravity in σ_8-λ (left panel), $m_{\nu,\,\mathrm{sterile}}$-$\lambda$ (right panel) planes assuming one massive sterile and three massless active neutrinos.

growth of matter density contrast using all data sets without galaxy clusters. In order to constrain the value of σ_8 (matter density contrast averaged by regions with the size $8h^{-1}$ obtained in liner regime) we take the cluster mass function measurements in our consideration. From the right panel of Fig. 2 we deduce that degeneracy between the mass of sterile neutrino $m_{\nu,\,\mathrm{sterile}}$ and the free parameter of $f(R)$ gravity λ is not so prominent as suggested before. Moreover, modified gravity consideration changes the constraint on the sterile neutrino mass insignificantly: $0.47 \text{ eV} < m_{\nu,\,\mathrm{sterile}} < 1 \text{ eV}$ (2σ) in $f(R)$ model in contrast to $0.45 \text{ eV} < m_{\nu,\,\mathrm{sterile}} < 0.92 \text{ eV}$ (2σ) in ΛCDM.

We can also explore the function form $f(R)$ by getting a constraint on the $f(R)$ gravity parameter λ. We find $\lambda > 9.4$ (2σ) in the case of the fourth massive sterile neutrino and others taken massless. When the systematic uncertainty in galaxy cluster mass measurements $\delta M/M \approx 0.09$[23] is included in the likelihood functions, the constraint is relaxed: $\lambda > 8.2$ (2σ).

Nevertheless, the Universe described by modified gravity with one massive sterile neutrino remains slightly more preferable by 1.3σ confidence level as compared to ΛCDM consideration. Moreover, if the sterile state of mass $\approx 1.5\,\mathrm{eV}$ really exists for explanation of various anomalies in neutrino oscillation experiments, the $f(R)$ gravity yields significant improvement that corresponds to $\chi^2 = 19.05$ with one degree of freedom.

5. Conslusion

We find that 1.5 eV sterile neutrino is much better consistent with the $f(R)$ gravity rather than with the standard ΛCDM model. Besides, if the mass of sterile neutrino is free parameter, modified gravity improves the maximum likelihood slightly in comparison with the standard cosmology.

References

1. A. S. Chudaykin, A. A. Starobinsky, D. S. Gorbunov and R. A. Burenin, Cosmology based on $f(R)$ gravity with $\mathcal{O}(1)$ eV sterile neutrino, *J. Cosmol. Astropart. P.* **05**, p. 004 (2015).
2. S. Nojiri and S. D. Odintsov, Introduction to modified gravity and gravitational alternative for dark energy, *Int. J. Geom. Meth. Mod. Phys.* **4**, p. 115 (2007).
3. A. A. Starobinsky, Disappearing cosmological constant in $f(R)$ gravity, *JETP Lett.* **86**, p. 157 (2007).
4. H. Motohashi, A. A. Starobinsky and J. Yokoyama, Matter power spectrum in $f(R)$ gravity with massive neutrinos, *Prog. Theor. Phys.* **124**, p. 541 (2010).
5. Y. I. Izotov, T. X. Thuan and N. G. Guseva, A new determination of the primordial He abundance using the HeI 10830A emission line: cosmological implications, *Mon. Not. Roy. Astron. Soc.* **445**, p. 778 (2014).
6. K. A. Olive *et al.*, Review of Particle Physics, *Chin. Phys. C* **38**, p. 090001 (2014).
7. P. Anselmann *et al.*, First results from the Cr-51 neutrino source experiment with the GALLEX detector, *Phys. Lett. B* **342**, p. 440 (1995).
8. D. N. Abdurashitov *et al.*, The Russian-American gallium experiment (SAGE) Cr neutrino source measurement, *Phys. Rev. Lett.* **77**, p. 4708 (1996).
9. C. Giunti and M. Laveder, Statistical Significance of the Gallium Anomaly, *Phys. Rev. C* **83**, p. 065504 (2011).
10. S. A. Appleby, R. A. Battye and A. A. Starobinsky, Curing singularities

in cosmological evolution of F(R) gravity, *J. Cosmol. Astropart. P.* **1006**, p. 005 (2011).
11. A. D. Felice and S. Tsujikawa, $f(R)$ theories, *Living Rev. Rel.* **13**, p. 3 (2010).
12. A. C. A. Lewis and A. Lasenby, Efficient Computation of Cosmic Microwave Background Anisotropies in Closed Friedmann-Robertson-Walker Models, *Astrophys. J* **538**, p. 473 (2000).
13. A. Lewis and S. Bridle, Cosmological parameters from CMB and other data: a Monte-Carlo approach, *Phys. Rev. D* **66**, p. 103511 (2002).
14. P. A. R. Ade *et al.*, Planck 2013 results. I. Overview of products and scientific results, *Astron. Astrophys.* **571**, p. A1 (2014).
15. C. L. Bennett *et al.*, Nine-year Wilkinson Microwave Anisotropy Probe (WMAP) Observations: Final Maps and Results, *Astrophys. J. Suppl.* **208**, p. 20 (2013).
16. S. Das *et al.*, The Atacama Cosmology Telescope: temperature and gravitational lensing power spectrum measurements from three seasons of data, *J. Cosmol. Astropart. P.* **1404**, p. 014 (2014).
17. R. Keisler *et al.*, A Measurement of the Damping Tail of the Cosmic Microwave Background Power Spectrum with the South Pole Telescope, *Astrophys. J.* **743**, p. 28 (2011).
18. J. K. Parejko *et al.*, The clustering of galaxies in the SDSS-III Baryon Oscillation Spectroscopic Survey: the low redshift sample, *Mon. Not. Roy. Astron. Soc.* **429**, p. 98 (2013).
19. C. Maraston *et al.*, Stellar masses of SDSS-III BOSS galaxies at z$\sim$0.5 and constraints to galaxy formation models, *Mon. Not. Roy. Astron. Soc.* **435**, p. 2764 (2013).
20. D. H. Jones *et al.*, The 6dF Galaxy Survey: final redshift release (DR3) and southern large-scale structures, *Mon. Not. Roy. Astron. Soc.* **399**, p. 683 (2009).
21. A. G. Riess *et al.*, A 3% Solution: Determination of the Hubble Constant with the Hubble Space Telescope and Wide Field Camera 3, *Astrophys. J.* **732**, p. 119 (2011).
22. P. A. R. Ade *et al.*, Planck 2013 results. XVII. Gravitational lensing by large-scale structure, *Astron. Astrophys.* **571**, p. A17 (2014).
23. A. A. Vikhlinin *et al.*, Chandra Cluster Cosmology Project II: Samples and X-ray Data Reduction, *Astrophys. J.* **692**, p. 1033 (2009).
24. R. A. Burenin *et al.*, The 400 Square Degree ROSAT PSPC Galaxy Cluster Survey: Catalog and Statistical Calibration, *Astrophys. J. Suppl.* **172**, p. 561 (2007).

Variations of fundamental constants and multidimensional gravity

K. A. Bronnikov[a,b,c,*] and M. V. Skvortsova[a]

[a] *Center for Gravitation and Fundamental Metrology, VNIIMS,*
46 Ozyornaya St., Moscow 119361, Russia
[b] *Institute of Gravitation and Cosmology, PFUR,*
6 Miklukho-Maklaya St., Moscow 117198, Russia
[c] *National Research Nuclear University "MEPhI"*
(Moscow Engineering Physics Institute), Kashirskoe sh. 31, Moscow 115409, Russia
**E-mail: kb20@yandex.ru*

We try to explain the recently reported large-scale spatial variations of the fine structure constant α, in agreement with other cosmological observations, in the framework of curvature-nonlinear multidimensional gravity. The original theory is reduced to a scalar-tensor theory in four dimensions, and the corresponding isotropic cosmologies are considered in both Einstein and Jordan conformal frames. In the Jordan frame one obtains simultaneous variations of α and the gravitational constant G, equal in magnitude. Long-wave small inhomogeneous perturbations of isotropic models allow for explaining spatial variations of α.

Keywords: Varying fundamental constants; multidimensional gravity; cosmological models; perturbations.

1. Introduction

The recently reported[1] spatial variations of the fine structure constant α are not yet confirmed by other research groups but are actively discussed. According to Ref. 1, joint data from the Keck and VLT telescopes have revealed a dipole type anisotropy (termed "the Australian dipole") in the deflections $\delta\alpha$ from the value α_0 measured on Earth, described as

$$\delta\alpha/\alpha_0 = (1.10 \pm 0.25) \times 10^{-6}\, r \cos\psi \qquad (1)$$

(c.l. 4.1σ), where r is the distance from Earth measured in billions of light years and ψ is the angle between the line of sight and the dipole axis, located at a declination of $-61 \pm 9\,\mathrm{deg}$ and right ascension of $17.3 \pm 0,6$ hours. It is of interest that such spatial variations could be detected by terrestrial instruments with accuracies raised by a factor of 100 from the present-day level.[2]

Laboratory measurements constrain today's variations of α by[3]

$$\dot{\alpha}/\alpha = (-1.6 \pm 2.3) \times 10^{-17}/\text{yr}. \tag{2}$$

Of the same order of magnitude are constraints from decay product analysis of the Oklo natural reactor that operated about 2×10^9 years ago:[4]

$$-0.35 \times 10^{-18}/\text{yr} \le \dot{\alpha}/\alpha \le 5 \times 10^{-18}/\text{yr} \tag{3}$$

(assuming that α changed uniformly in this period). These data are illustrated in Fig. 1: on Earth, at least since the Oklo times, α is constant up to $\sim 10^{-17}$/yr, whereas the quasar data indicate a variation rate within 10^{-15}/yr for about 10^{10} years. It is also possible that all variations are purely spatial and seem to be time-dependent due to finiteness of the velocity of light.

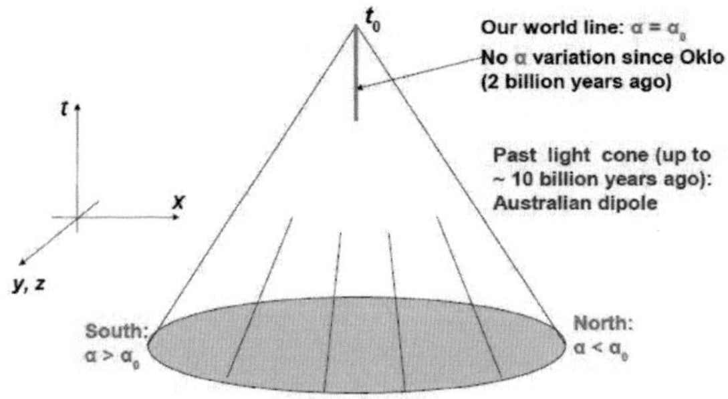

Fig. 1. Empirical data on variations of α.

Theoretical explanations of these data[5–9] are mostly based on general relativity with scalar fields introduced *ad hoc*. However, similar results naturally follow[10,11] from curvature-nonlinear multidimensional gravity.[12,13] This approach, without fixing the number of dimensions, deals with multidimensional Lagrangians containing different curvature-nonlinear terms (functions of the Ricci scalar, the Ricci tensor squared, the Kretschmann invariant), and employs the assumption of slow changes of the relevant functions on the Planck scale). Depending on further assumptions, this approach leads to a variety of models[12–14] useful for solving a wide range of problems. Thus, it has been applied to a description of early inflation and the present accelerated expansion of the Universe in a unified way[15] and to a possible explanation of the origin of the electroweak theory.[16] Concerning its application to FPC variations, one can note that a common advantage of multidimensional theories is their ability to describe variations of different constants in a unified way as a consequence of a varying size of extra dimensions.[17]

Multidimensional gravity, being reduced to four dimensions, takes the form of a (multi-)scalar-tensor theory which can be written in Einstein or Jordan conformal frame, and it is by now uncertain which of them corresponds to observations.[18] In Jordan's frame, in addition to α, the gravitational constant G also varies. Such variations have not yet been discovered, there are only upper bounds of which the tightest one is[10]

$$-4.2 \times 10^{-14}/\mathrm{yr} < \dot{G}/G < +7.5 \times 10^{-14}/\mathrm{yr}. \tag{4}$$

Since variations of α are at most $\sim 10^{-15}$ per year, variations of G of the same magnitude, if any, are in agreement with (4).

2. Basic Equations

Consider a curvature-nonlinear theory of gravity with the action

$$S = \int \sqrt{{}^D g}\, d^D x \,(L_g + L_m), \qquad L_g = F(R) + c_1 R^{AB} R_{AB} + c_2 \mathcal{K}, \tag{5}$$

where $A, B, \ldots = \overline{0, D-1}$, $F(R)$ is an arbitrary function of D-dimensional scalar curvature R, $\mathcal{K} = R^{ABCD} R_{ABCD}$ is the Kretschmann invariant, c_1, $c_2 = \mathrm{const}$, L_m is a matter Lagrangian, and ${}^D g = |\det(g_{AB})|$.

The metric in $(D = 4 + d_1)$-dimensional space-time is taken in the form

$$ds^2 = g_{\mu\nu} dx^\mu dx^\nu + \mathrm{e}^{2\beta(x)} b_{ab} dx^a dx^b \tag{6}$$

where x^μ are the observable four coordinates while b_{ab} is the metric of a compact d_1-dimensional space of constant curvature $K_1 = \pm 1$.

It is hard to solve the field equations of the full theory (5). According to the purpose of this study, we perform the following steps:

(a) Integrate out the extra dimensions and express everything in terms of 4D quantities and $\beta(x)$; thus we have

$$R = R_4 + \phi + 2d_1 \Box\beta + d_1(d_1+1) g^{\mu\nu} \partial_\mu\beta \partial_\nu\beta, \tag{7}$$

$$\phi(x) := K_1 d_1 (d_1 - 1)\, \mathrm{e}^{-2\beta(x)}, \tag{8}$$

where R_4 and $\Box$ are the 4D scalar curvature and d'Alembertian operator. The action in four dimensions has the form

$$S = \frac{1}{2}\mathcal{V}[d_1] \int \sqrt{{}^4 g}\, d^4 x\, \mathrm{e}^{d_1\beta}\, [L_g + L_m], \qquad {}^4 g = |\det(g_{\mu\nu})|, \tag{9}$$

$\mathcal{V}[d_1]$ being the volume of a compact d_1-dimensional space of unit curvature.

(b) Apply the slow-change approximation[12,13] (compared with the D-dimensional Planck scale), *i.e.*, consider each derivative ∂_μ to be proportional to a small parameter ε, and neglect anything of order higher than $O(\varepsilon^2)$. As a result, we obtain a 4D scalar-tensor theory in its Jordan frame.

(c) Apply the standard conformal mapping leading to the Einstein frame, that is, reduce the original theory to general relativity with a minimally coupled scalar field with a nonzero potential.

(d) Construct isotropic cosmological models with the scalar field ϕ defined in (8) (or equivalently β) as the only source. Estimate possible variations of the fine structure constant α assuming that the electromagnetic field $F_{\mu\nu}$ appears in the original D-dimensional theory with the Lagrangian

$$L_{\rm em} = \alpha_1^{-1} F_{\mu\nu}F^{\mu\nu}, \qquad \alpha_1 = \text{const.} \tag{10}$$

In four dimensions emerges the factor $\mathrm{e}^{d_1\beta}$ because $\sqrt{{}^D g} = \sqrt{{}^4 g}\,\mathrm{e}^{d_1\beta}$, which ultimately leads to variations of the effective fine structure constant:

$$\alpha/\alpha_0 = \mathrm{e}^{d_1(\beta_0-\beta)}, \tag{11}$$

where, as usual, the index 0 marks the values of any quantities taken where and when the observation is carried out. We assume that the field $F_{\mu\nu}$ is weak enough and does not affect the gravitational field.

(e) Since isotropic models are unable to explain a spatial dependence of any physical quantity, we consider small perturbations of β varying from point to point in space, which in turn lead to spatial variations of α.

At step (c) we obtain the following Einstein-frame Lagrangian:[10,12]

$$L = \widetilde{R}_4 + K_{\rm E}(\phi)\widetilde{g}^{\mu\nu}\partial_\phi\partial_\nu\phi - 2V_{\rm E}(\phi) + \widetilde{L}_m, \tag{12}$$

$$K_{\rm E}(\phi) = \frac{1}{4\phi^2}\left[6\phi^2\left(\frac{F''}{F'}\right)^2 - 2d_1\phi\frac{F''}{F'} + \frac{1}{2}d_1(d_1{+}2) + \frac{4(c_1{+}c_2)\phi}{F'}\right], \tag{13}$$

$$-2V_{\rm E}(\phi) = (\operatorname{sign} F')\frac{\mathrm{e}^{-d_1\beta}}{F'(\phi)^2}[F(\phi) + c_*\phi^2], \tag{14}$$

where the tilde marks quantities obtained from or with the Einstein-frame metric $\widetilde{g}_{\mu\nu}$, $F = F(\phi)$, $F' = dF/d\phi$, etc., and $c_* = c_1/d_1 + 2c_2/[d_1(d_1-1)]$.

Further on we suppose that the function $F(R)$ has the simple form $F(R) = R^2 - 2\Lambda_D$, where Λ_D is the initial D-dimensional cosmological constant, and consider model I with $\Lambda_D > 0$ and model II with $\Lambda_D = 0$.

3. Isotropic Cosmology

Depending on the choice of $F(R)$, the constants c_1 and c_2 and the matter Lagrangian, the theory (5) can lead to a great variety of cosmological models.[12,13] Here we briefly describe some particular models[10,11] related to the generic behavior of the potential V_E near $\phi = 0$. If $\phi \to 0$ at late times, the extra dimensions grow: $b = e^{\beta} \sim 1/\sqrt{|\phi|} \to \infty$. Such models can be viable if the growth is sufficiently slow and the size $b(t)$ does not reach detectable values. The admissible range of $b(t)$ comprises as many as 16 orders of magnitudes if the D-dimensional Planck length l_D coincides with its 4D counterpart: the upper bound corresponds to lengths $\sim 10^{-17}$ cm or energies of a few TeV. This estimate changes if l_D is larger.

We try to describe the modern stage of the Universe evolution on the basis of the theory (12) assuming the 4D Einstein-frame metric in the spatially flat FRW form $d\widetilde{s}_4^2 = dt^2 - a^2(t)d\vec{x}^2$ with the scale factor $a(t)$. It is convenient to rewrite the Lagrangian (12) in terms of β instead of ϕ:

$$L = \widetilde{R}_4 + 2K_0(\partial\beta)^2 - 2V(\beta) + \widetilde{L}_m, \tag{15}$$

$$K_0 = \frac{1}{4}\big[d_1^2 - 2d_1 + 12 + 4(c_1 + c_2)\big]. \tag{16}$$

The Einstein-scalar equations with the unknowns $\beta(t)$ and $a(t)$ are

$$3\frac{\dot{a}^2}{a^2} = K_0\dot{\beta}^2 + V(\beta), \qquad 2K_0\Big(\ddot{\beta} + 3\frac{\dot{a}}{a}\dot{\beta}\Big) = -\frac{dV}{d\beta}. \tag{17}$$

The form of the potential $V(\beta)$ is different in models I and II.

Model I: $F(R) = R^2 - 2\Lambda$, $\Lambda_D > 0$. In this case,[10]

$$V(\beta) = V_{10}e^{-2\bar{d}\beta}, \qquad V_{10} = \frac{\Lambda_D}{4d_1^2(d_1-1)^2}, \qquad \bar{d} = \frac{d_1 - 4}{2}. \tag{18}$$

and $d_1 > 4$ is required. Assuming slow rolling ($|\ddot{\beta}| \ll 3H\dot{\beta}$, $K_0\dot{\beta}^2 \ll V(\beta)$, $H = \dot{a}/a$ is the Hubble parameter), we obtain a solution describing power-law inflation in our space and much slower expansion in the extra one:

$$a(t) = a_1(t + t_1)^p, \qquad a_1 = \text{const}, \qquad p = K_0/\bar{d}^2. \tag{19}$$

$$b(t) = e^{\beta} = b_1(t + t_1)^{1/\bar{d}}, \tag{20}$$

where t_1 is an integration constant. Our accelerating Universe is fairly well described if $p \gg 1$ (and then the slow-rolling conditions hold), $t_* = t_1 + t_0 = pt_H \gg t_H = 1/H_0$, t_0 is the present time and t_H is the Hubble time. With $p \gg 1$, the power-law expansion is almost exponential: the effective equation-of-state parameter is $w = \text{pressure/density} = -1 + 2/(3p)$.

The smallness of the observed variations of α constrains the initial parameters c_1 and c_2. Indeed, according to (11),

$$\frac{\alpha}{\alpha_0} = \left(\frac{t+t_1}{t_0+t_1}\right)^{-2d_1/(d_1-4)} \approx 1 - \frac{2d_1}{d_1-4}\frac{t-t_0}{t_*}, \tag{21}$$

so that $\dot{\alpha}/\alpha \sim 10^{-10}/p$ per year. To obtain 10^{-17} per year according to (2), one needs $p \gtrsim 10^7$ and, by (19) and (16), $c_1 + c_2 \gtrsim 10^6$. This inequality is, however, substantially relaxed in the perturbed model considered below.

If we assume that our observations take place in Jordan's frame where the metric $g_{\mu\nu}$ is simply the 4D part of (6), it turns out that model I is not viable: the acceleration is too large, the parameter $-w \gg 1$, and the expansion ends with a Big Rip singularity ($a \to \infty$ at finite cosmic time).[11]

Model 2: $\Lambda_D = 0,\ F(R) = R^2$. In this case,

$$V(\beta) = V_{20}\mathrm{e}^{-d_1\beta}, \qquad V_{20} = -\frac{1+c_*}{8} = -\frac{1}{8}\left[1 + \frac{c_1}{d_1} + \frac{c_2}{d_1(d_1-1)}\right]. \tag{22}$$

We assume $V_{20} > 0$, which is possible under a proper choice of c_1 and c_2. The slow-rolling solution similar to (19), (20) is[11]

$$a(t) = a_2(t+t_2)^q, \qquad q = \frac{1}{d_1^2}[d_1^2 - 2d_1 + 12 + 4(c_1+c_2)]; \tag{23}$$

$$b(t) \equiv \mathrm{e}^{\beta} = b_0\left(\frac{t+t_2}{t_{**}}\right)^{2/d_1}, \quad b_0 = \left(\frac{t_{**}\sqrt{V_{20}}}{\sqrt{3q}}\right)^{2/d_1}, \tag{24}$$

where a_2 and t_2 are integration constants and $t_{**} = t_0 + t_2 = 1/H_0 = qt_H$. The slow-rolling conditions hold as long as $q \gg 1$, and $w = -1 + 2/(3q)$, so the model is good enough for the present state of the Universe. A sufficiently small size b_0 of the extra dimensions requires $d_1 > 8$.

For time variations of α we now have

$$\frac{\alpha}{\alpha_0} = \left(\frac{b}{b_0}\right)^{-d_1} \approx 1 - 2\frac{t-t_0}{t_*}, \qquad \frac{\dot{\alpha}}{\alpha_0} \approx -\frac{2H_0}{q} = -\frac{2}{qt_H}. \tag{25}$$

The empirical bounds on $\dot{\alpha}$ require $q \gtrsim 10^7$, leading, as in model I, to the constraint $c_1 + c_2 \gtrsim 10^6$.

This model turns out to be equally viable in Einstein and Jordan frames.[11] The latter predicts variations of the gravitational constant G: being constant by definition in the Einstein frame, it changes in Jordan's by the same law as α,

$$G/G_0 = \alpha/\alpha_0 = \mathrm{e}^{d_1(\beta_0-\beta)}. \tag{26}$$

Since the observational constraints on G are much weaker than those on α, see (4), if our model well describes the behavior of α, the observational bounds on G variations hold automatically.

4. Perturbed Isotropic Models and Spatial Variations of α

Let us now try to describe spatial variations of α by assuming a more general form of the Einstein-frame metric,

$$ds_{\rm E}^2 = {\rm e}^{2\gamma}dt^2 - {\rm e}^{2\lambda}dx^2 - {\rm e}^{2\eta}(dy^2 + dz^2), \tag{27}$$

where γ, λ, η aro functions of x and t. The conditions that the metric (27) only slightly differs from homogeneity and isotropy are

$$\gamma = \delta\gamma(x,t), \qquad \lambda = \ln a(t) + \delta\lambda(x,t), \qquad \eta = \ln a(t) + \delta\eta(x,t),$$

where all "deltas" are small and are considered in the linear approximation. We also replace the field $\beta(t)$ with $\beta(t)+\delta\beta(x,t)$. Then, starting from (12), using the Einstein equations and the gauge $\delta\eta = 0$, we can exclude $\delta\gamma$ and $\delta\lambda$ and arrive at the wave equation for $\delta\beta$:

$$\delta\ddot\beta + \frac{3\dot a}{a}\delta\dot\beta - \frac{1}{a^2}\delta\beta'' + \delta\beta\left[\frac{2\dot\beta^2}{H^2}VK_0 + \frac{2\dot\beta}{H}V_\beta + \frac{1}{2K_0}V_{\beta\beta}\right] = 0 \tag{28}$$

with an arbitrary constant K_0 and an arbitrary potential $V(\beta)$. Assuming that the background quantities $a(t)$ and $\beta(t)$ are known, it remains to find a solution for $\delta\beta$ which, being added to the background $\beta(t)$, would be able to account for the observed picture of variations of α.

Since the background is x-independent, we can separate the variables by assuming $\delta\beta = Y(t)\sin k(x+x_0)$ where k has the meaning of a wave number, and the constant x_0 is an arbitrary "phase". Then, instead of solving the emerging equation for $Y(t)$, since we are describing time intervals much shorter than the characteristic times t_* or t_{**} of the above isotropic models, we can restrict ourselves to the initial data taken at the present moment t_0, i.e., suppose simply $Y = Y_0 + Y_1(t - t_0)$, $Y_{0,1} = \text{const}$.

Considering, for certainty, model I, in this approximation we obtain the following expression for variations of α:

$$\alpha/\alpha_0 \approx 1 + d_1[\beta_0 - \beta(t) - (Y_0 + Y_1(t - t_0))\sin k(x + x_0)] + O(\epsilon^2), \tag{29}$$

where $O(\epsilon^2)$ means $O((t - t_0)^2/t_*^2)$ and $\beta(t)$ is the background solution given by (20). Requiring $\alpha/\alpha_0 = 1 + O(\epsilon^2)$ at $x = 0$ (that is, no variation of α on Earth in the first-order approximation), we obtain the condition

$$Y_1 \sin(kx_0) = -1/(\bar d\, t_*). \tag{30}$$

On the past light cone of the point $t = t_0$, $x = 0$, assuming a very long wave of perturbations, $kx \sim (t - t_0)/t_* \ll 1$, we obtain[10,11]

$$\alpha/\alpha_0 \approx 1 - d_1 Y_0 \sin(kx_0) + d_1 Y_0\, kx \cos(kx_0) + O(\epsilon^2), \qquad (31)$$

without time dependence up to $O(\epsilon^2)$ while the x dependence is linear. The expression (31) should be fit with (1) to explain the spatial variations of α in agreement with observations[1] by choosing proper relations between the input parameters d_1, c_1, c_2 and the constants k, x_0, Y_0, Y_1, which can be ascribed to slightly inhomogeneous initial conditions at primordial inflation. A small constant shift in the α value at $x = 0$ in (31) is unimportant.

The condition $c_1 + c_2 \gtrsim 10^6$ is now cancelled since small (if any) variations of α on Earth are provided by the equality (30), actually due to a proper phase in the perturbation wave. Instead, to provide the validity of our condition $p \gg 1$, it is sufficient to require $c_1 + c_2 \gtrsim 1000$.

It can also be verified[11] that the perturbed model II, in both Einstein and Jordan pictures, can be used to account for the "Australian dipole" equally well in terms of purely spatial variations of α. In Jordan's picture, the same variations are predicted for the gravitational constant G.

The above models are very simple, approximate and tentative but still work fairly well at times not too far from the present epoch, explaining both the accelerated expansion of the Universe and variations of α. They do not take into account other kinds of matter than the scalar field β (which may be called a quintessence of multidimensional origin); their inclusion must lead to a better description on a wider time interval.

A possible discovery of parallel evolution of constants like α and G would be a strong argument in favor of multidimensional physics.

In conclusion, it is appropriate to mention another approach to multidimensional gravity that also does not *a priori* fix the number of dimensions, it deals with product space-times consisting of a number of factor spaces and such sources of gravity as fields of antisymmetric forms of different ranks interacting with sets of scalar fields.[17,20–24] In this apporach, a large number of exact solutions were obtained, and FPC variations are also predicted. More details and further references can be found in the contribution by V.D. Ivashchuk and V.N. Melnikov to the present volume.

Acknowledgments

The work of KB was performed in part within the framework of the Center FRPP supported by MEPhI Academic Excellence Project (contract No. 02.a03.21.0005, 27.08.2013).

References

1. J. K. Webb et al., *Phys. Rev. Lett.* **107**, 191101 (2011).
2. V. A. Dzuba, M. S. Safronova, U. I. Safronova and V. V. Flambaum, *Actinide ions for testing the spatial α variation hypothesis.* ArXiv: 1508.07681.
3. T. Rosenband et al., *Phys. Rev. Lett.* **98**, 220801 (2007).
4. M.S. Onegin, *Mod. Phys. Lett. A* **27** (40), 1250232 (2012).
5. T. Chiba and M. Yamaguchi, *JCAP* **1103**, 044 (2011); ArXiv: 1102.0105.
6. K. A. Olive, M. Peloso, and A. J. Peterson, *Where are the walls?* ArXiv: 1204.4391.
7. K. Bamba, S. Nojiri, S.D. Odintsov, *Phys. Rev. D* **85**, 044012 (2012).
8. J. D. Barrow and S. Z. W. Lip, *Phys. Rev. D* **85**, 023514 (2012).
9. A. Mariano and L. Perivolaropoulos, *Phys. Rev. D* **87**, 043511 (2013).
10. K. A. Bronnikov, V. N. Melnikov, S. G. Rubin, and I. V. Svadkovsky, *Gen. Rel. Grav.* **45**, 2509 (2013).
11. K.A. Bronnikov and M.V. Skvortsova. *Grav. Cosmol.* **19**, 114 (2013).
12. K. A. Bronnikov and S. G. Rubin, *Phys. Rev. D* **73**, 124019 (2006).
13. K. A. Bronnikov and S. G. Rubin, *Black Holes, Cosmology and Extra Dimensions* (World Scientific, Singapore, 2012).
14. K. A. Bronnikov, R. V. Konoplich and S. G. Rubin, *Class. Quantum Grav.* **24**, 1261 (2007).
15. K. A. Bronnikov, S. G. Rubin and I. V. Svadkovsky, *Phys. Rev. D* **81**, 084010 (2010).
16. S. V. Bolokhov, K. A. Bronnikov and S. G. Rubin, *Phys. Rev. D* **84**, 044015 (2011).
17. V. N. Melnikov, in: *Albert Einstein Century Int. Conf.* (Eds. J.-M. Alimi and A. Fuzfa, AIP Conf. Proc. 861, 2006), p. 109.
18. K. A. Bronnikov and V. N. Melnikov, *Gen. Rel. Grav.* **33**, 1549 (2001).
19. E. V. Pitjeva, *Solar System Research* **47**, 386 (2013).
20. V. D. Ivashchuk and V. N. Melnikov, *Class. Quantum Grav.* **18**, R87 (2001).
21. K. A. Bronnikov, V. D. Ivashchuk and V. N. Melnikov, *Grav. Cosmol.* **3**, 203 (1997).
22. K. A. Bronnikov, *Grav. Cosmol.* **4**, 49 (1998).
23. K. A. Bronnikov and V. N. Melnikov, *Nucl. Phys. B* **584**, 434 (2000).
24. V. D. Ivashchuk and V. N. Melnikov, *Grav. Cosmol.* **20**, 182 (2014).

TianQin mission concept*

Jun Luo[†]

Sun Yat-sen University, 135 West Xingang Rd., Guangzhou, 510275, China
[†]*E-mail: junluo@sysu.edu.cn*

Jianwei Mei, Chenggang Shao, Yan Wang, Hsien-Chi Yeh, and Ze-bing Zhou

MoE Key Laboratory of Fundamental Quantities Measurement, School of Physics, Huazhong University of Science and Technology, 1037 Luoyu Rd., Wuhan 430074, P.R. China

Vadim Milyukov and Michail Sazhin

Sternberg Astronomical Institute, Moscow State University, Moscow 119992, Russia

We introduce the mission concept of TianQin, a spaceborne gravitational waves detector. TianQin will be consisted of three drag-free satellites, forming an equilateral-triangle constellation and orbiting Earth with an altitude of about 10^5 km. The major scientific goal of TianQin is to detect possible gravitational radiation from Hm Cnc (RX J0806.3+1527), a candidate ultracompact white dwarf binary with a putative orbital period of about 321.5 s. Based on current best models of the binary system, we expect SNR $\geq$10 after 3 month of observation, given that the noise is dominated by 1 pm $\mathrm{Hz}^{-1/2}$ in position noise and 3×10^{-15} m s^{-2} $\mathrm{Hz}^{-1/2}$ in residual acceleration. The progress of preliminary study will be presented in the conference.

Keywords: TianQin; spaceborne gravitational wave detector.

1. Introduction

The theory of general relativity has been 100 years old. During the past 100 years, the theory has passed many non-trivial tests, such as orbital precession of Mercury, deflection of light by Sun, gravitational redshift, Shapiro delay, frame dragging, gravitational lensing and so on. Despite such great success, however, there is one notable missing piece in the experimental test of general relativity — the direct detection of gravitational wave. Indirect

*This talk is based on Ref. 1. TianQin in Chinese means a zither in space. So the TianQin instrument is metaphorically seen as a zither that is being played by the nature itself through gravitational waves.

evidence for the existence of gravitational wave has been provided by looking at the orbital decay of binary systems, such as the famous Hulse-Taylor binary pulsar. But a direct detection of gravitational wave is still lacking. There are several ground based gravitational wave experiments that are currently operating. People are also looking at pulsars and the cosmological microwave background for evidence of gravitational waves. Many proposals for detecting gravitational waves in space have been prepared. See[1] for a list of references.

We propose a new and realistic gravitational wave detection mission, the TianQin experiment. TianQin will be consisted of three satellites, forming an equilateral-triangle constellation and orbiting Earth with an altitude of about 10^5 km. The inter-satellite distance will be monitored using laser interferometry and the disturbance reduction system will be used to reduce the effect of non-gravitational forces. The major scientific goal of TianQin is to detect possible gravitational radiation from Hm Cnc (RX J0806.3+1527), a candidate ultracompact white dwarf binary with an orbital period of about 321.5 s. Based on current best models of the binary system, we expect signal-to-noise ratio SNR $\geq$10 after 3 month of observation, given that the noise is dominated by 1 pm $\mathrm{Hz}^{-1/2}$ in position noise and 3×10^{-15} m s^{-2} $\mathrm{Hz}^{-1/2}$ in residual acceleration.

Here we introduce the preliminary mission concept and explain some basic considerations went into the construction of such a mission concept.

2. The Preliminary Concept of TianQin

All existing proposals for spaceborne gravitational wave missions aim to build a gravitational wave observatory, despite the fact that no direct detection of gravitational wave in space has ever been achieved. As a result, all of them face significant technological challenges. So far the only spaceborne gravitational wave mission that has some chance of getting launched is eLISA from ESA.[2] The eLISA pathfinder will be launched by the end of this year. The mission will last 9 months, and will test essential technologies such as the dragfree control and high precision laser ranging technology.

In view of this fact, we want our proposal to be very realistic, focusing on making a reliable detection first.

To find an easiest way to detect gravitational waves, we try to identify one most accessible source, and then optimize the design of the experiment towards the chosen source. Three identical satellites in equilateral triangle formation will be used for TianQin, so as to make the experiment as simple as possible. The detector plane (the plane formed by the three satellites) is chosen to face the chosen source, so that the response can be maximized. By focusing on a most accessible source, the required sensitivity of TianQin can be significantly reduced, which then allows us to use shorter armlengths and geocentric orbits. The latter can help cut the cost significantly and lower the demand on satellite technology.

2.1. *The reference source*

TianQin will be tuned toward a specially chosen source. Apart from being able to produce strong gravitational waves, the source should also: (1) be constantly available, (2) have a relatively short orbital period, and (3) lay close to the ecliptic.

A lot of work has been done on studying the sources for gravitational waves. In particular, a subset of ultra-compact galactic binaries (known as the LISA verification binaries[3,4]) has been identified to verify the performance of eLISA instruments. TianQin is interested in the same frequency band as LISA/eLISA, and the LISA verification binaries are very good targets for TianQin.

Since the primary goal of TianQin is to detect gravitational waves from one particular source, we simply pick the most accessible among the LISA verification binaries. This leads us naturally to the binary system RX J0806.3+1527 (also called HM Cnc, we will call it J0806 for short hereafter).

It is possible that a better source could be found in the near future. In that case, we might want to modify the design of TianQin in accordance with the newly found source. But at present, we will use J0806 as an example to map out the details of the experiment. The experience from working with J0806 can also make the possible transition to a new source a lot smoother. With this said, we will focus on the example J0806 in the following.

RX J0806.3+1527 The system has an orbital period of about 5.4 minutes, corresponding to an orbital frequency 3.11 millihertz, and the corresponding frequency of the radiated gravitational wave is about 6.22 millihertz. The mass of the primary star is about 0.55 solar mass, while the companion mass is about 1/3 to 2/3 that of the primary one. The binary orbit has an inclination of about 38 degree. There is large uncertainty in the distance to J0806, ranging from 50 pc to 5 kpc. The galactic latitude of J0806 is about 20 degree, which means that the distance cannot be too much greater than 5 kpc. J0806 also has the useful feature of being less than 5 degree away from the ecliptic.

With detailed parameters for the source J0806, a rough estimation shows that an armlengths at the order 10^5 km will be enough for an acceptable detection in a few months, assuming reasonable noise levels of various components of the experiment.

2.2. *Orbit*

TianQin will use geocentric orbit to lower the demand on cost and on satellite technology. Choosing geocentric orbits make it possible for us to focus most resource on other key technologies, most importantly the laser interferometer and the disturbance reduction system.

Geocentric orbits has a significant thermal issue with uneven and time varying thermal load on the satellites due to the Sun. But since the source J0806 is very close to the ecliptic and the detector plane is chosen to face the source, the detector plane is nearly vertical to the ecliptic. There are two times each year when the Sun light to the detector is at large angles to the detector plane. During such times, direct Sun shining on the telescopes can be avoided and thermal fluctuation is also minimal. These are quieter time windows where reliable detection can be carried out.

A possible set of TianQin orbits has been identified. They are nearly circular orbits with radius about 10^5 km. The relative velocity (range rate) between each pair of satellites varies no more than 10m/s in five years. Its compensation is well within the reach of present technologies. The variation of subtended angles can be separated into two parts short term fluctuations and long term shifts in the averaged value of the variation. The short term fluctuation has periods at the order of a few days and magnitudes at about 0.1°. This part can be corrected using laser beam pointing control with fast steering mirror. On the other hand, the attitude control of telescope with gimbal mechanism might be needed to deal with the larger variations accumulated through the long term shift. But this can be done at intervals in the order of several months.

2.3. *Requirement on key technologies*

The magnitude of the gravitational waves from J0806 is found to be at the order $10^{-23} \sim 10^{-22}$ (assuming the distance D =5 kpc). Our goal is to have a confident detection within a few months (say 90 days — corresponding to an integrated magnitude $10^{-19}/\text{Hz}^{1/2}$).

The main sources of noise include the position noise and the acceleration noise. For position measurement, we set the noise goal to be 1 pm/Hz$^{1/2}$. To have some design margin, we look for a signal-to-noise ration SNR=10 after 90 days of observation. Depending on the true distance to J0806, the requirement on the residual acceleration is found to be $(34 \sim 3.1) \times 10^{-15}$ m/s^2/Hz$^{1/2}$, corresponding to the distance $D=$ (50 pc $\sim$ 5 kpc).

A preliminary error budget has been given for the laser interferometer and the disturbance reduction system in Ref. 1.

3. Conclusion

The proposed TianQin mission is a space-based detector of gravitational waves in the mHz frequency band. To have a realistic experiment, we choose the most accessible GW source as the primary target and optimize the experiment toward detecting the gravitational waves from this particular source. This makes the experiment more a detector, rather than an observatory.

To reduce mission complexity, we choose geocentric orbits. This makes it possible to focus most of our resource on the development of the laser interferometer and the disturbance reduction system, which are the two key components showing a gap to what is required by the experiment. Nevertheless, we are confident that these two critical systems will reach the required maturity in the near future.

Ultimately, TianQin is expected to serve as a spring board for a fleet of future gravitational wave observatories.

References

1. J. Luo, *et al*, "TianQin: a space-borne gravitational wave detector", Submitted to Class. Quant. Grav.
2. eLISA homepage: `https://www.elisascience.org`.
3. A. Stroeer and A. Vecchio, Class. Quant. Grav. **23**, S809 (2006) [astro-ph/0605227].
4. G. Nelemans, *LISA Verification Binaries*, web interface to database, Radboud University. URL: `http://www.astro.ru.nl/nelemans/dokuwiki/doku.php?id=verification\_binaries:intro`.

Relational-statistical theory of space-time and interactions

Yu. S. Vladimirov

Faculty of Physics, M.V. Lomonosov State University
Leninskie Gory, Moscow 119991 Russia
E-mail: yusvlad@rambler.ru

The relational-statistical approach to the nature of space-time and physical interactions is proposed.

Keywords: Relational approach; space-time; gravitation; physical interactions.

1. Introduction

1.1. For the last 100 years, the principles underlying the basics of general relativity have already been analyzed, and the opportunities of the theory have been apparently exhausted. Invoking new ideas is necessary.

The long-term unsuccessful attempts to create a quantum theory of gravitation, i.e. a combination of the principles of general relativity and quantum theory, leads the author to a strong belief: the solution of this problem is possible only with a solution of a deeper problem, derivation of classical space-time relations from an independent system of more elementary physical factors and concepts of a microcosm instead of traditional acceptance of a priori space-time for all theoretical constructs.

1.2. The analysis of the foundations of conventional ideas of the nature, space-time and physical interactions show that now it is necessary to distinguish three main metaphysical approaches (paradigms) within which researches in the field of fundamental physics: 1) the field-theoretic one, on the basis of which the quantum theory is under construction, 2) the geometric one, within which the general theory of relativity and a number of its natural generalizations is constructed, and 3) the relational one, whose ideological bases were written in the works of G. Leibnitz, E. Mach and some other scientists.

1.3. Within the framework of these paradigms (approaches), there are different interpretations of the three key concepts (categories) of modern physics: space-time, interactions, and matter (particles)

The main basic reason of failures of the attempts of quantization of gravitation consists in a radical difference of concepts on which the synthesized theories are based.

1.4. It has been specified in the works of many authors that the physics of a microcosm in classical space-time representations lose its force, for example, in the book of D.I. Blokhintsev "Space-time in a microcosm". Moreover, in Louis de Broglie, L.I. Mandelshtam, G.F. Chew, P. K. Rashevsky, R. Penrose's works and others, the problem of searching for an independent system of the concepts inherent to physics of a microcosm (including quantum mechanics) were set out, of which it is necessary to bring classical concepts of space-time and classical physics.

1.5. This problem cannot be solved either within the field-theoretic paradigm or within the geometric paradigm since in the the field-theoretic paradigm the existential background, within the geometrical paradigm, which is only about change of properties (metric, topological, etc.), an already set existential continuum is already necessary for a definition of the key concept of a field. This problem can be solved only within the relational paradigm.

2. Relational Approach to Geometry and Physics

2.1. The relational paradigm consists of two inseparably linked parts: 1) the relational approach to the nature of space-time and 2) descriptions of physical interactions on the basis of the concept of action-at-a-distance. The latter is caused by the relational approach with the standard concept of short-range interaction, because in the absence of an *a priori* set existential background fields have no ability to propagate.

2.2. For research within the relational paradigm it was necessary to find a suitable mathematical formalism which for a long time was not sufficient for supporters of the relational approach. The foundation of such mathematical formaslism was laid in the so-called theory of physical structures offered in the late sixties in the works of Yu.I. Kulakov and G.G. Mikhailichenko. The foundation of the universal theory of relations was laid, the relation, the key concept of the whole relational approach. I.E. Tamm positively spoke of the advantages of this direction of research.

2.3. In the works of Kulakov's group, two types of new fundamentals in the theory of relations were developed: 1) on one set of elements (the unary systems of real relations) and 2) on two sets of elements (the binary systems of real relations). It was shown that, first, there are no substantial theories on three and more sets of elements and, second, that from binary systems of relations, by peculiar gluing together of elements of two sets, it is possible to pass on to unary systems of relations.

2.4. We will briefly characterize the basic concepts and an essence of the theory of systems of relations. It is set on one or two sets of certain elements (of abstract nature). Between couples of any elements of one set or any elements of two opposite sets, numbers — pair relations are set. It is postulated that there is an algebraic law connecting all pair relations between r elements of one set (for unary systems) or pair relations between s elements of one and r elements of the other set (for binary systems of relations). The letters r and (r, s) are called a rank of the system of relations. The existence of fundamental symmetry. feasibility of the law for any selection of elements is postulated. On the basis of these and some other auxiliary postulates, a set of functional and differential equations is constructed, from which follow all possible laws of systems of relations of various ranks.

2.5. In the works of Kulakov's group, the laws of unary systems of the real relations (USRR) of ranks 3, 4 and 5 were found, which were generalized to any ranks. It was shown that they correspond to known geometries, and their dimension n is connected with the rank r of system of the relations by the ratio $n = r/2$. In particular, it was shown that for $r = 5$ there are 10 and only 10 types of geometries which are Euclidean, pseudo-Euclidean, Lobachevsky, Riemann (constant curvature), quasisymplectic and some exotic geometries. It is shown that the laws of these USRR exist in the form of equality to zero of determinants like the Cayley-Menger or Gram determinants. All basic geometrical concepts (coordinates of elements-points, angles, areas, volumes etc.) are defined by minors of the determinants in the law. The Minkowski geometry corresponds to the law in the form of equality to zero of the Cayley-Menger determinant constructed on 6 points-events.

2.6. But for the solution of the problem discovered by Kulakov, theories of binary systems of relations on two sets of elements appeared as the most interesting ones. As these systems of relations are constructed by the same principles as unary systems of relations corresponding to usual geometry, it is natural to suppose that with the advent of binary systems, a new class geometries, binary ones, was discovered. Their theory was simpler. Since from binary systems of relations by identifying elements of two sets it is possible to pass to unary systems (geometries), the binary geometry can be considered to be more elementary. And therefore it is natural to raise a question of creation of a theory of physical interactions on the basis of binary rather than unary geometry.

2.7. Binary geometry can be considered as a pre-geometry. It is suitable for the description of physics of a microcosm and a quantum theory since two sets of elements can be considered as those describing the initial and final states of a quantum system, and the relations between elements of opposite sets as a prototype of S-matrices. But for realization of this idea it was required to generalize the Kulakov–Mikhailichenko theory to the case of complex binary relations. Such a theory was developed in our works, called the theory of binary systems of complex relations (BSCR). It was shown that for creation of basic provisions of the theory of classical space-time and interactions, the BSCR of minimumally the following ranks should be used: (2,2), (3,3) and (4,4).

3. Consequences of the Relational Approach

Without having an opportunity to develop in detail the content of the constructed theory, we will only list the main, already obtained results of principle character. It is possible to examine their detailed statement in the enclosed list of publications.

3.1. The stratified structure of classical space has been proved. It is known that in the classical theory we deal with a fiber bundle on the base and a layer. The base is the coordinate space, and a fiber is the space of velocities (the momentum space). Since the parameters of elements of any of the two sets of BSCR are determined by relations to elements of the other set, each element is characterized by two sets of parameters determined by a basis of the previous state or a basis of the subsequent state. It is suggested to consider the parameters introduced by a basis of the past a prototype of the coordinate space and those determined by a basis from the future a prototype of the space of velocities.

3.2. It is shown that the BSCR elements of the minimum nondegenerate rank (3,3) are characterized by a 2-component spinor, and massive particles are determined by a couple of elements and are described as a bispinor. Thus a key nature of BSCR of the rank (3,3) explains why elementary particles are described as spinors, and no more habitual vectors or scalars admit that. We will note that R. Penrose, in his twistor theory, actually postulated 2-component spinors whereas in our approach spinors arise automatically.

3.3. The spinor character of the BSCR elements of rank (3,3) dictates the existence in this theory of $SL(2, C)$, and this in turn gives a theoretical justification of dimension four and signature (+ - - -) classical space-time. Actually it solves the fundamental problem posed by E. Mach: "Why is space 3-dimensional?" Already in the XX century A. Einstein and A. Eddington reflected over this question.

3.4. In BSCR of the higher rank (4,4) (in the elementary binary multidimensionality), elements are described by a 3-component Finsler spinor. Thereby a new channel of generalizations of the theory of 2-component spinors is opened. We must remind that in the standard generalizations of the spinor based on Clifford's algebras over a field of real numbers there are only spinors with numbers components 4, 8, 16, ... , 2^n. In this theory a Finsler spinor with any number of components is possible.

3.5. BSCR of rank (3,3) actually shows the quadratic nature of a metric in the unary geometry received from it. A transition from BSCR of rank (4,4) to a unary system of real relations in the most general case leads to a 9-dimensional Finsler geometry with a cubic definition of the metric. For the description of the standard Minkowski geometry it is necessary to use a reduced option of BSCR of rank (4,4).

3.6. In the most general version of the theory of BSCR of rank (4,4) the $SL(3, C)$ group can be used for the description of electroweak and strong interactions. The electromagnetic interaction appears to be the simplest option described by the reduced BSCR of rank (4,4). Proceeding from it, it is possible to claim that the classical space-time relations are caused by the electromagnetic interactions.

4. The Relational-Statistical Approach

4.1. A solution of the above problems needs a relational view of geometry and physics with added ideas of a macroscopic (statistical) nature of space-time. This idea has been discussed in the literature for a long time. According to this idea, the classical complex of concepts of space-time is valid only for sufficiently large systems of elementary particles, macrosystems, and results from a peculiar summing of a huge number of certain factors inherent to physics of a microcosm. Observations in favor of this idea were made by E. Zimmerman, P. K. Rashevsky, R. Penrose, Van Danzig and other authors.

4.2. The relational principles, together with the idea about a macroscopic nature of space-time, make a relational-statistical approach to the solution of the above problem of deriving the classical space-time concepts.

Here, an extremely important role is played by Mach's principle treated in the broadest understanding as dependence of local properties of physical systems on the properties of the whole world.

5. Consequences of the Relational-Statistical Approach

Without going into details, we will specify a number of basic problems which can be solved by the relational-statistical approach.

5.1. On the basis of the relational-statistical approach, the theory of atoms without using the concepts of classical space-time and postulation of the equations of quantum mechanics is under construction. It rests on the so-called basic 44-relation possessing $SL(3,C)$ invariance and on the conditions of connection of particles forming an atom. It results in the $O(4)$ property — symmetry of atomic system that, in accord with the principles of the used approach, allows us to obtain the well-known quantum numbers characterizing the structure of atoms. This result can be regarded as a justification of the works of the 1930s by V.A. Fock and E.A. Hilleraas which discovered the $O(4)$ symmetry of H-like atoms.

5.2. The physics of a microcosm is adequately described by complex numbers due to the absence of the concept "more-less" in the microcosm, unlike classical physics described by real numbers. A transition from complex numbers to real numbers (even integers) is realized in the relational-statistical structure of atoms and other bound states of elementary particles. This statement actually resolves the correspondence dispute of R. Penrose with Kronecker on which numbers are "divine": the complex ones or integers.

5.3. On the basis of relational-statistical approach a new interpretation of quantum mechanics is proposed, substantially related to the Feynman formulation, however with other interpretations of the method of continual integration. It is based not on summation of deposits from classical trajectories, but on summation of deposits from huge "seas" of the pair relations.

5.4. It is known that the equations of quantum and classical physics have solutions which are both retarded and advanced. The latter are excluded in a wilful way, that means postulation of the principle of causality. Within the relational-statistical approach, the result obtained by R. Feynman and J. Wheeler in 1945 remains in force: the absence of advanced solutions is caused by the influence of matter around the world (Mach's principle).

5.5. The duality, specified by Feynman and Wheeler from results connected with the possibility of replacement of the condition of absoluteness of a world absorber by absoluteness of a world emitter in the past has been eliminated. It is because their reasoning was based on ready classical space-time whereas in the relational-statistical approach space-time is caused by a system of world emitters in the past.

5.6. Within the relational-statistical theory, the result of Feynman and Wheeler consisting in that the emergence of bremsstrahlung force in the equations of the motion of charges in classical electrodynamics is caused by the Universe response, i.e., it is one more manifestation of Mach's principle.

5.7. In the relational-statistical approach it has been shown that gravitational interaction has a derivative (secondary) nature, it is caused by the existence of other types of physical interactions, mainly the electromagnetic one. It allows us to look at the problem of theoretical justification of gravitational waves in a new way, as well as unsuccessful attempts to quantize gravity as a special (geometric) type of matter and unsuccessful attempts of experimental detection of gravitational radiation.

It must be reminded that the idea of a derivative nature of gravitation is not new. Earlier on the basis of other reasons this idea was expressed in the works of A.D. Sakharov, Adler and some other authors.

5.8. Since in the relational-statistical theory Mach's principle takes place, in this approach natural are A. Eddington's and P. Dirac's hypotheses of connection of physical constants of the micro- and mega-world. Subsequently these hypotheses were developed in K.P. Staniukovich and 5. N. Melnikov's works.

5.9. The relational-statistical approach to geometry and physics allows us to look from a new point of view at the global properties of the Universe and at a number of already stated hypotheses and unresolved problems of cosmology. In particular, it is already possible to give a relational-statistical justification of the cosmological redshift.

6. Conclusion

In conclusion, we must emphasize the most important conclusion at which the relational-statistical approach is aimed, clarification of a mechanism of formation of the classical space-time concepts. Having solved this problem, it will be possible to put the following, not less global and practically important problem: are there are new opportunities to influence the space-time relations and is it possible to do so. Frankly speaking, the scientific community hoped to find a solution of this problem by means of the general theory of relativity or its natural generalizations (within the geometric paradigm), however, in the last century it has not been done.

References

1. Yu.I. Kulakov, Yu.S. Vladimirov, A.V. Karnaukhov, *Introduction to the theory of physical structures and binary geometrophysics* (Archimedes Publishing, Moscow, 1991) (In Russian).
2. Yu.S. Vladimirov, *Metaphysics* (BINOM: Laboratory of knowledge, Moscow, 2009) (In Russian).
3. Yu.S. Vladimirov, *Relational theory of space time and interactions. Part 1. Theory of systems of relations* (MSU Publishing, Moscow, 1996) (In Russian).
4. Yu.S. Vladimirov, *Relational theory of space time and interactions. Part 2. Theory of physical interactions* (MSU Publishing, Moscow, 1998) (In Russian).
5. Yu.S. Vladimirov, *Physics of long-range action. Part 1. Nature of space-time* (LIBROKOM Book house, Moscow, 2012) (In Russian).
6. Yu.S. Vladimirov, Binary Geometrophysics: Space-Time, Gravitation, *Grav. Cosmol.* **1**(3), 184 (1995).
7. Yu.S. Vladimirov, A.V. Solovyov, Finsler N-spinors with complex components, *Hypercomplex numbers in geometry and physics* **5**(2), 90 (2009) (In Russian).

"Spring theory of relativity" originating from quantum mechanics

Alexander P. Yefremov

Peoples' Friendship University of Russia
Miklukho–Maklaya st. 6, Moscow, 117198, Russia
E-mail: a.yefremov@rudn.ru

Compact derivation of mathematical equations similar to those of quantum and classical mechanics is given on the base of fractal decomposition of a three-dimensional space. In physical units the equations become Shrödinger and Hamilton-Jacobi equations, the wave function of a free particle associated with a virtual ring. Locally uniform motion of the ring in the physical space provides an original helix (or regular cylindrical spring) model of a relativistic theory equivalent in results with special relativity, the free particle's relativistic Lagrangian emerging automatically. Irregular spring model generates theory similar to general relativity.

Keywords: Hypercomplex numbers; fractal space; quantum mechanics; classical mechanics; virtual ring; relativity.

1. Mathematical Derivation of Equations of Quantum and Classical Mechanics

It is fact (though surprising) that the equations of quantum and classical mechanics can be derived from pure mathematical considerations. Detailed deduction is given in the paper Ref. 1; below a sketch of the procedure is outlined.

A local area of three-dimensional (3D) space is fully described by Cartan's frame initiating the Cartesian system of coordinates. In its turn the frame, an arbitrarily moving axial vector triad, may be perfectly imaged by vector quaternion units represented by adequately parameterized three 2×2-matrices $\mathbf{q}_k$ with common properties* (see Ref. 2): $\det \mathbf{q}_k = 1$, $\operatorname{Tr} \mathbf{q}_k = 0$, and postulated product law

$$\mathbf{q}_k \mathbf{q}_m = -\delta_{km} + \varepsilon_{kmj} \mathbf{q}_j. \tag{1}$$

*Small Latin indices are 3D, δ_{km}, ε_{kmj} are Kronecker and Levi-Civita tensors, summation in repeated indices is implied.

The 4th (real) unit I completing the quaternion basis is represented by the unit 2×2-matrix that may be regarded as a metric g of locally flat surface (2D fractal space) endowed with a vector basis ψ^+, ψ^- satisfying the properties of a dyad

$$g\psi^{\pm}\psi^{\pm} = 1, \quad g\psi^{\pm}\psi^{\mp} = 0. \tag{2}$$

With introduction of the covectors $\varphi^{\pm} \equiv g\psi^{\pm}$ the unit I is represented as the sum of vector-covector tensor products

$$I = \psi^+\varphi^+ + \psi^-\varphi^-. \tag{3a}$$

One discovers that the vector units $\mathbf{q}_k$ also can be structured by the same dyad as

$$\mathbf{q}_1 = -i\,(\psi^+\varphi^- + \psi^-\varphi^+),\ \mathbf{q}_2 = \psi^+\varphi^- - \psi^-\varphi^+,\ \mathbf{q}_3 = i\,(\psi^+\varphi^+ - \psi^-\varphi^-), \tag{3b}$$

the decomposition Eq. (3b) together with the dyad properties Eq. (2) making the triad $\mathbf{q}_k$ to identically satisfy the law Eq. (1). The primitive dyad $\tilde{\psi}^+ = \begin{pmatrix} 0 \\ 1 \end{pmatrix}$, $\tilde{\psi}^- = \begin{pmatrix} 1 \\ 0 \end{pmatrix}$, $\tilde{\varphi}^+ = \begin{pmatrix} 0\,1 \end{pmatrix}$, $\tilde{\varphi}^- = \begin{pmatrix} 1\,0 \end{pmatrix}$ forms the 2D metric $I = \begin{pmatrix} 1\,0 \\ 0\,1 \end{pmatrix}$, and due to Eq. (3a), Eq. (3b) generates the constant vector triad expressed through Pauli matrices (the tilde marks out this special case) $\tilde{\mathbf{q}}_1 = -i\begin{pmatrix} 0\,1 \\ 1\,0 \end{pmatrix}$, $\tilde{\mathbf{q}}_2 = -i\begin{pmatrix} 0 & -i \\ i & 0 \end{pmatrix}$, $\tilde{\mathbf{q}}_3 = -i\begin{pmatrix} 1 & 0 \\ 0 & -1 \end{pmatrix}$. The simplest $SU(2)$ transformation $\psi'^{\pm} = e^{\pm i\alpha}\tilde{\psi}^{\pm}$, $\varphi'^{\pm} = e^{\mp i\alpha}\tilde{\psi}^{\pm}$ preserves the law Eq. (1) though forcing triad $\mathbf{q}'_k$ [computed as in Eq. (3b)] to rotate about vector $\mathbf{q}'_3 = \tilde{\mathbf{q}}_3$ at angle 2α. Next mapping, a conformal stretching $\psi^{\pm} = \sigma\psi'^{\pm} = \sigma\, e^{\pm i\alpha}\tilde{\psi}^{\pm}$, $\varphi^{\pm} = \sigma\varphi'^{\pm} = \sigma\, e^{\mp i\alpha}\tilde{\varphi}^{\pm}$, $\sigma \in \mathbb{R}$, $\sigma \neq 0$ causes the defect $\sigma^2 \neq 1$ of the metric Eq. (3a) and the triad Eq. (3b). To avoid it, i.e. to save the algebra units, one can regard the complex-number function (comprising both variables of the mapping) $\lambda \equiv \sigma\, e^{i\alpha}$ as a function of coordinates ξ_N of an abstract M-dimensional space ($N = 1, ..., M$) and of a free parameter θ. Then the normalizing integral can be defined

$$f(\theta) = \int_{V_\xi} \lambda\lambda^* dV_\xi = \int_{V_\xi} \sigma^2 dV_\xi = 1; \tag{4}$$

when used as a factor in Eq. (3a), Eq. (3b) it restores unity of the metric and triad. The algebra is saved "forever" (in the sense of parameter θ) if the integral in Eq. (4) is a constant: $\partial_\theta f = 0$; due to the Gauss theorem this produces the continuity-type equation

$$\partial_\theta(\lambda\lambda^*) + \nabla_\xi(\lambda\lambda^*\mathbf{k}) = 0 \tag{5}$$

where $\nabla_\xi \equiv \partial/\partial\xi_N$, $\mathbf{k}$ is an arbitrary ("propagation") vector. If this vector points in the space direction of increase of the phase (and of the angle of $\mathbf{q}_k$ rotation) $k = \nabla_\xi\,\alpha = (i/2)\nabla_\xi \ln(\lambda^*/\lambda)$ then Eq. (5) is reduced to the square-root format

$$\left(i\,\partial_\theta + \frac{1}{2}\nabla_\xi^2 - W\right)\lambda = 0, \tag{6}$$

where $W(\xi,\,\theta)$ is an arbitrary function. Separated into a real and imaginary parts Eq. (6) forms the set

$$\partial_\theta\sigma + \nabla_\xi\,\sigma\,\nabla_\xi\,\alpha + \frac{1}{2}\sigma\,\nabla_\xi^2\alpha = 0, \tag{7}$$

$$\partial_\theta\,\alpha + \frac{1}{2}(\nabla_\xi\,\alpha)^2 + W - \frac{1}{2}\frac{\nabla_\xi^2\sigma}{\sigma} = 0, \tag{8}$$

where Eq. (7) is a square root of the continuity-type Eq. (5), while Eq. (8) is a vanishing imaginary square-root part of Eq. (5) defined as function W.

Now let the abstract space be associated with the 3D geometric space; then we endow the unit-less math magnitudes with properties of the space coordinates and time

$$x_k \equiv \varepsilon\,\xi_k, \quad t \equiv \tau\,\theta \tag{9}$$

using some physical standards of length $\varepsilon = \hbar/(mV)$ and duration $\tau = \varepsilon/V = \hbar/(mV^2)$ built out of appropriate (for some physical system) constant characteristics, where $\hbar$ is an angular momentum (e.g. the Planck constant), m is a mass, V is a velocity. In new variables from Eq. (9) the expression Eq. (6) becomes precisely the Shrödinger equation

$$\left(i\,\hbar\partial_t + \frac{\hbar^2}{2m}\nabla^2 - U\right)\lambda = 0, \tag{10}$$

[with $\nabla \equiv \partial/\partial x_k$ and $U \equiv (mV^2)W$ treated as a potential energy], while the set Eq. (7) and Eq. (8) converts into the Bohm equations,[3] in general, the full equivalent of Eq. (10)

$$\partial_t\sigma + \frac{\hbar}{m}\nabla\sigma\,\nabla\alpha + \frac{\hbar^2}{2m}\sigma\,\nabla^2\alpha = 0, \tag{11}$$

$$\hbar\,\partial_t\alpha + \frac{\hbar^2}{2m}(\nabla\alpha)^2 + U - \frac{\hbar^2}{2m}\frac{\nabla^2\sigma}{\sigma} = 0. \tag{12}$$

One can notice though that under the assumption that the phase is proportional to action function S

$$S \equiv \hbar\alpha, \tag{13}$$

with $\sigma = 1$ (no stretching of the fractal dyad), Eq. (11) reduces to equation of solenoidal (tube-like) motion of incompressible fluid $\nabla \mathbf{v} = 0$ where $\mathbf{v} \equiv (1/m)\nabla S$, while Eq. (12) becomes the Hamilton-Jacoby equation of classical mechanics for a particle of mass m in an exterior field U

$$\partial_t S + \frac{1}{2m}(\nabla S)^2 + U = 0. \tag{14}$$

Thus, indeed, we can state that pure math equations Eq. (6) – Eq. (8), playing a fundamental role in algebra's support, are in fact the full equivalent of equations of quantum and classical mechanics (Eq. (10), Eq. (14)) provided they are written in appropriate physical units. This fact together with understanding that here we operate with "square-root" entities from 2D fractal space prompts original models of the wave function and of 3D particle.

2. Geometric Models of a Particle: Fractal 2D-Cell, Rotating Massive Point, and Virtual Ring

Function $\lambda = \sigma\, e^{i\alpha}$ is a factor of the constant vector $\tilde{\psi}^+$, so the couple $\psi^+ = \lambda\tilde{\psi}^+$, $\psi^- = \lambda^*\tilde{\psi}^-$ forms a plane area changing its value with variation of the phase; the process can be conceived as the area's "pumping over" from real to imaginary sector of the fractal space. This picture of a "flickering 2D-cell" fully determined by properties of λ gives in fact a geometric (or rather "pregeometric", since fractal) image of the wave functions of a particle. This assumption is supported by analyses of normalizing integral Eq. (4) in physical variables Eq. (9) having the form $(1/\varepsilon^3)\int_V \sigma^2 dV = 1$ where $dV = \varepsilon^3 dV_\xi$. If $\sigma \equiv \sqrt{\rho(x,t)/\rho_{\text{mean}}}$ then the integral becomes a definition of particle's mass

$$\int_V \rho\, dV = \rho_{\text{mean}}\varepsilon^2 = m, \tag{15}$$

the integration volume (a cube with edge equal to the characteristic length) containing the mass density $\rho(x,t)$. In fact the integral Eq. (15) just squeezes a mass distributed in a small volume into a point-like particle, but now this "material point" drastically differs from that of classical mechanics since it is endowed with "frozen-in" triad rotating about one axis at angle equal to the doubled phase of its 2D-cell's flickering. The option to detect proper rotation of a moving material point entails emerging of quantum features of the motion, and leads to model of a *virtual ring* linked with question of measurement; this as well concerns classical objects, e.g. planets of the Solar system.

Consider an example of a particle moving with velocity $V = \text{const}$ while rotating with angular frequency $\Omega = \text{const}$; then one can ascribe to the particle a "wave length" $l = VT = 2\pi V/\Omega$, i.e. its path for period of rotation $T = 2\pi/\Omega$, what yields the velocity-frequency link $V = \Omega r$ where $r \equiv l/(2\pi)$ is radius of a length l of some circumference. So apart from a linearly transported point mass one can model the motion either by a massive point uniformly revolving about a center at radius r or by a rotating rigid (infinitesimally thin) ring of mass m, evidently virtual. Both models have the same moment of inertia $I = mr^2$, but the ring looks better for symmetry reasons. The ring's proper angular momentum is computed as $L = I\Omega = mr^2\Omega = mrV = \text{const}$, and together with velocity $V$and angular frequency Ω it can be chosen as a basic magnitude to express the energy through ΩL (e.g. kinetic energy $E_{\text{kin}} = \Omega\, mrV/2 \sim \Omega L$), and the momentum through kL (if $k = 1/r$ is the "wavenumber") $p = (1/r)\, mrV = kL$. So mechanical characteristics of classical particles may be measured (with certain dimensional coefficients) in values of some standard angular momentum provided this standard is properly determined for the considered mechanical system (as it is done in quantum mechanics where the standard is $\hbar$). Respective mechanical equation of the type Eq. (10) can be written for this case

$$\left(i\, L\partial_t + \frac{L^2}{2m}\nabla^2 - U\right)\lambda = 0, \tag{16}$$

(similarly the substitution $\hbar \to L$ gives respective Bohm-type equations) while the virtual ring producing a "wave" is associated with the De Broglie-type wave function

$$\lambda(x,t) = e^{i\alpha} = e^{\frac{i}{L}(\mathbf{pr}-Et)}. \tag{17}$$

If $U = 0$ then the function Eq. (17) describes a solenoidal motion $\boldsymbol{\nabla}\mathbf{p} = 0$ of the particle inside a tube of radius r with energy $E = p^2/(2m)$. Ring's phase in Eq. (17) is

$$\alpha = \frac{S}{L} = \frac{\mathbf{pr} - Et}{L} = \frac{mV^2}{2L}t \equiv \omega t, \tag{18}$$

where $\omega = \Omega/2$, as noted above, is the 2D-cell flickering frequency, twice as less as Ω, frequency of rotation of the triad attached to the particle and imaged by rotation of the virtual ring. From Eq. (18) one finds $mV^2/L = \Omega$, i.e. we return back to the definition $L = mrV$. On the other hand, it is known that the full time derivative of the mechanical action gives the Lagrange function $dS/dt = \Lambda$, so for the stationary motion the phase can be formulated as following

$$\alpha = \omega t = (\Lambda/L)t. \tag{19}$$

Equation (18) and Eq. (19) mean that the 2D-cell's phase flicker (the virtual ring rotates), hence the "wave function" exists, due to the fact that $\Lambda \neq 0$. If a particle moves in a field of force then both constituents of the phase, kinetic and potential energy, are present; a freely moving particle ($U = 0$) has in the phase just the kinetic energy term $E = \omega L$, its value equal to that of proper energy of the virtual ring. But if a free particle is immobile then its Lagrangian is zero, there is no flickering or rotation, hence no virtual ring model and wave properties.

This observation by the way seems to be of importance for quantum mechanics where the phase factor $e^{i\omega t}$ (with $\omega =$ const) is usually associated with constant energy of the quantum system. But certain systems with constant energy have variable function Λ; so if it is indeed a Lagrangian that should be coupled to the phase, then some known solutions of Shrödinger equations ought to be reanalyzed. But here we just mention the problem; main purpose of the rest of this study is to show that the virtual ring conjecture bringing some quantum features to classical mechanics in the case of extreme velocities leads to different understanding of the relativity theory.

3. "Spring Theory of Relativity"

In fact the suggested above virtual ring model of a moving rotating body is not sufficiently good even for classical mechanics since one fails to picture the particle of mass m moving with velocity V by similarly moving though rotating ring of mass m because the energy of such a ring is twice as great as that of the particle. So this ring model "has no trajectory" and to describe the motion we'll have to instantly and inertialessly (wasting no energy) transport the ring at its wavelength along the particle trajectory after each period of rotation. Of course this "quantum" displacement is obviously at variance with traditional description of a classical object, so a "material" ring model in Newtonian mechanics should be adequately improved; we delay the task to a next study (see Sec. 4). But the privilege of the discussed model is its virtual character, and we employ it to discover an extra liaison between pure math and physics.

Now we associate the virtual ring with *ultimate particle's border* where linear velocity of proper rotation is equal to the fundamental velocity c. This means that for a standard particle, e.g. electron with the mass m, proper rotational frequency $\Omega = 2\omega$, and size $\varepsilon = 2r$, the linear velocity is $c = r\Omega = \varepsilon\omega$. For such a model we define the standard proper angular momentum as$\hbar \equiv m\varepsilon^2\omega = m\varepsilon c$; by the way the characteristic length (which as mentioned above can be chosen arbitrary) in this case coincides with the Compton wave length

$$\varepsilon = \hbar/(mc). \tag{20}$$

Moreover, the ring describing such a particle is permitted to move smoothly in the physical space (as a real particle's border) with a velocity $u = \text{const}$, the angle between $\mathbf{\Omega}$ and $\mathbf{u}$ denoted as β. Then any particle's "bordering point" depicts a helix-type line; let velocity of this point be always equal to c, then the helix square line element has the form

$$dl^2 = c^2dt^2 = \varepsilon^2 d\alpha^2 + 2\varepsilon \sin 2\alpha \sin\beta\, d\alpha\, udt + u^2dt^2 \tag{21}$$

with the phase $\alpha = \omega t$. All terms in Eq. (21) are constant apart from $\sin 2\alpha$, hence $\sin\beta = 0$, or $\beta = 0, \pi$, i.e. $\mathbf{\Omega} \uparrow\uparrow \mathbf{u}$ (or $\mathbf{\Omega} \uparrow\downarrow \mathbf{u}$), and the helix has the shape of a cylindrical spring (Fig. 1).

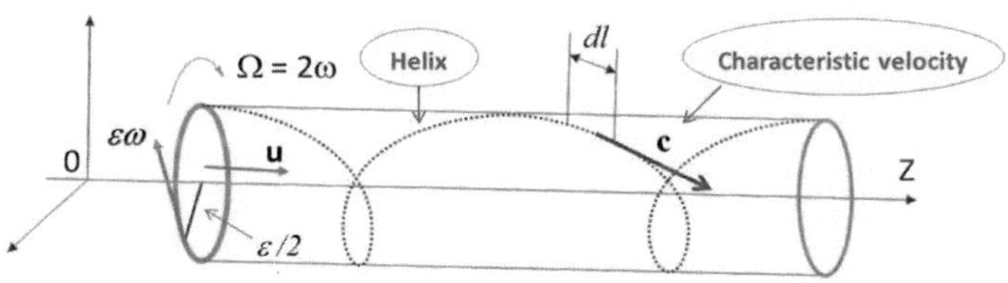

Fig. 1. Solenoidal tube-like motion: cylindrical "spring".

From Eq. (21) the rotational square path of the ring's point is expressed as

$$\varepsilon^2 d\alpha^2 = c^2dt^2 - dr^2, \tag{22}$$

and one can find the "wave" phase value as an integral $\alpha = \pm(c/\varepsilon)\int_{\Delta t}\sqrt{1 - u^2/c^2}dt$. Substitution of Eq. (20) with the choice of sign "minus" (clockwise rotation) yields precisely the action functional for a free relativistic particle

$$S \equiv \alpha\,\hbar = -mc^2 \int_{\Delta t} \sqrt{1 - \frac{u^2}{c^2}}\,dt \tag{23}$$

previously only postulated. The correlation Eq. (22), describing in special relativity invariance of a space-time interval, in this model has specific geometrical meaning illustrated by Fig. 2.

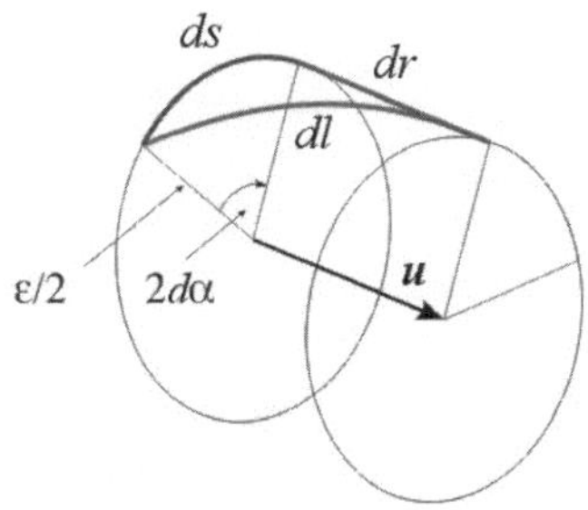

Fig. 2. Geometrical sense of SR line-element.

Here the elements of Eq. (22) form a cylindrical triangle in 3D space. The time part of the interval $cdt = dl$ is an element of the helix line, the particle's path dr is an element of the cylinder generatrix, and the line element of special relativity $ds = \varepsilon\, d\alpha$ has the sense of an arc length of the immobile particle's "ultimate border" which for a free particle in proper frame of reference is a definite number (invariant of special relativity). When unrolled in a plane this curved triangle becomes an equivalent of the Minkowski diagram.

Now we notice that Eq. (23) in the differential form in the case of weak-relativity

$$\hbar d\alpha = \hbar\,\partial_t \alpha dt + \hbar \partial_n \alpha dx_n = -Edt + p_n dx_n \cong -(mc^2 + mu^2/2)dt + mu_n dx_n$$

automatically yields De Broglie's energy-frequency and momentum-wave vector ratios

$$E = (mc^2 + mu^2/2) = |\omega|\,\hbar, \quad p_n = k_n\hbar. \tag{24}$$

So the relativistic ring conception leads to description not only of kinetic energy $mu^2/2$, but as well of the rest energy $E_{\text{rest}} \equiv mc^2 = |\omega|\,\hbar$ in this model generated by proper rotation. Thus the "spring" theory based on the rotating ring model in a way unites relativity (space-time interval invariance) and quantum properties of a particle.

Apart from imaging an inertial relativistic particle the ring model may be used to describe a particle moving in a force field when the regular spring geometry is distorted. To develop the theory we consider a Frenet triad $\mathbf{q}_n(t, x)$ attached to the ring so that $\mathbf{q}_3$ is tangent to a trajectory $x_k(t)$, $\mathbf{q}_1$, $\mathbf{q}_2$ permanently rotating about $\mathbf{q}_3$ (see Fig. 3).

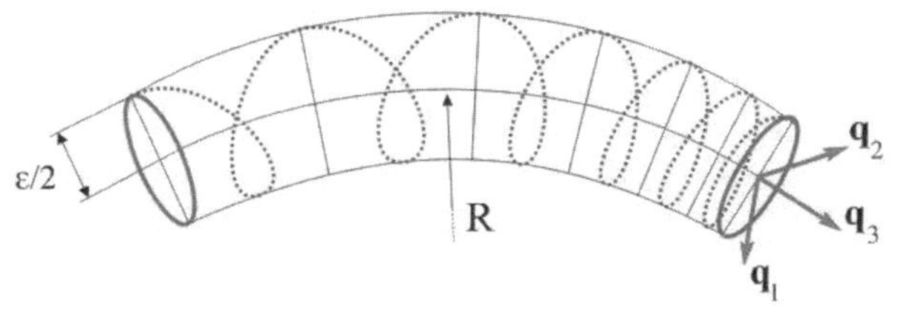

Fig. 3. Irregular "spring".

The 1-form of the helix line-element is a sum of transverse and lengthwise differentials $d\mathbf{l} = d\boldsymbol{\rho} + d\mathbf{r}$. The transverse vector-quaternion $\boldsymbol{\rho} \equiv (\varepsilon/2)\,\mathbf{q}_1(t, x)$ determines a bordering point on $\mathbf{q}_1$; when translated along $x_k(t)$ it generates the 1-form $d\boldsymbol{\rho} = (\varepsilon/2)[\omega_{t\,12}\mathbf{q}_2 dt + \omega_{r13}\mathbf{q}_3 dr]$ expressed through the components of the quaternion connection $\omega_{\xi kn}$ whose time-derivative component $\omega_{t\,12}dt = 2d\alpha$ has meaning of the angle of rotation (or double phase)[4], while the path-derivative component $\omega_{r13}dr \equiv 2dr/R$ describes the trajectory's local curvature of radius $R(x)/2$. The lengthwise 1-form is $d\mathbf{r} = \mathbf{q}_3 dr$ so that square interval of the curved cylindrical spring is

$$dl^2 = d\mathbf{l}\, d\bar{\mathbf{l}} = \varepsilon^2 d\alpha^2 + (1 + \varepsilon/R)^2\, dr^2; \tag{25}$$

here the quaternion conjugation is formally used to provide positive length value. Further on we'll use the simplifying notation $e^{2G(x)} \equiv [1 + \varepsilon/R(x)]^2$, and in addition we admit that the line-element Eq. (25) can be irregularly squeezed, a function $e^{-2W(x)}$ describing the compression $dl^2 = e^{-2W(x)}(\varepsilon^2 d\alpha^2 + e^{2G(x)} dr^2)$; then respective "space-time" interval has the form

$$\varepsilon^2 d\alpha^2 = ds^2 = e^{2W(x)} c^2 dt^2 - e^{2G(x)}\, dr^2 \equiv g_{\mu\nu} dx^\mu dx^\nu, \tag{26}$$

with $g_{\mu\nu}$ having sense of a variable diagonal metric of 4D space-time ($\mu, \nu =$ 0, 1, 2, 3). If the spring is not cylindrical [see Eq. (21)] then variable mixed space-time terms arise, and the interval becomes identical to that of general relativity.

For non-relativistic particle-weak-field system $dr/(cdt) = u/c \ll 1$, $e^{2W} \cong 1 + 2W$ with neglected curvature of the trajectory $e^{2G} \approx 1$ Eq. (26) yields the 1-form (with the Lagrangian in the right-hand-side)

$$dS \equiv \hbar d\alpha = -mc^2 dt\sqrt{e^{2W} - e^{2G}(u/c)^2} \approx -[mc^2(1+W) - mu^2/2]\,dt,$$

hence (since $dS/dt = \partial_t S + u_n \partial_n S$) the Hamilton-Jacobi equation $\partial_t S + \partial_n S\, \partial_n S/(2m) + U = 0$, the potential $U \equiv mc^2 W + E_{\text{rest}}$ containing in this case proper energy of the virtual ring. So the theoretical model of non-uniformly moving ring imaged by an irregular spring and described by the compression function $e^{2W} \equiv g_{00}$ involves in the game a certain mechanical potential, in this approach meaning any particle-field interaction, not necessarily gravitational.

4. Conclusion

Discovered recently interconnection between pure algebraic correlations and a logically confined cluster of mechanical theories of a particle in this study is supplemented with the virtual ring conjecture. Reflecting a plausible physical fact that a quantum particle really possesses rotational degrees of freedom, this model as well makes it possible to endow classical objects, such as a star system planets, with specific quantum-like wave properties and to equivalently consider (under some restrictions) the Shrödinger-type equation to be adequately describing moving and rotating "material points". Of course this fresh idea is far from being exhaustively elaborated (see note at the beginning of the Sec. 3). It is quite possible that an extended version of the ring model should be investigated uniting point-and-ring properties, perhaps in the shape of a "Saturn-type" object. Distribution of the mass and mechanical energy between the central point and the ring can make feasible simultaneous presence of rotational and translational energy of such a classical object its fractal equivalent obviously possessing wave properties. This model will be presented in the subsequent study.

References

1. A. P. Yefremov, General theory of particle mechanics arising from a fractal surface, *Grav. Cosmol.* **21**, 19 (2015).

2. A. P. Yefremov, Quaternions: algebra, geometry and physical theories, *Hypercomplex Numbers in Geometry and Physics* **1**, 104 (2004); arXiv:math-ph/0501055v1 (2005).
3. D. Bohm, A Suggested Interpretation of the Quantum Theory in Terms of "Hidden" Variables. I, *Phys. Rev.* **85**, 166 (1952).
4. A. P. Yefremov, *Quaternion spaces, frames of reference and fields* (Moscow, PFUR Publ., 2005).

Mathematical models of cosmological evolution of statistical system of scalar charged particles

Yu. G. Ignat'ev*, A. A. Agathonov, M. L. Mikhailov and D. Yu. Ignatyev

Lobachevsky Institute of Mathematics and Mechanics,
Kazan Federal University,
Kazan, Russian Federation
**E-mail: ignatev_yu@rambler.ru*
www.kpfu.ru

We consider the macroscopic model of plasma of scalar charged particles obtained by statistical averaging of the microscopic equations of particle dynamics in a scalar field. On the basis of kinetic equations and their rigorous integral consequences a self-consistent set of equations was formulated which describes the self-gravitating plasma of scalar charged particles. The corresponding closed cosmological model was obtained and it was numerically simulated for the case of one-component degenerated Fermi gas and two-component Boltzmann system.

Keywords: Physics of the early universe; particle physics-cosmology connection; inflation; phantom scalar interaction.

1. Introduction

In the recent years there emerged a large number of articles devoted to scalar fields, which are necessary for an explanation of the secondary acceleration of the Universe. It is reasonable to single out models with a minimal interaction, where the scalar field interacts with "ordinary matter" only gravitationally, and models with non-minimal interactions, where the scalar field interacts with "ordinary matter" by means of some coupling realized through a factor depending on the scalar field potential in the Lagrange matter function or on a factor depending on the scalar field derivatives.

In contrast to mentioned nonminimal models of scalar interaction, we consider statistical systems of scalar charged particles in which some sort of particles can directly interact with a scalar field by means of a fundamental scalar charge. On the other hand, a statistical system possessing a non-zero scalar charge and being itself a source of a scalar field, can efficiently influence the scalar field, managing its behavior. Such a scalar interaction was introduced in the general-relativistic theory in 1982 by Yu.G. Ignat'ev[1–4] and a bit later by G.G. Ivanov.[5] In Refs. 2, 3, on the basis of kinetic theory, a self-consistent set of equations describing a statistical system of particles with scalar interaction was obtained. In the recent papers Refs. 6–8, the macroscopic theory of statistical systems with a scalar interaction was significantly improved and extended to the case of phantom scalar fields.

2. Mathematical Model of a Self-Gravitating Plasma of Scalar Charged Particles

We consider a rigorous dynamic model of interaction of a scalar field with elementary particles, based on the Hamilton microscopic equations of motion and subsequent statistical averaging procedure.[9,10] In the case of the flat Friedman model the complete set of macroscopic equations, which describe a self-gravitating plasma of scalar charged particles, consists of:

(1) Scalar field equation:

$$\ddot{\Phi} + 3\frac{\dot{a}}{a}\Phi + \epsilon_2 m_s^2 \Phi = -4\pi\epsilon_1\sigma; \tag{1}$$

(2) Einstein equation:

$$3\frac{\dot{a}^2}{a^2} = 8\pi\mathcal{E}; \tag{2}$$

(3) Energy conservation law:

$$\dot{\mathcal{E}}_p + 3\frac{\dot{a}}{a}(\mathcal{E}_p + \mathcal{P}_p) = \sigma\dot{\Phi}; \tag{3}$$

(4) Particle number conservation law:

$$\dot{n}_a + 3\frac{\dot{a}}{a}n_a = 0, \tag{4}$$

where $a(t)$ is a scale factor; $\mathcal{E} = \mathcal{E}_p + \mathcal{E}_s$; $\epsilon_2 = 1$ corresponds to a classical and $\epsilon_2 = -1$ to a phantom scalar field; for a field with repulsion of equally charge particles $\epsilon_1 = 1$ and for a field with attraction of equally charge particles $\epsilon_1 = -1$. Macroscopic scalars:

$$n_a = \frac{2S+1}{2\pi^2} m_*^3 \int_0^\infty \frac{\sinh^2 x \cosh x dx}{e^{-\gamma_a + \lambda_* \cosh x} \pm 1}; \tag{5}$$

$$\mathcal{E}_p = \sum_a \frac{2S+1}{2\pi^2} m_*^4 \int_0^\infty \frac{\sinh^2 x \cosh^2 x dx}{e^{-\gamma_a + \lambda_* \cosh x} \pm 1}; \tag{6}$$

$$\mathcal{P}_p = \sum_a \frac{2S+1}{6\pi^2} m_*^4 \int_0^\infty \frac{\sinh^4 x dx}{e^{-\gamma_a + \lambda_* \cosh x} \pm 1}; \tag{7}$$

$$T_p = \sum_a \frac{2S+1}{2\pi^2} m_*^2 \int_0^\infty \frac{\sinh^2 x dx}{e^{-\gamma_a + \lambda_* \cosh x} \pm 1}; \tag{8}$$

$$\sigma = \sum_a \frac{(2S+1) m_*^3}{2\pi^2} q \int_0^\infty \frac{\sinh^2 x dx}{e^{-\gamma_a + \lambda_* \cosh z} \pm 1}; \tag{9}$$

where $x = \sinh t$ and two dimensionless scalar functions are introduced:

$$\lambda_* = \frac{m_*}{\theta}; \quad \gamma_a = \frac{\mu_a}{\theta}.$$

Energy density and pressure of the scalar field:

$$\mathcal{E}_s = \frac{\epsilon_1}{8\pi}(\dot{\Phi}^2 + \epsilon_2 m_s^2 \Phi^2), \quad \mathcal{P}_s = \frac{\epsilon_1}{8\pi}(\dot{\Phi}^2 - \epsilon_2 m_s^2 \Phi^2). \tag{10}$$

It is interesting to investigate the degenerated Fermi system of scalar charged particles due to maximum simplicity of the mathematical model and the possibility of interpretation of such a system as a cosmological dark (cold) matter. Total degeneracy condition $\gamma(t) = \dfrac{\mu_f}{\theta} \to \infty$ gives the infitite value of reduced chemical potential. In this case integration of the macroscopic densities can be expressed in elementary functions:

$$n = \frac{1}{\pi^2} p_F^3; \tag{11}$$

$$\mathcal{E}_p = \frac{m_*^4}{8\pi^2}\left[\psi\sqrt{1+\psi^2}(1+2\psi^2) - \ln(\psi + \sqrt{1+\psi^2})\right]; \tag{12}$$

$$\mathcal{P}_p = \frac{m_*^4}{24\pi^2}\left[\psi\sqrt{1+\psi^2}(2\psi^2 - 3) + 3\ln(\psi + \sqrt{1+\psi^2})\right]; \tag{13}$$

$$\sigma = q\frac{m_*^3}{2\pi^2}\left[\psi\sqrt{1+\psi^2} - \ln(\psi + \sqrt{1+\psi^2})\right]; \tag{14}$$

where dimensionless function ψ is a ratio of Fermi momentum to effective fermion mass:

$$\psi = \frac{p_F}{|m_*|}.$$

We also considered a highly non-degenerate Boltzmann plasma. That means that reduced chemical potentials are small functions: $\gamma_a(t) = \dfrac{\mu_a}{\theta} \to 0$. Carrying out the integration we obtain following expressions of macroscopic scalars for Boltzmann plasma:

$$n_a = \frac{2S+1}{\pi^2} m_*^3 \frac{\mathrm{K}_2(\lambda_a)}{\lambda_a} e^{\gamma_a}; \tag{15}$$

$$\mathcal{E}_p = \frac{1}{2\pi^2} \sum_a (2S+1) m_*^4 \left(\frac{\mathrm{K}_3(\lambda_a)}{\lambda_a} - \frac{\mathrm{K}_2(\lambda_a)}{\lambda_a^2} \right); \tag{16}$$

$$\mathcal{P}_p = \frac{1}{2\pi^2} \sum_a (2S+1) m_*^4 \frac{\mathrm{K}_2(\lambda_a)}{\lambda_a^2}; \tag{17}$$

$$\sigma = \sum_a \frac{2S+1}{2\pi^2} q_a (m + q_a \Phi)^3 \frac{\mathrm{K}_1(\lambda_a)}{\lambda_a}; \tag{18}$$

where K_n is a Bessel function of imaginary order. We considered the case of symmetry between scalar charged particles and antiparticles:

$$m_- = m_+ = m_* \Rightarrow \begin{cases} 1. \, m = 0; \; m_* = |q\Phi| \\ 2. \, q = 0; \; m_* = m. \end{cases} \tag{19}$$

3. The Numerical Simulation

The numerical integration of the set of equations was carried out in the applied mathematics package "Mathematica".

We can state that at large times $\sim 10^5 t_{Pl}$ the cosmological evolution of our model does not vary from the evolution of matter with a minimal scalar interaction. However at smaller times $10^5 t_{Pl}$ the evolution of matter with a nonminimal scalar interaction is characterized by a large number of the behaviour types with respect to matter with a minimal scalar interaction. The results of numerical simulation depend insignificantly on the particle statistics: in fact, almost identical results were obtained for one-component degenerate Fermi systems and for two-component Boltzmann plasma.

Numerical simulation of the cosmological evolution of the statistical system of scalar charged particles revealed two brand new features of such cosmological model behavior with respect to minimal models. The first difference is in the appearance of giant bursts (up to 10^{10}) of the invariant cosmological acceleration Ω at the time interval $8 \cdot 10^3 \div 2 \cdot 10^4 t_{Pl}$. As we noted, the presence of such bursts requires significant modification of quantum theory for generation of cosmological gravitational perturbations. The second characteristic feature of the cosmological model with a scalar particle interaction is strong heating ($3 \div 8$ orders of magnitude) of a statistical system at these times of evolution.

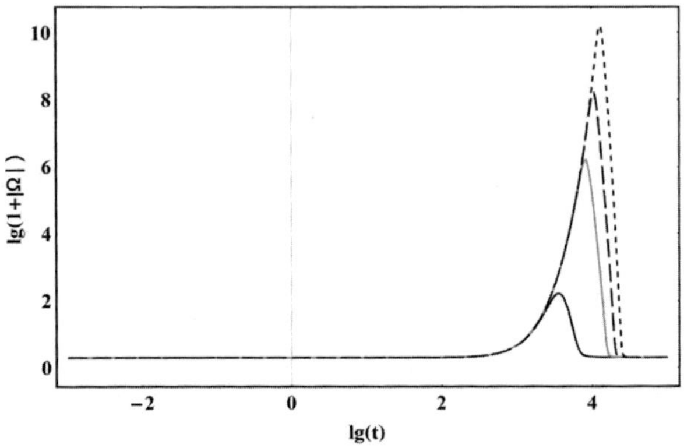

Fig. 1. The dependency of the invariant cosmological acceleration ($\Omega = a\ddot{a}/\dot{a}^2$) evolution on the value of the scalar charge. Bottom to top: $q = 0.001; 0.1; 1; 10$.

References

1. Yu. G. Ignat'ev, *Russian Physics Journal* **25**, No. 4, 92 (1982).
2. Yu. G. Ignat'ev, *Russian Physics Journal* **26**, No. 8, 15 (1983)).
3. Yu. G. Ignat'ev, *Russian Physics Journal* **26**, No. 8, 19 (1983).
4. Yu. G. Ignat'ev, *Russian Physics Journal* **26**, No. 12, 9 (1983).
5. G. G. Ivanov, *Russian Physics Journal* **26**, No. 1, 32 (1983).
6. Yu. G. Ignat'ev, *Russian Physics Journal* **55**, No. 2, 166 (2012).
7. Yu. G. Ignatiev, *Russian Physics Journal* **55**, No. 5, 550 (2012).
8. Yu. G. Ignatyev, *Russian Physics Journal* **55**, No. 11, 1345 (2013).
9. Yu. G. Ignat'ev, A. A. Agathonov and D. Yu. Ignatyev, *Grav. Cosmol.* **20**, 304 (2014);
10. Yurii Ignat'ev, Alexander Agathonov, Mikhail Mikhailov and Dmitry Ignatyev. *Astrophysics and Space Science*, **357**, Issue 1, 21 (2015).

Electromagnetic and gravitational radiation from massless particles

D. V. Gal'tsov

Moscow State University, Faculty of Physics, Department of Theoretical Physics
E-mail: galtsov@phys.msu.ru

We demonstrate that full description of both electromagnetic and gravitational radiation from massless particles lies outside the scope of classical theory. Synchrotron radiation from the hypothetical massless charge in quantum electrodynamics in external magnetic field has finite total power while the corresponding classical formula diverges in the massless limit. We argue that in both cases classical theory describes correctly only the low-frequency part of the spectra, while the total power diverges because of absence of the UV frequency cutoff. Failure of description of gravitational radiation from massless particles by classical General Relativity may be considered as another appeal for quantization of gravity apart from the problem of singularities.

The limit of zero mass of radiating charge in classical electrodynamics is non-trivial. As was discussed recently,[1,2] the Lienard-Wiechert potentials, due to the factor $(1 - \mathbf{rv}/r)$ in the denominator, diverge in the direction of the instantaneous velocity for $|\mathbf{v}| = 1$. Attempts to regularize this singularity had not led to reasonable results, so it was argued in Ref. 2 that the truth is that the massless charge does not radiate at all. Similar conjecture was promoted before on different grounds.[3] This, however, apparently contradicts to infiniteness of the massless limit $\mu \to 0$ of the well-known formula for synchrotron radiation[4]:

$$P_{\rm cl} = \frac{2e^4 H^2}{3\mu^2}\left(\frac{E}{\mu}\right)^2, \qquad (1)$$

where E denotes the energy of the charge, and H – the magnetic field. This discrepancy led us to investigate this problem in the quantum theory.[5]

Radiation from massless charges in quantum electrodynamics is also a non-trivial problem. Within the perturbation theory one encounters, apart from the usual infrared divergencies, the *collinear singularities*, occurring when the photon is emitted from the massless legs of the Feynmann diagrams in the direction of the momentum of a charge.[6] This is manifestation of degeneracy of states of the charge and the photon moving along the same line. Elimination of collinear divergences is achieved using Kinoshita-Lee-Nauenberg[7,8] prescription of averaging over an ensemble of degenerate states. Note that the mentioned above line singularities of classical retarded potentials look similar to collinear singularities of quantum theory. Another complication is the screening of the electromagnetic field of the massless charge due to vacuum polarization.[9] However, these problems are evaded, if interaction of the charge with an external electromagnetic field is treated non-perturbatively, using the exact operators in the external classical field. In the case of magnetic field one deals with the bound states of the charge at Landau levels, and the momentum in the plane orthogonal to magnetic field is not conserved. This removes collinear divergences and modifies the propagation function, leading, in particular, to non-zero quantum correction to mass in the one-loop order. In this approach radiation from the massless charge is non-zero and finite.[5]

This analysis also reveals that classical theory still can be applied if only one computes not the Lienard-Wiechert potentials, but their spectral expansion. Indeed, in view of the correspondence principle, it can be expected that classical theory describes correctly the low frequency part of radiation whatever the mass of the radiating charge is. In the massive theory[4] one has for the low frequencies:

$$\frac{dP}{d\omega} = \frac{e^2 \omega_H 3^{1/6} \Gamma(2/3)}{\pi} \left(\frac{\omega}{\omega_H} \right)^{1/3}, \tag{2}$$

where $\omega_H = eH/E$. This expression depends only on the particle energy and it is unchanged for zero mass. The difference between massive and massless particles, however, is that in the former case the formation length $l \sim 1/(\gamma\omega_H)$ of radiation in a given direction is finite, leading to the frequency cutoff at $\omega_{\rm cr} \sim l^{-1}\gamma^2 \sim \omega_H \gamma^3$, where γ is the Lorentz-factor $\gamma = E/\mu = 1/\sqrt{1-v^2}$. In the massless case $\omega_{\rm cr} \to \infty$, so there is no frequency cutoff. The Lienard-Wiechert potentials in coordinate representation account for the total field, so they are inappropriate for description of radiation from the massless charge indeed. But performing the spectral decomposition, one can give a reasonable estimate using classical theory with the quantum cutoff $\omega_{\rm quant} = E/\hbar$. Integrating (2) up to this cutoff one obtains:

$$P_{\rm cut} = \int_0^{\omega_{\rm quant}} \frac{dP}{d\omega} d\omega = \frac{e^2 \sqrt{3}\Gamma(2/3)}{4\pi\hbar^2} (3e\hbar H E)^{2/3}. \tag{3}$$

Transition to the massless limit in the quantum theory of synchrotron radiation of massive charge[4,10] is subtle, since the results of the latter depend on two dimensionless parameters (in the units $\hbar = c = 1$):

$$f = \frac{H}{H_0} = \frac{eH}{\mu^2}, \qquad \chi = \frac{H}{H_0}\frac{E}{\mu} = \frac{eHE}{\mu^3}, \tag{4}$$

diverging as $\mu \to 0$. Because of this, the standard approximations used in the quasiclassical case (high Landau levels in both initial and final sates) in terms of the Macdonald or Airy functions fail, and one has to develop an alternative approximation scheme. We used[5] the mass-operator in the Schwinger formalism[10,11] (see also Ref. 12) setting the charge mass to zero *ab initio* and obtaining the spectral power as the integral

$$\frac{dP}{d\omega} = \frac{e^2 v}{4\pi E}\int_0^{\infty}\left(E^2(8-v^2)(1-v)^2 x\sin\psi + \frac{eHv}{x^2}(1-\cos\psi)\right)dx, \tag{5}$$

where $v = \omega/E$ and $\psi = \frac{x^3 E^2}{3eH}v(1-v)^2$. Evaluating the integral over x we get

$$\frac{dP}{d\omega} = \frac{2e^2\,\Gamma(2/3)}{27\hbar E}\,(3e\hbar HE)^{2/3}\,\mathcal{P}(\hbar\omega/E), \tag{6}$$

where the Planck's constant is restored, and the normalized spectral function is introduced

$$\mathcal{P}(v) = \frac{27}{2\pi\sqrt{3}}\,v^{1/3}(1-v)^{2/3}, \qquad \int_0^1 \mathcal{P}(v)\,dv = 1. \tag{7}$$

This spectrum is smooth, exhibiting maximum at

$$\hbar\omega_{\max} = \frac{1}{3}E. \tag{8}$$

The average photon energy is

$$\langle\hbar\omega\rangle = E\int_0^1 v\mathcal{P}(v)\,dv = \frac{4}{9}E, \tag{9}$$

while the total energy loss per unit time reads

$$P = \int_0^{E/\hbar} P(\omega)d\omega = \frac{2e^2\,\Gamma(2/3)}{27\hbar^2}\,(3e\hbar HE)^{2/3}. \tag{10}$$

It differs from the estimate (3) only by a numerical coefficient. The expression (6) has the following unusual features. It is non-perturbative in the fine structure constant e^2/hc, and it has no classical limit $\hbar \to 0$, being essentially quantum. This could be expected in view of the Eq. (9).

Now we are going to show that gravitational radiation form massless particles exhibits similar features, though, as could be expected from the Weinberg's theorems on soft photons and gravitons,[6] the corresponding divergencies are weaker (logarithmic). Note that, while consideration of massless charges in electrodynamics appeals rather to the limiting case of the realistic theory, in gravity it is not so, since all particles, including truly massless ones (photons) are subject to gravitational interaction and thus are entitled to emit gravitational radiation. In the textbooks it is tacitly assumed that General Relativity describes correctly gravitational radiation from any classical sources. This turns out not to be true. Namely, the spectrum of gravitational radiation from massless particles moving along null geodesics in curved space-time, being computed classically, is UV divergent. This can be considered as the second argument (apart from the problem of classical singularities) appealing to quantization of gravity.

Gravitational radiation from massive bodies moving along ultrarelativistic geodesics around black holes (gravitational synchrotron radiation, GSR) was considered in early 1970-ies as possible mechanism of enhancement of the flux of gravitational waves from the center of Galaxy to explain the (unconfirmed) Weber's results. Calculations were performed in the Schwarzschild[13,14] and Kerr[15,16] geometries. Here we discuss the massless limit in the GSR theory, which is interesting not only in view of the above conceptual problem, but also in view of discovery of high energy astrophysical sources involving strong fluxes of photons and neutrinos.

Consider for simplicity the Schwarzschild case. Timelike geodesics parameterized by the proper time τ obey the radial equation $(dr/d\tau)^2+U(r) = 0$ with the effective potential

$$U(r) = \left(1 - \frac{2M}{r}\right)\left(\frac{L^2}{r^2} + 1\right) - \gamma^2\,,$$

where $\gamma = E/\mu$ as before and L is the angular momentum. The most interesting are the winding orbits which perform many turns around black hole before being scattered back, or absorbed by the hole. To estimate gravitational radiation one considers circular orbits, whose radii $r_{\rm p}$ correspond to two conditions $U(r_{\rm p}) = 0 = U'(r_{\rm p})$. Solving these equations with respect to γ and L, one finds $\gamma = (1 - 2M/r_{\rm p})\,(1 - 3M/r_{\rm p})^{-1/2}$, $L/\gamma = (Mr_{\rm p})^{1/2}\,(1 - 2M/r)^{-1}$, while the rotation frequency in terms of the orbit radius $r_{\rm p}$ reads $\omega_0 = d\phi/dt = (M/r_{\rm p})^{1/2}$. Relativistic time-like circular orbits with $3M < r_{\rm p} < 6M$ are unstable. Those which lie in the interval $3M < r_{\rm p} < 4M$ become unbound under small perturbations, they correspond to large angle scattering with the impact parameter $b = L/\left(\gamma^2 - 1\right)^{1/2} = r_{\rm p}\,(4M/r_{\rm p} - 1)^{-1/2}$ In the ultrarelativistic case $\gamma \gg 1$ the unbound orbits with the impact parameter close to the critical value $b = 3\sqrt{3}M$ scatter on the black hole with multiple revolutions, radiation from these orbits can be estimated using the GSR power. The limit $\gamma = \infty$ corresponds to the massless particles (photon orbits), the corresponding rotation radius being $r_{\rm ph} = 3M$. Its dislocation with respect to timelike ultrarelativistic orbits can be characterized by a small parameter δ via $r_{\rm p} = (3+\delta)M$. Two useful relations then follow in the leading order in δ: $dt/d\tau = \sqrt{3/\delta}$, $\gamma^2 = 1/3\delta$.

Radiation field is expanded in terms of the appropriate angular harmonics, labeled by two integers l, m, $|m| \leqslant l$, the main contribution for large γ coming from $|m| \gg 1$ and l differing from $|m|$ by 0, 1, depending on polarization. The total intensity can be presented as a sum

$$P_{\rm GSR} = \sum_{m=1}^{\infty} \frac{E^2\,\omega_0}{M}\, F_m(r_p, M)\, \frac{m_{\rm cr}}{m}\, {\rm e}^{-m/m_{\rm cr}}\,, \tag{11}$$

where each term corresponds to radiation with the frequency $\omega = m\omega_0$, F_m is a smooth function of the parameters, and the critical frequency is

$$m_{\rm cr} = \frac{12}{\pi}\gamma^2\,. \tag{12}$$

The spectrum is therefore a falling function of the harmonic number m, and it is cut off classically at the frequency γ times smaller than in the case of flat space synchrotron radiation (because of the increasing formation length due to closeness of ultrarelativistic timelike geodesics of the radiating particle and the null geodesics followed by gravitons [14]).

The main contribution (96%) to the total power comes from the polarization commonly denoted as $\otimes$. This quantity was computed in Ref. 16 (in earlier papers only the distribution over the harmonics was given) and reads:

$$P_{\rm GSR} = \frac{6{\rm e}^{-\pi/4}(r_{\rm p}-M)\,|\Gamma(1/4+i/4)|^2}{\pi^{3/2} r_{\rm p}^2 (r_{\rm p}+3M)} E^2 \ln(E/\mu)\,. \tag{13}$$

The last factor reflects the logarithmic divergence of the sum (11) if the frequency cutoff (12) is removed, what happens in the limit of zero mass $\gamma = \infty$. We therefore conclude that General Relativity fails to provide full description of gravitational radiation form massless particles. Quantum theory of gravitational synchrotron radiation is not developed yet, but one can hope to get a correct estimate of the total power replacing the divergent logarithm by $\ln(E/\hbar\omega_0)$ as we tested in the electromagnetic case.

Radiation of scalar $s = 0$ and vector $s = 1$ waves is described similarly to Eq. (11) with different m-dependence, namely $(m/m_{\rm cr})^{1-s}$, with $s = 2$ standing for gravitational waves. Thus the divergence of the total power as

$\mu \to 0$ is γ^2 for $s = 0, 1$. Softer divergence in the gravitational case is due to the fact that the effective gravitational coupling is proportional to the energy, as was noted by Weinberg long ago.[6]

This work was supported by RFBR under the project 14-02-01092-a.

References

1. F. Azzurli and K. Lechner, *Phys. Lett. A* **377**, 1025 (2013) [arXiv:1212.3532]; F. Azzurli and K. Lechner, *Annals Phys.* **349**, 1 (2014) [arXiv:1401.5721 [hep-th]].
2. K. Lechner, *J. Math. Phys.* **56**, no. 2, 022901 (2015) [arXiv:1405.4805 [hep-th]].
3. L. Xiang and Y. G. Shen, *Int. J. Theor. Phys.* **46**, 576 (2007); B. P. Kosyakov, *J. Phys. A* **41**, 465401 (2008) [arXiv:0705.1228 [hep-th]]; Y. Yaremko, *Electron. J. Theor. Phys.* **9**, no. 26, 153 (2012).
4. A. A. Sokolov and I. M. Ternov, Synchrotron Radiation, Pergamon, New York, 1968).
5. D. V. Gal'tsov, *Phys. Lett. B* **747**, 400 (2015) [arXiv:1505.06775 [hep-th]].
6. S. Weinberg, *Phys. Rev.* **140**, B516 (1965).
7. T. Kinoshita, *J. Math. Phys.* **3**, 650 (1962).
8. T. D. Lee and M. Nauenberg, *Phys. Rev.* **133**, B1549 (1964).
9. V. G. Vaks, *Soviet Physics JETP* **13** 556-561 (1961); V. N. Gribov, *Nucl. Phys. B* **206**, 103 (1982).
10. J. Schwinger, Particles, Sources, and Fields (Addison- Wesley, Heading, Mass.), Vol. III, Chap. 5, Sec. 6; *Phys. Rev. D* 7, 1969 (1973).
11. W. Y. Tsai, *Phys. Rev. D* **8**, 3460 (1973).
12. V. N. Baier, V. M. Katkov and V. M. Strakhovenko, *Sov. Phys. JETP* **71**, 657 (1990) [*Zh. Eksp. Teor. Fiz.* **98**, 1173 (1990)].
13. C. W. Misner, R. A. Breuer, D. R. Brill, P. L. Chrzanowski, H. G. Hughes and C. M. Pereira, *Phys. Rev. Lett.* **28**, 998 (1972); R. A. Breuer, R. Ruffini, J. Tiomno and C. V. Vishveshwara, *Phys. Rev. D* **7**, 1002 (1973).
14. I. B. Khriplovich and E. V. Shuryak, *Zh. Eksp. Teor. Fiz.* **65**, 2137 (1973).
15. P. L. Chrzanowski and C. W. Misner, *Phys. Rev. D* **10**, 1701 (1974).
16. D. V. Gal'tsov and A. A. Matyukhin, *Yad. Fiz.* **45**, 894 (1987).

Schwinger effect in (A)dS and charged black hole

Sang Pyo Kim

Department of Physics, Kunsan National University,
Kunsan 573-701, Korea
E-mail: sangkim@kunsan.ac.kr

In an Anti-de Sitter space and a charged black hole the Schwinger effect is either enhanced by the Hawking radiation or suppressed by the negative curvature. We use the contour integral method to calculate the production of charged pairs in the global (A)dS space. The charge emission from near-extremal black hole is found from the AdS geometry near the horizon and interpreted as the Schwinger effect in a Rindler space with the surface gravity for the acceleration as well as the Schwinger effect in AdS space.

Keywords: Schwinger effect; Hawking radiation; (Anti-) de Sitter space; charged black hole, Unruh effect.

1. Introduction

A charged Reissner-Nordström (RN) black hole emits not only the Hawking radiation of all species of particles due to the horizon but also the Schwinger emission of charged particles due to the electric field near the horizon, and thus provides an arena where one may explore the intertwinement between quantum electrodynamics (QED) and quantum aspect of black holes. The pair production by a uniform electric field in an (A)dS space also exhibits the interplay between QED and quantum gravity effect. As summarized in Fig. 1, vacuum fluctuations produce pairs from the vacuum and the horizon of a black hole or dS space and the electric field physically separate pairs. The black hole radiation has a thermal distribution with the Hawking temperature as a Unruh temperature with the surface gravity on the horizon.[1] Cai and Kim show that the Schwinger effect in the (A)dS space has also a thermal interpretation with a Unruh temperature of accelerating charges in that space.[2] Using the geometry $AdS_2 \times S^2$ near the horizon of a near-extremal black hole, Chen et al have explicitly found the Schwinger effect from the charged black hole.[3,4]

In this paper, after briefly reviewing the recently introduced contour integral method,[5] we apply it to the pair production by a uniform electric field in the global (A)dS_2 geometry. Then, using the near-horizon geometry $AdS_2 \times S^2$ of a near-extremal black hole, we find the Schwinger effect from the charged black hole and propose a thermal interpretation of the charge emission.[6,7] In the in-in formalism, the quantum field theory is extended to the complex plane of time or space and the out-vacuum transported along a loop starting from the past infinity is distinguished from the in-vacuum by simple poles of the frequency or momentum in the background spacetime and/or electromagnetic field.[5] Hence, particle production is characterized and determined by polons, that is, contour integrals of all homotopy classes of winding number one encircling poles.[5,8,9] Differently from the planar coordinates without any finite pole, a massive scalar field in the global coordinates has simple poles both at the north and south pole of the global geometry, whose contributions interfere and result in the Stokes phenomenon of the dS radiation depending on the spacetime dimension.[8] It is investigated whether the Schwinger effect may have such a Stokes phenomenon in the global dS_2 geometry. The holographic Schwinger effect is also studied in the dS space[10] and AdS space.[11]

2. Pair Production via Contour Integrals

In the functional Schrödinger picture and the in-in formalism, the in-vacuum in the past infinity evolves to the future infinity and return to

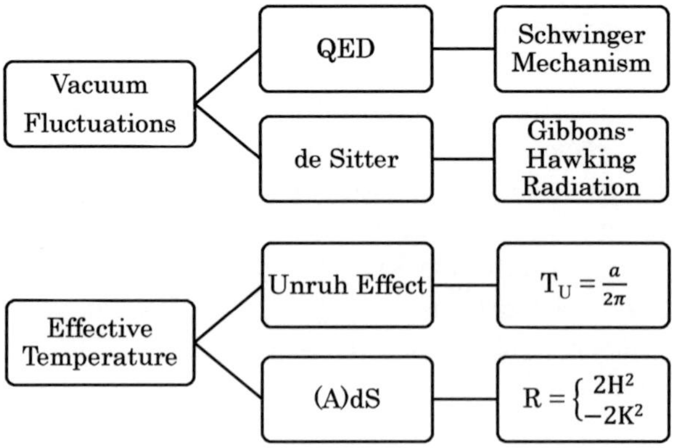

Fig. 1. The Schwinger effect and the Gibbons-Hawking radiation in dS from quantum fluctuations. The temperature for the Unruh effect in (A)dS_2 is $T_{\rm eff} = \sqrt{T_U^2 + \frac{R}{8\pi^2}}$.[12]

the past infinity, which gives the mean number of pairs[5]

$$\mathcal{N}_\kappa = \Big|\sum_J \langle 0_\kappa, \text{in}|0_\kappa, C_J^{(1)}\rangle^2\Big| = \Big|\sum_J e^{-i\oint_{C_J^{(1)}} \omega_\kappa(z)dz}\Big|. \quad (1)$$

Here, the quantum evolution is extended to the complex plane and $C_J^{(1)}$ denotes distinct contours of winding number one. For instance, the global dS geometry has simple poles not only at infinity but also at the north and south poles, which contribute constructively or destructively depending on whether the spacetime dimension is even or odd.[8] On the other hand, the planar dS space has the simple pole at infinity only and has a nondestructive radiation with the Gibbons-Hawking temperature. A physical reasoning may be that the propagator and hence the Hyugens principle exhibits a different nature. In the in-in formalism, the scattering matrix between the in-vacuum and another in-vacua tranported along contours $C_J^{(n)}$ of all winding numbers gives the vacuum persistence

$$2\text{Im}\mathcal{L}_{\text{eff}} = \pm\sum_\kappa\Big|\sum_J\sum_n \ln(\langle 0_\kappa, \text{in}|0_\kappa, C_J^{(n)}\rangle^2)\Big| = \pm\sum_\kappa \ln(1 \pm \mathcal{N}_\kappa), \quad (2)$$

where the upper (lower) sign is for scalar (spinor) QED and the mean number $\mathcal{N}_\kappa$ is determined by the contours of winding number one in Eq. (1).

3. Schwinger Effect in (A)dS

The production of charged pairs by a uniform electric field in a dS space exhibits both the Schwinger effect and the Gibbons-Hawking radiation as shown in Fig. 1.[13,14] An AdS space confines virtual pairs and limits the Schwinger pair production through the Breitenlohner-Friedman bound.[2,14,15] Now, we consider the Schwinger effect in the two-dimensional global dS geometry

$$ds^2 = -dt^2 + \cosh^2(Ht)dx^2, \quad (3)$$

and in the uniform electric field with a vector potential $A_1 = -E\sinh(Ht)/H$. We distinguish the weak-field limit ($qE \ll H^2/2$) from the strong-field limit ($qE \gg H^2/2$), in which the pair production has different characteristics.

First, we consider the weak-field limit ($\gamma > 0, \lambda > 0$ below). Following Ref. 8, a charged scalar has the time-dependent frequency

$$\omega^2(t) = (\gamma H)^2 + \frac{(\lambda H)^2}{\cosh^2(Ht)}, \quad (4)$$

where

$$\gamma = \sqrt{\left(\frac{qE}{H^2}\right)^2 + \frac{m^2}{H^2} - \frac{1}{4}}, \quad \lambda = \sqrt{\frac{1}{4} - \left(\frac{qE}{H^2}\right)^2}. \tag{5}$$

The frequency (4) has two poles at $\tau = \pm i\pi/(2H)$ corresponding to the north and south pole of the global geometry, which lead to the so-called Stokes phenomenon of dS radiation.[8] As will be shown below, the Schwinger effect has also the Stokes phenomenon in the weak-field limit. Let us introduce a conformal transformation $z = e^{Ht}$ and restrict to a single Riemann sheet. Then, the contour integral takes the form

$$\oint \omega(t)dt = \gamma \oint \frac{dz}{z(z^2+1)} \sqrt{(z-z_1)(z-z_1^*)(z-z_2)(z-z_2^*)}, \tag{6}$$

where

$$z_1 = i\Big(\sqrt{1 + \frac{\lambda^2}{\gamma^2}} + \frac{\lambda}{\gamma}\Big), \quad z_2 = i\Big(\sqrt{1 + \frac{\lambda^2}{\gamma^2}} - \frac{\lambda}{\gamma}\Big). \tag{7}$$

The integrand has finite simple poles at $z = \pm i$ corresponding to the north and south pole of the global geometry and another simple pole from $z = \infty$. The residues at $z = \pm i$ take the opposite signs due to the branch cuts. Summing over four distinct contours of winding number one, we obtain from Eq. (1)

$$\mathcal{N}_{\rm dS} = \big(e^{2i\pi\lambda} + 2 + e^{-2i\pi\lambda}\big)e^{-2\pi\gamma} = 4\cos^2(\pi\lambda)e^{-2\pi\gamma}. \tag{8}$$

The oscillatory factor is the characteristic of the Stokes phenomenon.

Second, in the strong field limit ($\gamma > 0, \bar{\lambda} > 0$) the integrand has different roots

$$z_1^{(\pm)} = \frac{\bar{\lambda}}{\gamma} \pm i\sqrt{1 - \frac{\bar{\lambda}^2}{\gamma^2}}, \quad z_2^{(\pm)} = -\frac{\bar{\lambda}}{\gamma} \pm i\sqrt{1 - \frac{\bar{\lambda}^2}{\gamma^2}}, \tag{9}$$

where

$$\bar{\lambda} = \sqrt{\left(\frac{qE}{H^2}\right)^2 - \frac{1}{4}}. \tag{10}$$

Then, the mean number of pairs from four distinct contours, for instance Fig. 2, is given by

$$\mathcal{N}_{\rm dS} = \big(e^{2\pi\bar{\lambda}} + 2 + e^{-2\pi\bar{\lambda}}\big)e^{-2\pi\gamma} = 4\cosh^2(\pi\bar{\lambda})e^{-2\pi\gamma}. \tag{11}$$

The mean number (11) is consistent with the exact result from the Bogoliubov transformation using the exact solution.[16]

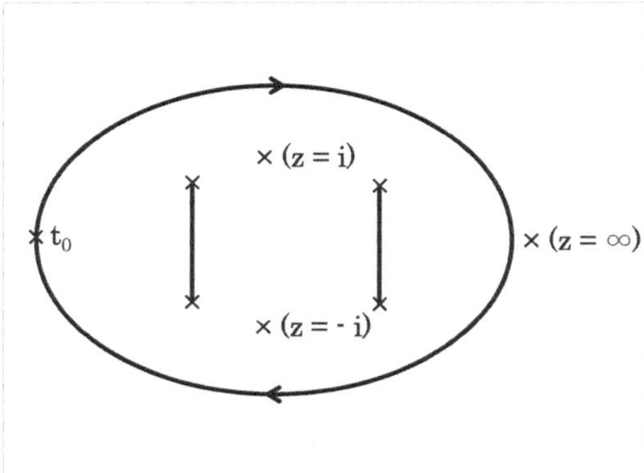

Fig. 2. The frequency (6) for a charged scalar in a uniform electric field and the global dS_2 has two pairs of branch cuts in the complex plane and a pair of simple poles at $z = \pm i$ and another simple pole at $z = \infty$. The contour includes both poles $z = \pm i$, whose residues sum vanishes but receives a contribution from $z = \infty$.

Finally, a charged scalar in the global AdS geometry

$$ds^2 = -\cosh^2(Kx)dt^2 + dx^2, \tag{12}$$

and in a vector potential $A_0 = -E\sinh(Kx)/K$ for the uniform electric field has the momentum

$$q^2(x) = (\gamma_{\mathrm{AdS}}K)^2 - \frac{(\lambda_{\mathrm{AdS}}K)^2}{\cosh^2(Kx)}, \tag{13}$$

where

$$\gamma_{\mathrm{AdS}} = \sqrt{\left(\frac{qE}{K^2}\right)^2 - \frac{m^2}{K^2} - \frac{1}{4}}, \quad \lambda_{\mathrm{AdS}} = \sqrt{\left(\frac{qE}{K^2}\right)^2 + \frac{1}{4}}. \tag{14}$$

As for the dS space, four distinct contours contribute to the mean number

$$\mathcal{N}_{\mathrm{AdS}} = 4\sinh^2(\pi\lambda_{\mathrm{AdS}})e^{-2\pi\gamma_{\mathrm{AdS}}}. \tag{15}$$

4. Schwinger Effect in Near-Extremal Charged Black Hole

The near-extremal black hole has the near-horizon geometry[3,4]

$$ds^2 = -\frac{\rho^2 - B^2}{Q^2}d\tau^2 + \frac{Q^2}{\rho^2 - B^2}d\rho^2 + Q^2 d\Omega_2^2, \tag{16}$$

where the coordinates are stretched as $\rho = (r - Q)/\epsilon, \tau = \epsilon t$ and the deviation of the charge from the mass is measured as $M - Q = (\epsilon B)^2/(2Q)$.

The Coulomb potential is $A_0 = -\rho/Q$ on the horizon. The geometry (16) is the planar coordinates of $AdS_2 \times S^2$. The radial momentum for the spherical harmonic Y_{lm} and energy ω is given by

$$S(\rho) = \int \frac{\sqrt{(q\rho - \omega Q)^2 Q^2 - \left(m^2 Q^2 + (l+1/2)^2\right)(\rho^2 - B^2)}}{\rho^2 - B^2} d\rho. \quad (17)$$

Note that pairs are produced when the BF bound ($Q < l/q$) for AdS is violated.[2] Taking a contour of figure eight counterclockwise enclosing $\rho = B$ and $\rho = -B$ gives the leading term $\mathcal{N} = e^{-(\mathcal{S}_a - \mathcal{S}_b)}$, whereas the exact formula from the field equation is given by[3,4]

$$\mathcal{N} = \left(\frac{e^{-(\mathcal{S}_a - \mathcal{S}_b)} - e^{-(\mathcal{S}_a + \mathcal{S}_b)}}{1 + e^{-(\mathcal{S}_a + \mathcal{S}_b)}}\right)\left(\frac{1 - e^{-(\mathcal{S}_c - \mathcal{S}_a)}}{1 + e^{-(\mathcal{S}_c - \mathcal{S}_b)}}\right), \quad (18)$$

where

$$\mathcal{S}_a = 2\pi q Q, \; \mathcal{S}_b = 2\pi q Q \sqrt{1 - \left(\frac{m}{q}\right)^2 - \left(\frac{l+1/2}{qQ}\right)^2}, \; \mathcal{S}_c = 2\pi \frac{\omega}{B} Q^2. \quad (19)$$

Interestingly, the mean number of emitted particles consists of three parts, the Boltzmann amplification factor, the Schwinger effect in AdS_2, and the Schwinger effect in a two-dimensional Rindler space:[6,7]

$$\mathcal{N} = e^{\frac{\bar{m}}{T_{\rm CK}}} \times \left\{\frac{e^{-\frac{\bar{m}}{T_{\rm CK}}} - e^{-\frac{\bar{m}}{\bar{T}_{\rm CK}}}}{1 + e^{-\frac{\bar{m}}{\bar{T}_{\rm CK}}}}\right\} \times \left\{\frac{e^{-\frac{\bar{m}}{\bar{T}_{\rm CK}}}\left(1 - e^{-\frac{\omega - qA(r_{\rm H})}{T_{\rm H}}}\right)}{1 + e^{-\frac{\omega - qA(r_{\rm H})}{T_{\rm H}}} e^{-\frac{\bar{m}}{\bar{T}_{\rm CK}}}}\right\}. \quad (20)$$

Here, $\bar{m} = \sqrt{m^2 + (l+1/2)^2/Q^2}$, $T_{\rm H}$ is the Hawking temperature, and $T_{\rm CK}$ and $\bar{T}_{\rm CK}$ are the effective temperature in AdS_2 introduced by Cai and Kim[2]

$$T_{\rm CK} = T_{\rm U} + \sqrt{T_{\rm U}^2 - \left(\frac{1}{2\pi Q}\right)^2}, \quad \bar{T}_{\rm CK} = T_{\rm U} - \sqrt{T_{\rm U}^2 - \left(\frac{1}{2\pi Q}\right)^2}, \quad (21)$$

where $T_{\rm U} = q/(2\pi \bar{m} Q)$ is the Unruh temperature for charge accelerated by the electric field on the horizon. The Schwinger effect and vacuum persistence in the Rindler space has been first calculated by Gabriel and Spindel.[17]

Acknowledgments

The author thanks Rong-Gen Cai and Chiang-Mei Chen for useful discussions. The participation of ICGAC12 was supported in part by APCTP. This work was supported by IBS (Institute for Basic Science) under IBS-R012-D1.

References

1. W. G. Unruh, Notes on black-hole evaporation, *Phys. Rev. D* **14**, 870 (1976).
2. R.-G. Cai and S. P. Kim, One-loop effective action and Schwinger effect in (anti-) de Sitter space, *JHEP* **09** (2014) 72.
3. C.-M. Chen, S. P. Kim, I.-C. Lin, J-R. Sun and M.-F. Wu, Spontaneous pair production in Reissner-Nordstrm black holes, *Phys. Rev. D* **85**, 124041 (2012).
4. C.-M. Chen, J.-R. Sun, F.-Y. Tang and P.-Y. Tsai, Spinor particle creation in near extremal Reissner–Nordstrm black holes, *Class. Quantum Grav.* **32**, 195003 (2015).
5. S. P. Kim, New geometric transition as origin of particle production in time-dependent backgrounds, *Phys. Lett. B* **725**, 500 (2013).
6. S. P. Kim, H. K. Lee, and Y. Yoon, Thermal Interpretation of Schwinger Effect in Near-Extremal Reissner-Nordström Black Hole [arXiv:1503.00218].
7. S. P. Kim, Schwinger Effect, Hawking Radiation and Gauge-Gravity Relation [arXiv:1506.03990].
8. S. P. Kim, Geometric origin of Stokes phenomenon for de Sitter radiation, *Phys. Rev. D* **88**, 044027 (2013).
9. S. P. Kim, Particle Production from Geometric Transition in Expanding Universe, *JPS Conf. Proc.* **1**, 013110 (2014).
10. W. Fischler, P. H. Nguyen, J. F. Pedraza, and W. Tangarife, Holographic Schwinger effect in de Sitter space, *Phys. Rev. D* **91**, 086015 (2015).
11. C.-M. Chen, Pair Production in Near Extremal Charged Black Holes, in this Proceedings (2015).
12. S. Deser and O. Levin, Accelerated detectors and temperature in (anti-) de Sitter spaces, *Class. Quantum Grav.* **14**, L163 (1997).
13. J. Garriga, Pair production by an electric field in (1+1)-dimensional de Sitter space, *Phys. Rev. D* **49**, 6343 (1994).
14. S. P. Kim and D. N. Page, Schwinger pair production in dS_2 and AdS_2, *Phys. Rev. D* **78**, 103517 (2008).
15. B. Pioline and J. Troost, *JHEP* **03** (2005) 043.
16. S. P. Kim, P. W.-Y. Hwang and T.-C. Wang, Schwinger mechanism in dS_2 and AdS_2 revisited [arXiv:1112.0885].
17. Cl. Gabriel and Ph. Spindel, Quantum Charged Fields in (1+1) Rindler Space, *Ann. Phys.* **284**, 263 (2000).

Pair production in near extremal charged black holes

Chiang-Mei Chen[1a], Jia-Rui Sun[2b] and Fu-Yi Tang[1c]

[1] *Department of Physics, National Central University, Chungli 320, Taiwan*
[2] *Institute of Astronomy and Space Science, Sun Yat-Sen University, Guangzhou 510275, China*
[a] *cmchen@phy.ncu.edu.tw*, [b] *sunjiarui@sysu.edu.cn*, [c] *foue.tang@gmail.com*

We study the spontaneous pair production, including the Schwinger mechanism and the Hawking radiation, of charged scalar and spinor particles from the near horizon region of (near) extremal charged black holes in the probe field limit. The pair production rate and the absorption cross section, as well as the retarded Green's functions of the probe fields are analytically computed. Moreover, the holographic description dual to the pair production is discussed.

Keywords: Pair production; black hole; Hawking radiation; Schwinger mechanism.

1. Introduction

The spontaneous pair production occurring in charged black holes mixes two independent processes, the Schwinger mechanism[1] and the Hawking radiation.[2] Since this effect is expected to happen in the near horizon region, we investigate the particle emissions (scalar and spinor particles) in the spacetime of the near-horizon region of the near extremal Reissner-Nordström (RN) black hole[3,4] without back-reactions. The spacetime has the geometric structure $AdS_2 \times S^2$ plus a constant electric field, which allows us to analytically solve the Klein-Gorden (KG) equation and the Dirac equation for the probe scalar and spinor fields, respectively and obtain the exact solutions in terms of the well-known hypergeometric functions. Further, by imposing the particle viewpoint boundary condition on the solutions, the physical quantities: the vacuum persistence amplitude, the mean number of produced pairs, the absorption cross section ratio and the retarded Green's function of the probe fields are gained. In addition, the physical quantities calculated from the gravity side are showed to match well with those results of the scalar and spinor operators in the dual boundary conformal field theories (CFTs) side, based on the RN/CFT correspondence.[5–10] Moreover, the corresponding thermal interpretation has been discussed.[11–13]

2. The Near Horizon Near Extreme RN Black Holes

The geometry of the near horizon near extreme RN black hole has the structure of an $AdS_2 \times S^2$ as[3]

$$ds^2 = -\frac{\rho^2 - B^2}{Q^2} d\tau^2 + \frac{Q^2}{\rho^2 - B^2} d\rho^2 + Q^2 d\Omega_2^2,$$
$$A = -\frac{\rho}{Q} d\tau; \quad F = \frac{1}{Q} d\tau \wedge d\rho, \tag{1}$$

where ρ is the radial coordinate of the AdS_2 section, B labels the deviation from extreme limit and acts as the new horizon radius in this geometry, and Q is the charge of the original RN black hole. Geometrically, the positive curvature of S^2 exactly compensates the negative part of the AdS_2.

3. Particle Creation

3.1. *Boundary conditions*

There is a potential barrier due to the electromagnetic and gravitational forces in the near horizon region, the pair production becomes a tunneling process by considering the fluxes in the scattering matrix theory. Two boundary conditions can be imposed (see Fig. 1):

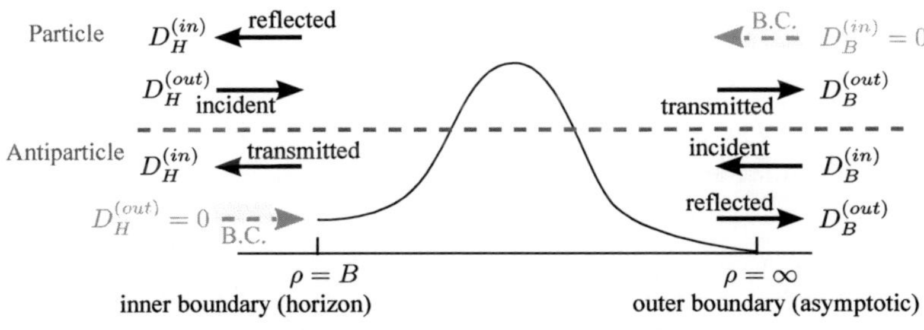

Fig. 1. Outer boundary condition (upper picture): no incoming flux at asymptotic. Inner boundary condition (lower picture): no outgoing flux at horizon.

The first one is the outer boundary condition (particle viewpoint), which restricts no incoming fluxes at the asymptotic outer boundary. In the Stückelberg-Feynman picture, the outgoing (transmitted) flux at the asymptotic represents the spontaneous produced *particle*, the outgoing (incident) flux at the horizon represents the total particles created by vacuum fluctuations, and the incoming (reflected) flux represents the re-annihilated.

The second one is the inner boundary condition (antiparticle viewpoint) that restricts no outgoing fluxes at the inner boundary, which means that the incoming (transmitted) flux at the horizon represents the spontaneous produced *antiparticle*, the incoming (incident) and the outgoing (reflected) fluxes at the asymptotic represents the total created antiparticles and the re-annihilated. Since the particle and anti-particle are always produced in pair due to the charge conservation and/or the energy-momentum conservation, these two boundary conditions are actually equivalent.[3]

3.2. *Physical quantities*

For bosonic and fermionic particles, there are two versions of flux conservations[14]

$$\begin{aligned} \text{bosons}: \quad & |D_{\text{incident}}| = |D_{\text{reflected}}| + |D_{\text{transmitted}}|, \\ \text{fermions}: \quad & |D_{\text{reflected}}| = |D_{\text{incident}}| + |D_{\text{transmitted}}|, \end{aligned} \tag{2}$$

and two versions of Bogoliubov relations

$$\text{bosons}: \quad |\mathcal{A}|^2 - |\mathcal{B}|^2 = 1, \qquad \text{fermions}: \quad |\mathcal{A}|^2 + |\mathcal{B}|^2 = 1. \tag{3}$$

where the vacuum persistence amplitude $|\mathcal{A}|^2$ and the mean number of produced pairs $|\mathcal{B}|^2$ are given by the ratio of the flux components in the Coulomb gauge

$$|\mathcal{A}|^2 \equiv \frac{D_{\text{incident}}}{D_{\text{reflected}}}, \quad |\mathcal{B}|^2 \equiv \frac{D_{\text{transmitted}}}{D_{\text{reflected}}}. \tag{4}$$

Moreover, from the viewpoint of scattering of an incident flux from the asymptotic boundary, we can define the absorption cross section ratio as

$$\sigma_{\text{abs}} \equiv \frac{D_{\text{transmitted}}}{D_{\text{incident}}} = \frac{|\mathcal{B}|^2}{|\mathcal{A}|^2}. \tag{5}$$

4. Scalar Production

The EoM of the probe scalar field Φ with mass m and charge q propagating in the near horizon geometry (1) is

$$D_\alpha D^\alpha \Phi - m^2 \Phi = 0, \tag{6}$$

where $D_\alpha \equiv \nabla_\alpha - iqA_\alpha$ with ∇_α being the usual covariant derivative. The corresponding radial flux is

$$D = i\sqrt{-g}g^{\rho\rho}(\Phi D_\rho \Phi^* - \Phi^* D_\rho \Phi). \tag{7}$$

Using the ansatz $\Phi = \exp(-i\omega\tau + in\phi)R(\rho)S(\theta)$, the KG equation can be separated into the standard spherical harmonic equation with eigenvalue $\lambda_\ell = \ell(\ell+1)$ (ℓ: integer) and the radial part

$$\partial_\rho[(\rho^2 - B^2)\partial_\rho R] + \left[\frac{(q\rho - \omega Q)^2 Q^2}{\rho^2 - B^2} - m^2 Q^2 - \lambda_\ell\right] R = 0, \tag{8}$$

which resembles the EoM of a scalar field with effective mass $m_{\text{eff}}^2 = m^2 - q^2 + \lambda_\ell/Q^2$ propagating in an AdS$_2$ geometry with radius $L_{\text{AdS}} = Q$. We find that the existence condition for the pair production is the violation of the Breitenlohner-Freedman (BF) bound[15,16] in the AdS$_2$ (or effectively AdS$_3$) spacetime, i.e.

$$(m^2 - q^2)Q^2 + (\ell + 1/2)^2 < 0, \tag{9}$$

which indicates that the mass of the created particle should be smaller than its charge.

Eq.(8) can be exactly solved by the following solution

$$\begin{aligned} R(\rho) &= c_1(\rho - B)^{(-i/2)(a-\tilde{a})}(\rho + B)^{(i/2)(a+\tilde{a})} F\Big(\frac{1}{2} + i\tilde{a} + ib, \frac{1}{2} + i\tilde{a} - ib; \\ &\quad 1 - ia + i\tilde{a}; \frac{1}{2} - \frac{\rho}{2B}\Big) + c_2(\rho - B)^{(i/2)(a-\tilde{a})}(\rho + B)^{(i/2)(a+\tilde{a})} \\ &\quad F\Big(\frac{1}{2} + ia + ib, \frac{1}{2} + ia - ib; 1 + ia - i\tilde{a}; \frac{1}{2} - \frac{\rho}{2B}\Big), \end{aligned} \tag{10}$$

with parameters $a \equiv qQ, \tilde{a} \equiv \omega Q^2/B, b \equiv \sqrt{(q^2 - m^2)Q^2 - (\ell + 1/2)^2}$. Then the Bogoliubov coefficients and the absorption cross section ratio in particle viewpoint can be obtained as

$$\begin{aligned} |\mathcal{A}|^2 &= \frac{|D_{\text{H}}^{(\text{out})}|}{|D_{\text{H}}^{(\text{in})}|} = \frac{\cosh(\pi a - \pi b)\cosh(\pi\tilde{a} + \pi b)}{\cosh(\pi a + \pi b)\cosh(\pi\tilde{a} - \pi b)}, \\ |\mathcal{B}|^2 &= \frac{|D_{\infty}^{(\text{out})}|}{|D_{\text{H}}^{(\text{in})}|} = \frac{\sinh(2\pi b)\sinh(\pi\tilde{a} - \pi a)}{\cosh(\pi a + \pi b)\cosh(\pi\tilde{a} - \pi b)}, \\ \sigma_{\text{abs}} &= \frac{|\mathcal{B}|^2}{|\mathcal{A}|^2} = \frac{\sinh(2\pi b)\sinh(\pi\tilde{a} - \pi a)}{\cosh(\pi a - \pi b)\cosh(\pi\tilde{a} + \pi b)}. \end{aligned} \tag{11}$$

where $D_H^{(\text{out})}, D_H^{(\text{in})}$ are the outgoing and ingoing fluxes at horizon, respectively and $D_\infty^{(\text{out})}$ is the outgoing flux at asymptotic boundary (results for antiparticle viewpoint can be found in Ref. 3). In the extremal limit $B \to 0\,(\tilde{a} \to \infty)$, the leading term of $|\mathcal{B}|^2$ leads to the Schwinger formula $|\mathcal{B}|^2 \simeq e^{-\pi m^2 Q/q} \simeq e^{-\pi m^2 r_H^2/qQ}$. We can compare the ratio $|\mathcal{B}(B=0)|^2/|\mathcal{B}(B \neq 0)|^2 \geq 1$, which means the production rate in the extremal limit (Schwinger) is greater than that in the near extremal limit (Schwinger + Hawking). In other words, as the geometry changes from the extremal to near extremal black hole, the increasing gravitational force will reduce the electromagnetic repulsive force for the Schwinger mechanism[3].

5. Spinor Production

The EoM for charged spinor particles is the Dirac equation

$$[\gamma^a e_a{}^\mu (\partial_\mu + \Gamma_\mu - iqA_\mu) + m]\,\Psi = 0, \tag{12}$$

where $e_a{}^\mu$ is the tetrad and $\Gamma = \Gamma_\mu dx^\mu = \frac{1}{4}\gamma^a\gamma^b\omega_{ab}$ is the spin connection one-form. Choosing the ansatz $\Psi = \exp(-i\omega\tau)[R_+(\rho)\Phi^+_{\kappa,n}(\theta,\phi) + R_-(\rho)\Phi^-_{\kappa,n}(\theta,\phi)]$, where $\Phi^\pm_{\kappa,n}$ are the spherical spinors correspond to the eigenvalues, $\kappa = \mp(j+1/2)$ (j: half-integer) and the projection $n(-\ell < n < \ell)$, eq.(12) reduces to two first order coupled equations[4]

$$\frac{\sqrt{\rho^2 - B^2}}{Q}\partial_\rho \mathcal{R}_\pm \mp i\frac{\omega Q - q\rho}{\sqrt{\rho^2 - B^2}}\mathcal{R}_\pm + \left(\frac{\kappa}{Q} \pm im\right)\mathcal{R}_\mp = 0, \tag{13}$$

where $\mathcal{R}_\pm = R_+ \pm R_-$. Using the new coordinate $z = (\rho + B)/2B$ and two re-scaled functions $\tilde{\mathcal{R}}_\pm = \Sigma^{\pm 1}\mathcal{R}_\pm$ where $\Sigma = (2B)^{ia} z^{i(\tilde{a}+a)/2}(z-1)^{-i(\tilde{a}-a)/2}$ with the parameters $a \equiv qQ, \tilde{a} \equiv \omega Q^2/B, \bar{b} \equiv \sqrt{(q^2 - m^2)Q^2 - (j+1/2)^2}$, eq.(13) can be further reduced to the hypergeometric differential equations

$$z(1-z)\partial_z^2\tilde{\mathcal{R}}_\pm + [\gamma_\pm - (\alpha_\pm + \beta_\pm + 1)z]\,\partial_z\tilde{\mathcal{R}}_\pm - \alpha_\pm\beta_\pm\tilde{\mathcal{R}}_\pm = 0, \tag{14}$$

with $\alpha_\pm = i(\bar{b} \mp a), \beta_\pm = -i(\bar{b} \pm a)$ and $\gamma_\pm = \frac{1}{2} \mp i(\tilde{a} + a)$. The solutions are

$$\begin{aligned}\tilde{\mathcal{R}}_\pm &= C_\pm F\left(i(\bar{b} \mp a), -i(\bar{b} \pm a); \frac{1}{2} \mp i(\tilde{a}+a); z\right) \\ &\quad + \bar{C}_\pm z^{\frac{1}{2} \pm i(\tilde{a}+a)} F\left(\frac{1}{2} - i(\bar{b} \mp \tilde{a}), \frac{1}{2} + i(\bar{b} \pm \tilde{a}); \frac{3}{2} \pm i(\tilde{a}+a); z\right).\end{aligned} \tag{15}$$

Also, the condition for the pair production to happen requires the parameter $\bar{b}$ to be real, namely,

$$(m^2 - q^2)Q^2 + (j + 1/2)^2 < 0. \tag{16}$$

which breaks the BF bound of spinor fields in the AdS_2 (or effectively AdS_3) spacetime. Moreover, according to the charge and energy conservation, the black holes lose their charge more than mass in the process of pair production, which ensures the cosmic censorship conjecture when the back-reaction is not included.

Likewise, the Bogoliubov coefficients and the absorption cross section ratio in the particle picture are given by

$$\begin{aligned} |\mathcal{A}|^2 &= \frac{|D_{\rm H}^{(\rm out)}|}{|D_{\rm H}^{(\rm in)}|} = \frac{\sinh(\pi a - \pi\bar{b})\cosh(\pi\tilde{a} + \pi\bar{b})}{\sinh(\pi a + \pi\bar{b})\cosh(\pi\tilde{a} - \pi\bar{b})}, \\ |\mathcal{B}|^2 &= \frac{|D_{\infty}^{(\rm out)}|}{|D_{\rm H}^{(\rm in)}|} = \frac{\sinh(2\pi\bar{b})\cosh(\pi\tilde{a} - \pi a)}{\sinh(\pi a + \pi\bar{b})\cosh(\pi\tilde{a} - \pi\bar{b})}, \\ \sigma_{\rm abs} &= \frac{|\mathcal{B}|^2}{|\mathcal{A}|^2} = \frac{\sinh(2\pi\bar{b})\sinh(\pi\tilde{a} - \pi a)}{\sinh(\pi a - \pi\bar{b})\cosh(\pi\tilde{a} + \pi\bar{b})}. \end{aligned} \tag{17}$$

It can be seem that results in eq.(17) and eq.(11) are simply related via $\sinh(\pi a \pm \pi\bar{b}) \leftrightarrow \cosh(\pi a \pm \pi b)$ and $\cosh(\pi\tilde{a} - \pi a) \leftrightarrow \sinh(\pi\tilde{a} - \pi a)$, except for ℓ is integer, while j is half-integer.

6. Holographic CFT Description

The near horizon geometry of the near extreme RN black hole is dual to a 2d CFT with left- and right-hand central charges and temperatures

$$c_L = c_R = \frac{6Q^3}{\ell}, \qquad T_L = \frac{\ell}{2\pi Q}, \qquad T_R = \frac{\ell B}{\pi Q^2}, \tag{18}$$

where ℓ is a parameter that can be interpreted as a measure of the U(1) bundle of the background spacetime.[5–10]

For the bulk scalar and spinor fields, the conformal dimensions of their dual operators are

$$\begin{aligned} &\text{scalar operator}: \qquad h_L = h_R = \frac{1}{2} \pm ib,, \\ &\text{spinor operator}: \qquad h_R = \frac{1}{2} \pm i\bar{b}, \quad h_L = 1 \pm i\bar{b}. \end{aligned} \tag{19}$$

Also note that the first law of thermodynamics of the black hole is identical to that of the dual CFT, i.e.

$$\frac{\delta M}{T_H} - \frac{\Omega_H \delta Q}{T_H} = \frac{\tilde{\omega}_L}{T_L} + \frac{\tilde{\omega}_R}{T_R}, \tag{20}$$

in which the black hole Hawking temperature and chemical potential are $T_H = \frac{B}{2\pi Q^2}$, $\Omega_H = A_\tau(B) = -B/Q$. Then

$$\tilde{\omega}_L = -q\ell \quad \text{and} \quad \tilde{\omega}_R = 2\omega\ell. \tag{21}$$

Thus the absorption cross sections in Eqs.(17), (11) match with their CFT's results, respectively, only up to some numerical factors. Consequently, the mean number of pairs $|\mathcal{B}|^2$ also match with the CFT 2-point functions, both for the scalars and fermions, via the relation $|\mathcal{B}|^2 = -\sigma_{\rm abs}(b \to -b)$. Furthermore, the retarded Green's functions and quasinormal modes of the bulk probe scalar and spinor fields also consistent with the results $G_R(\omega_L, \omega_R)$ from the dual CFT_2 side in which the frequencies take discrete values of the Matsubara frequencies, respectively.

7. Conclusions

We study the scalar and spinor pair production for the near extremal RN black hole without back-reaction. The KG equation and Dirac equation are solved in the near horizon region where the geometry is $\mathrm{AdS}_2 \times S^2$ and the background electric field is constant in the radial direction. The near horizon region contains causal horizon and dominated electric field which capture both contributions: the Hawking radiation and the Schwinger mechanism. The exact solutions were obtained in terms of the hypergeometric functions. By imposing the particle viewpoint boundary condition, the physical quantities associated the pair production can be derived by the ratio of boundary fluxes. In particular, the expressions of the vacuum persistence amplitude, the mean number of pairs and the absorption section ratio were obtained. The existence condition for the pair production is corresponding to the instability of the probe fields in the AdS_2, i.e. violating the BF bound, leads to the black hole losing their charge more than their mass in pair production process. This consequence is in agreement with the cosmic censorship conjecture that a naked singularity cannot be evolved form the complete gravitational collapse when the matter fields satisfy appropriated energy conditions. We also present the holographic dual IR CFT_2 description for the pair production process, it would be more interesting to study the asymptotically AdS black holes and further study the dual UV CFT description for the pair production on the AdS boundary.

Acknowledgments

We are grateful to Sang Pyo Kim for useful discussions. C.M.C. and F.Y.T. were supported by the Ministry of Science and Technology of the R.O.C. under the grant MOST 102-2112-M-008-015-MY3. J.R.S. was supported by the NSFC under Grant No. 11205058 and the Open Project Program of State Key Laboratory of Theoretical Physics, Institute of Theoretical

Physics, Chinese Academy of Sciences, China (No. Y5KF161CJ1).

References

1. J. S. Schwinger, *Phys. Rev.* **82**, 664 (1951).
2. M. K. Parikh and F. Wilczek, *Phys. Rev. Lett.* **85**, 5042 (2000).
3. C.-M. Chen, S. P. Kim, I.-C. Lin, J.-R. Sun and M.-F. Wu, *Phys. Rev. D* **85**, 124041 (2012) [arXiv:1202.3224 [hep-th]].
4. C.-M. Chen, J.-R. Sun, F.-Y. Tang and P.-Y. Tsai, *Class. Quant. Grav.* **32**, 195003 (2015) [arXiv:1412.6876 [hep-th]].
5. C.-M. Chen, J.-R. Sun and S.-J. Zou, *JHEP* **1001**, 057 (2010) [arXiv:0910.2076 [hep-th]].
6. C.-M. Chen, Y.-M. Huang and S.-J. Zou, *JHEP* **1003**, 123 (2010) [arXiv:1001.2833 [hep-th]].
7. C.-M. Chen and J.-R. Sun, *JHEP* **1008**, 034 (2010) [arXiv:1004.3963 [hep-th]].
8. C.-M. Chen, Y.-M. Huang, J.-R. Sun, M.-F. Wu and S.-J. Zou, *Phys. Rev. D* **82**, 066003 (2010) [arXiv:1006.4092 [hep-th]].
9. C.-M. Chen, Y.-M. Huang, J.-R. Sun, M.-F. Wu and S.-J. Zou, *Phys. Rev. D* **82**, 066004 (2010) [arXiv:1006.4097 [hep-th]].
10. C.-M. Chen and J.-R. Sun, *J. Phys. Conf. Ser.* **330**, 012009 (2011) [arXiv:1106.4407 [hep-th]].
11. S. P. Kim, H. K. Lee and Y. Yoon, "Thermal Interpretation of Schwinger Effect in Near-Extremal RN Black Hole," arXiv:1503.00218 [hep-th].
12. S. P. Kim, "Schwinger Effect, Hawking Radiation and Gauge-Gravity Relation," arXiv:1506.03990 [hep-th].
13. S. P. Kim, "Schwinger Effect in (A)dS and Charged Black Hole," arXiv:1509.05532 [hep-th].
14. S. P. Kim and D. N. Page, *Phys. Rev. D* **73**, 065020 (2006) [hep-th/0301132].
15. P. Breitenlohner and D. Z. Freedman, *Annals Phys.* **144**, 249 (1982).
16. P. Breitenlohner and D. Z. Freedman, *Phys. Lett. B* **115**, 197 (1982).

New experimental results on cosmic gravitational effects and the centenary of Einstein's equation

C. S. Unnikrishnan

Tata Institute of Fundamental Research, Homi Bhabha Road, Mumbai 400005, India
E-mail: unni@tifr.res.in, URL: www.tifr.res.in/~GraviMetro

As one celebrates the centennial triumph of the general theory of relativity (GTR), Einstein's theory of gravity, it is appropriate to review and assess the experimental foundations of the theory in the context of what we know about matter in the universe, a knowledge that did not exist when GTR was formulated. A worldview consistent with and rigorously based on experimental evidence indeed demands a paradigm change and theoretical revision. The new theory of dynamics and relativity is Cosmic Relativity, with both absolute space and absolute time of a preferred cosmic frame, and it implies a modified GTR on its new and deeper foundation. Supporting results from laboratory experiments on the relative one-way velocity of light and the direct gravitational action of the cosmic matter on dynamics are discussed, demanding a modified equation of relativistic gravitation, aptly called the Centenary Einstein's equation.

Keywords: Cosmic relatvity; speed of light; Einstein's equation.

1. Gravitational Environment of Dynamics

All fundamental theories of physics, most prominently theories of relativity and dynamics and the quantum theory, were completed with empty space as their background arena well before we had any significant knowledge about cosmological distribution of matter and hence its gravity. In contrast, all experimental tests are invariably done in the eternal gravitational presence of this vast amounts of matter. Therefore, empirical results naturally include all possible effects of the gravity of cosmic matter in physical phenomena, while the present theoretical structure does not incorporate any such effect. The main theme of this paper is to highlight this discrepancy and establish, based on several empirical evidence, that the true picture requires a paradigm change in which gravity of cosmic matter determines dynamics, relativistic effects and the very basis of physics.[1,2]

In this context, we need to stress that one of the unfinished aspects of fundamental physics is the real understanding of pseudo-forces.[3] Though the Machian speculation that pseudo-forces of physics might be induced by a true interaction with matter was a significant inspiration to Einstein while formulating the general theory of relativity, the theory itself turned out to be anti-Machian because Einstein reduced even true gravity to a geometrical pseudo-force, locally equivalent to an inertial force arising in coordinate transformations!

Today, after giant strides in observational cosmology, we know the average matter density and the distribution of matter in the universe, and the level of homogeneity and isotropy, fairly well. One of the first implications for fundamental theories is that this gravitational background should be explicitly included in the theories, permanently and inseparably, because it cannot be removed or assumed insignificant anywhere in the universe. Indeed, the Newtonian gravitational potential of all the matter in the universe, with average density close to the critical density up to the Hubble size, is about a billion (10^9) more than the gravitational potential due to the earth and amounts to 10^{17} m^2/s^2. Noting that this is numerically the same as the quantity c^2, all of relativity and dynamics can be phenomenologically reformulated as arising in the gravity of the universe.[2] More seriously, for any observer moving though the universe at a constant velocity $\vec{v}$, the matter flow is the same as the large current of the charge of gravity (mass and energy) in the direction $-\hat{v}$, and it is clear that the resulting space and its metric should be anisotropic, while remaining homogenous. Therefore, the *Lorentz transformations that preserves the isotropy of the metric is physically inconsistent in non-empty space.* In contrast, the Galilean transformation gives the correct anisotropic metric and also, the correct factor for time dilation due to the change in the metric factor g_{00}. Under the coordinate transformation

$$x' = x - vt; \quad t' = t \tag{1}$$

the metric (2D for brevity) changes as

$$\begin{bmatrix} -1 & 0 \\ 0 & 1 \end{bmatrix} \rightarrow \begin{bmatrix} -(1 - v^2/c^2) & v/c \\ v/c & 1 \end{bmatrix} = \begin{bmatrix} g_{00} & g_{0i} \\ g_{i0} & g_{11} \end{bmatrix} \tag{2}$$

Therefore, the physical time of a moving clock becomes $(-g_{00})^{1/2}\,dt' = dt'(1-v^2/c^2)^{1/2}$. Here, v is the velocity in the cosmic frame and hence without usual paradoxical ambiguity associated with relative velocities. Since the cosmic frame is the absolute preferred frame in this frame work, this *theory of cosmic relativity predicts that relative velocity of light relative to a moving observer is Galilean to first order*, $(v \pm c)$. This in fact is the crucial falsifiable prediction of the theory that directly contradicts the fundamental untested postulate of special relativity that the relative one-way speed of light is an invariant constant. Combined with this feature, the time dilation factor also implies the correct length contraction.

2. The Relative One-Way Speed of Light

It should be well-known that the null result of the Michelson-Morley experiment that compared the two-way propagation of light along two perpendicular directions is consistent with both Lorentz-Poincare relativity, with Galilean relative velocity for light, as well as with Einstein's special relativity, with absolute invariance of the velocity of light. It is generally believed that the true one-way speed of light can never be measured because of the impossibility to synchronize distant clocks in a way that is independent of the theory. However, we have found and validated a way to compare the genuine one-way velocity of light relative to an inertially moving observer without the need for synchronization of distant clocks,[2] or without using any clock at all!

To fix the ideas, let us consider propagation of a pulse of sound, which we know for sure is Galilean. If the pulse starts at the location of the inertially moving observer O', the separation ΔL between O' and the pulse increases linearly with his time t',

$$\Delta L = v_r \Delta t' = (c_s - v_o)\, \Delta t \tag{3}$$

where t refers to time in the stationary laboratory O.The relative velocity is denoted v_r and v_o and c_s are the velocities of the observer and the pulse of sound in the fixed frame O. In the experimental configuration in figure 1 with a round trip distance of L, the pulse has travelled a distance $\Delta L = L$ relative to O' when it catches up with the observer after one round trip. Therefore $v_r = L/\Delta t'_r$ where $\Delta t'_r$ is the round trip time relative to O'. This did not require any synchronization because the same clock is maintained. It is trivial to confirm that this gives the correct relative velocity for sound, as can be further confirmed by a real experiment. Since $\Delta t' = \Delta t$ to first order, independent of theory, measured relative velocity is $v_r = (c_s \pm v_o)$ depending on the direction of propagation of the pulse (much smaller second order $\sqrt{1 - v^2/c^2}$ effect is not relevant for this first order experiment that measures a v/c effect). A plot of $\Delta t'$ vs. v_o gives a linear graph with slope v_o/c_s *if and only if the relative velocity is Galilean.*

It is therefore obvious that if the relative velocity of sound were an invariant constant, $v_r = c_s$, then $\Delta t'$ would not depend on the velocity of the

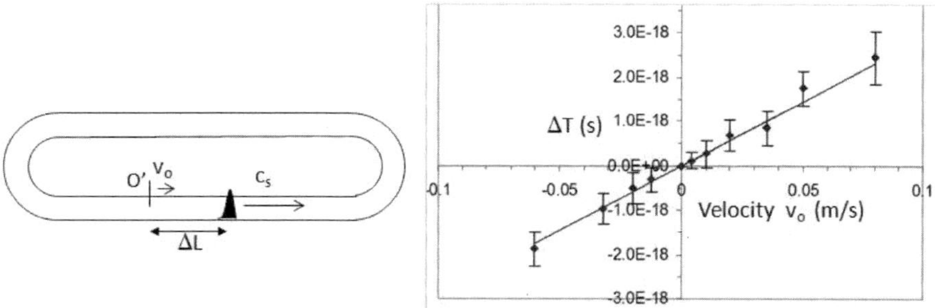

Fig. 1. Left: Validated method for comparing relative one-way velocity of sound or light. Right: Results from a run of the experiment with light ($L = 3$ m).

observer. If the wave is split and send in opposite directions the comparison of delay can be done interferometrically, without any real clock. This experiment with sound validates the experimental technique for measuring one-way relative speed without the need for synchronizing spatially separated clocks. With this important conceptual advance in the experimental technique, we have performed the same experiment with light, in exactly the same configuration.[2] One set of results is summarized in figure 1.

The results show unambiguously that the true one-way relative speed of light is Galilean, $c \pm v_o$, providing support for the frame work of cosmic relativity, while falsifying the core postulate of the theory of special relativity.

3. Demonstration of a Cosmic Gravitational Effect

Clinching evidence for cosmic relativity and the gravitational action of the matter in the universe is obtained from a direct experiment that resembles Ampère's foundational experiments on the interaction of a current with another, which we call the magnetic interaction in modern formalism. Magnetism is the real evidence for relativistic features of electrodynamics. In Ampère's experiments, it was established that parallel currents experience an attractive force and opposite currents repel. The Ampère force between two such current elements is

$$F = -\frac{i_1 dl_1 i_2 dl_2}{4\pi\epsilon_0 c^2 r^2} = -\frac{\mu_0 i_1 dl_1 i_2 dl_2}{4\pi r^2} \tag{4}$$

and the force is of order v^2/c^2. If the currents are in a loop, one can rewrite this in terms of a magnetic moment of the small loop interacting with a uniform magnetic field due to the other, and the energy of interaction is

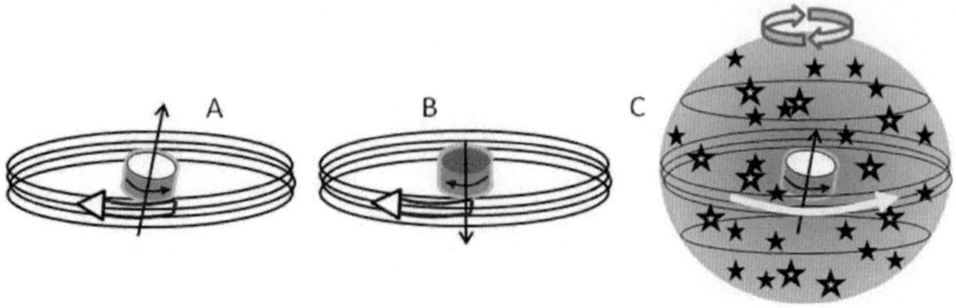

Fig. 2. Panels A and B indicate how a small current loop or magnet flips its direction when the current in the larger loop is in a direction opposite to that of the test loop current. Panel C: When a spinning disk is taken into a frame slowly rotating relative to the matter in the universe (indicated by dual arrows on top), the spinning disk flips due to the large cosmic gravitomagnetic torque.

$E = -\mu \cdot B$. The torque on the current loop due to the magnetic field $\tau = \mu \times B$ will flip the loop (or a small magnet) to the lower energy configurtion of parallel currents.

If the fundamental paradigm of cosmic relativity is correct, then one should be able to easily see the large gravitational effects of the universe in local dynamics by performing an Ampère experiment for gravity. Creating the analogue of a small current loop in gravity is easy because setting a piece of the charge of gravity – mass itself – is achieved by rotating a small massive disk. Einstein's general relativity applied to the environment near the earth predicts nearly unobservably small relativistic gravitomagnetic torques on such a rotating disk (gyroscope) arising from rotation of the earth (Kerr metric) or when rotated in an orbit relative to the earth in its Schwartzchild metric. These require special efforts in space-based experiments to even detect. But, cosmic relativity insists that a large external current can easily be generated by rotating the spinning disk relative to all the matter in the universe. The magnitude of such a current is enormous, of the order of $I_u \simeq M_u R_u \Omega$, where M_u and R_u denote some averaged mass and size of the universe.

As demonstrated directly ('on stage', so to say) the spinning disk indeed flips its direction promptly when a large matter current of cosmic matter is also generated by slow rotation relative to the cosmic frame, fully consistent with the relativistic interaction with the matter in the universe (figure 2).[4] The measured torque shows that parallel currents repel in gravity. The interaction is similar to the Ampère force, with coupling G/c^2. Since there is frame-independent transfer of energy and angular momentum due to the action of an external force, we have no choice but to trace it to one of the only two known long range interactions in physics. The disk being neutral, the only explanation possible is that the interaction responsible is gravity. However, such an interaction of the right magnitude is not contained in the vacuum solutions of Einstein's equation, and therefore our experiment demonstrates two important facts: (1) the large relativistic gravitomagnetism of cosmic matter is directly observable in simple terrestrial experiments and it validates the paradigm of cosmic relativity, (2) Einstein general relativity needs an important modification to permanently incorporate the gravitational effects of the matter in the universe because the equation is to be operative always in the presence of all this matter without exception. The proposed modification is straighforward and does not alter other tested predictions of general relativity and I call the new equation the Centenary Einstein's Equation,[4]

$$R_{ik} - \frac{1}{2} g_{ik} + \Lambda_{ik} g - \frac{8\pi G}{c^4} T_{ik(u)} = \frac{8\pi G}{c^4} T_{ik} \tag{5}$$

which does not allow a vacuum Einstein equation, $R_{ik} = 0$, except as an approximation that cannot predict the result of the experiment we just discussed.

4. Summary of Other Features

The fundamental metric that we derived from the Galilean transformation can be used as the basis for calculating physical effects instead of depending on the full modified Einstein equation.[2,5,6] The metric component $g_{0i} = v/c$ is like a vector potential A_i/c even though it is one component of a space-time tensor. For an observer in uniform motion in a circular path, the direction of velocity and hence A_i is time dependent resulting in the centrifugal force, $F_{cf} = m_g dA/dt = m_g v^2/r$. In frames of pure rotation, $dA/dt = 0$, but $\nabla \times A = \nabla \times v = 2\Omega$ and this is exactly the cosmic gravitomagnetic field in the fully relativistic theory. The Coriolis-Lorentz force $F_C = m_g v \times 2\Omega$ on a moving mass is an immediate consequence. It can be shown that the Newton's law of dynamics itself is a relativistic consequence of cosmic gravity and it is indeed a gravitomagnetic law, similar to Lenz's law in electrodynamics.[6] This also reveals the origin of the equivalence principle, both the weak equivalence principle and the Einstein equivalence principle because in an accelerated frame, $g_a = dA/dt = -dv/dt$.

One very important consequence is the cosmic gravitomagnetic effect on spins, both classical and quantum, when the relevant particle is in an orbital motion or in a rotating frame. We get for the energy and torque of interaction,

$$E_s = s \cdot (\nabla \times A)/2 = s \cdot \Omega; \quad \tau_s = s \times \Omega \tag{6}$$

the effect of which we already saw in the experiment with the spinning disk. Fundamental physical effects involving this large field g_{0i} is being explored in scenarios of spin transport in chiral molecules, quantum Hall effect etc. For example, it seems that the mysterious gauge field that dresses electrons to make the composite fermions required for a coherent treatment of the factional quantum Hall effect is indeed cosmic gravity!

5. Concluding Remarks

We have discussed the fundamental paradigm of cosmic relativity and its direct experimental evidence. The Galilean nature of the one-way relative speed of light and the direct gravitomagnetic action of cosmic matter on a gyro-disk in a rotating frame are demonstrated. A modified 'centenary' Einstein's equation that is consistent with known tests as well as these new experimental evidence is presented, which also incorporates cosmic gravity in the Machian sense and beyond.

Acknowledgments

I thank Valentin Rudenko for an initial invitation and brotherly attention during the visit to Moscow. I am thankful to V. N. Melnikov, S. V. Bolokhov, and V. Silin for their generous friendly help during the conference.

References

1. C. S. Unnikrishnan, Cosmic Relativity: The fundamental theory of relativity, its implications and experimental tests, arXiv:gr-qc/0406023.
2. C. S. Unnikrishnan, Physics in the 'Once-Given' Universe, in *Recent Developments in Theoretical Physics*, (Eds. Subir Ghosh and Guruprasad Kar, World Scientific, 2010), pp. 99-120.
3. F. Wilczek, Total Relativity, *Physics Today*, **57** (4), 10 (2004).
4. C. S. Unnikrishnan, Dynamics, relativity and the equivalence Principle in the 'once-given' universe, to appear in the *Proceedings of the 50th Moriond meeting on Gravitation* (La Thuile, 2015).
5. D. Sciama, On the origin of inertia, *MNRAS* **113**, 34 (1953).
6. C. S. Unnikrishnan, True dynamical tests of the equivalence principle, *Int. Jl. Mod. Phys (Conf. series)* **30**, 1460267 (2014).

Laboratory and astrophysical constraints on an axion

G. L. Klimchitskaya and V. M. Mostepanenko*

Department of Astrophysics, Central Astronomical Observatory at Pulkovo of the Russian Academy of Sciences, Saint Petersburg, 196140, Russia
and
Institute of Physics, Nanotechnology and Telecommunications, Peter the Great Saint Petersburg Polytechnic University, Saint Petersburg, 195251, Russia
**E-mail: vmostepa@gmail.com*

We review astrophysical, cosmological and laboratory constraints on different types of axion-like particles. Special attention is paid to constraints on the axion-nucleon coupling constants following from measurements of the Casimir interaction. These constraints are compared with the most recent results obtained from the Casimir-less experiment. The possibility of obtaining stronger constraints by using the polarized test bodies is discussed.

Keywords: Axion; Cavendish-type experiments; Casimir force; dark matter.

1. Introduction

Axion is the light pseudoscalar particle. It was introduced[1,2] in order to avoid strong CP violation and large electric dipole moment of a neutron. Axions were also considered[3] as possible constituents of dark matter. The originally introduced axions are pseudo-Nambu-Goldstone bosons arising due to violation of the Peccei-Quinn symmetry. After they were constrained to a narrow band in the parameter space,[4] many types of axion-like particles were proposed in different models. The most of them belong to one of the two groups: the hadronic (or QCD) axions[5,6] and the GUT axions.[7,8]

Axion-like particles interact with photons, electrons and nucleons. These interactions are used to constrain the respective coupling constants. Below we focus our attention on the coupling constants of axions with nucleons which are constrained by now not very strongly. The GUT axions a interact with nucleons ψ by the pseudoscalar Lagrangian density[3]

$$\mathcal{L}_{\rm ps} = -ig_{ak}\bar{\psi}\gamma_5\psi a, \tag{1}$$

where g_{ak} is the dimensionless coupling constant of an axion a with a neutron ($k = n$) or a proton ($k = p$) and γ_5 is the Dirac matrix. As to the

QCD (adronic) axions, they are the Goldstone bosons and interact with nucleons via the pseudovector Lagrangian density

$$\mathcal{L}_{\rm pv} = \frac{g_{ak}}{m_a}\bar{\psi}\gamma_5\gamma_\mu\psi\partial^\mu a, \tag{2}$$

where γ_μ are the Dirac matrices, m_a is the axion mass. Note that (1) results in a renormalizable quantum field theory, whereas the theory with the Lagrangian (2) is nonrenormalizable, because the effective coupling constant g_{ak}/m_a is dimensional. What is more, for the GUT axions g_{ak} and m_a are the independent parameters, whereas for adronic axions there is some relationship between the coupling constants and m_a.

Both Lagrangians (1) and (2) lead to common spin-dependent effective potentials of the nucleon-nucleon interaction due to exchange of one axion. This process cannot be used, however, for obtaining constraints on g_{ak} from force measurements between two unpolarized test bodies because the resulting interaction averages to zero. From (1) we can also find[10] the effective potential due to two-axion exchange between two nucleons with masses m_k and m_l situated at the points $\boldsymbol{r}_1 \neq \boldsymbol{r}_2$:

$$V_{kl}(|\boldsymbol{r}_1 - \boldsymbol{r}_2|) = -\frac{g_{ak}^2 g_{al}^2\, m_a}{32\pi^3 m_k m_l}\,\frac{K_1(2m_a|\boldsymbol{r}_1 - \boldsymbol{r}_2|)}{|\boldsymbol{r}_1 - \boldsymbol{r}_2|^2}, \tag{3}$$

where $K_1(z)$ is the modified Bessel function of the second kind. Similar potential for the Lagrangian density (2) with $m_a \neq 0$ remains unknown.

2. Cosmological and Astrophysical Constraints

Here, we list a few constraints obtained mostly from different processes with the QCD axions. The information below is based on the reviews in Refs. 11 and 12. We start from the cosmological constraints.

A lower bound for the axion mass derived from the cosmological arguments depends on an assumption whether the Peccei-Quinn symmetry breaks before the inflation stage or after it. In the former case one obtains $m_a > 6\,\mu$eV, whereas in the latter case the result depends on an assumption whether the axion strings decay rapidly or slow. In the former case it follows $m_a > 15\,\mu$eV, whereas in the latter case the result depends on whether the model of local or global strings is used in the analysis. For the local strings one arrives at the constraint $m_a > 100\,\mu$eV, but for the global ones to $m_a > 30\,\mu$eV. Thus, the cosmological constraints depend on a model.

Concerning the astrophysical constraints, stellar evolution places limits on the coupling constants of the QCD axions to photons, electrons and nucleons, and only indirectly on m_a. Thus, the axion-photon coupling constant is constrained from the study of axion cooling of some stars within globular clusters. This results in $m_a < 0.3\,\text{eV}$ or $m_a < 0.7\,\text{eV}$ depending on the model used.

Constraining the axion-electron coupling constant from the study of energy losses in globular cluster stars and white dwarfs leads to $m_a > 0.01\,\text{eV}$. Similar constraints on the axion-electron and axion-nucleon coupling constants, following from the study of neutrino emission from Supernova SN 1987a, result in $0.01\,\text{eV} < m_a < (2 \div 10)\,\text{eV}$ and $15\,\text{eV} < m_a < 10\,\text{keV}$.

Instead of stellar energy losses, one can look for axion fluxes from the Sun, detected at the Earth by axion telescopes, e.g., by CERN Axion Solar Telescope (CAST). This already resulted in the excluded region $0.3\,\text{eV} < m_a < 1\,\text{eV}$. The resonant conversion of dark matter axions to photons in a microwave cavity in strong magnetic field, i.e., Axion Dark Matter eXperiment (ADMX), is also prospective for constraining axions.

3. Constraints from Magnetometer and Gravitational Experiments

In this and following sections we list the strongest laboratory constraints on the masses and coupling constants of axions to nucleons considered as two independent parameters. The constraint on g_{an} obtained[13] by a magnetometer using spin-polarized K and ^{3}He atoms are shown by the line 1 in Fig. 1. The regions above each line are prohibited, and below each line are allowed. Atoms in this experiment are polarized, and, thus, the constraints of line 1 are equally applicable to the QCD and GUT axions.

The line 2 in Fig. 1 shows the strongest constraints obtained[14] from the most recent Cavendish-type experiment.[15] In this case (and in all other experiments discussed in the next section) the test bodies were unpolarized, and the additional force due to two-axion exchange was calculated by the additive summation of the effective potentials (3). As a result, the constraints of the lines 2–5 are applicable only to the GUT axions. We present the constraints of lines 2–5 under the natural assumption $g_{an} = g_{ap}$.[16]

Fig. 1. Constraints on the axion-nucleon coupling constant obtained from the magnetometer measurements (line 1), Cavendish-type experiment (line 2), Casimir-less experiment (line 3), measurements of the effective Casimir pressure (line 4), and from measurements of the lateral Casimir force (line 5).

4. Constraints from Measuring the Casimir Interaction and from the Casimir-Less Experiment

During the last few years a lot of precise experiments on measuring the Casimir interaction have been performed.[17] In several experiments the additional force due to two-axion exchange between nucleons of the test bodies was computed using Eq. (3). This force was not observed, and the theoretical predictions for the Casimir interaction were confirmed in the limits of some measurement errors. Taking this into account, the masses and coupling constants of the GUT axion-like particles have been constrained[18–21] (see also review in Ref. 22). In Fig. 1 we show the most strong constraints obtained in such a way from measurements of the effective Casimir pressure by means of micromachined oscillator[20] (line 4) and of the lateral Casimir force between corrugated surfaces by means of atomic force microscope[21] (line 5). As is seen in Fig. 1, experiments on measuring the Casimir interaction lead to much stronger constraints on axion-like particles of larger masses than the Cavendish-type experiments.

Very recently the improved constraints on the axion-nucleon interaction were obtained[23] from the new Casimir-less experiment.[24] This is a differential force measurement where the contribution of the Casimir force is subtracted, and the total signal is determined solely by the contribution from some hypothetical force. The obtained results are shown by the line 3 in Fig. 1. They are stronger than the previously known ones up to a factor of 60 within the wide region of axion masses from 1.7×10^{-3} to $0.9\,\mathrm{eV}$.

5. Conclusions and Discussion

As is seen from the foregoing, the laboratory constraints on an axion are more model-independent than the cosmological and astrophysical ones. It can be concluded also that measurements of the Casimir interaction, used for this purpose only recently, have already brought rather competitive constraints on the axion-nucleon coupling constants and axion masses. In future the constraints on axion-like particles, following from experiments on measuring the Casimir interaction, could be significantly improved by using the polarized test bodies. In this case the exchange of one axion between nucleons will contribute to the effective potential. The obtained results will be applicable to all types of axions and axion-like particles.

References

1. S. Weinberg, A new light boson? *Phys. Rev. Lett.* **40**, 223 (1978).

2. F. Wilczek, Problem of strong P and T invariance in the presence of instantons, *Phys. Rev. Lett.* **40**, 279 (1978).
3. J. E. Kim, Light pseudoscalars, particle physics and cosmology, *Phys. Rep.* **150**, 1 (1987).
4. K. Baker *et al.* The quest for axions and other new light particles, *Ann. Phys. (Berlin)* **525**, A93 (2013).
5. J. E. Kim, Weak-interaction singlet and strong CP invariance, *Phys. Rev. Lett.* **43**, 103 (1979).
6. M. A. Shifman, A. I. Vainstein and V. I. Zakharov, Can confinement ensure natural CP invariance of strong interactions? *Nucl. Phys. B* **166**, 493 (1980).
7. A. P. Zhitnitskii, On possible suppression of the axion hadron interactions, *Sov. J. Nucl. Phys.* **31**, 260 (1980).
8. M. Dine, F. Fischler and M. Srednicki, A simple solution to the strong CP problem with a harmless axions, *Phys. Lett. B* **104**, 199 (1981).
9. M. Kawasaki and K. Nakayama, Axions: Theory and cosmological role, *Annu. Rev. Nucl. Part. Sci.* **63**, 69 (2013).
10. S. D. Drell and K. Huang, Many-body forces and nuclear saturation, *Phys. Rev.* **91**, 1527 (1953).
11. L. D. Duffy and K. van Bibber, Axions as dark matter particles, *New J. Phys.* **11**, 105008 (2009).
12. J. Beringer *et al.* (Particle Data Group), G. G. Raffelt and L. J. Rosenberg, Axions and other similar particles, *Phys Rev. D* **86**, 010001 (2012).
13. G. Vasilakis, J. M. Brown, T. R. Kornack and M. V. Romalis, Limits on new long range nuclear spin-dependent forces set with a K–^{3}He comagnetometer, *Phys. Rev. Lett.* **103**, 261801 (2009).
14. E. G. Adelberger, B. R. Heckel, S. Hoedl, C. D. Hoyle, D. J. Kapner and A. Upadhye, Particle-physics implications of a recent test of the gravitational inverse-square law, *Phys. Rev. Lett.* **98**, 131104 (2007).
15. D. J. Kapner, T. S. Cook, E. G. Adelberger, J. H. Gundlach, B. R. Heckel, C. D. Hoyle and H. E. Swanson, Tests of the gravitational inverse-square law below the dark-energy length scale, *Phys. Rev. Lett.* **98**, 021101 (2007).
16. E. G. Adelberger, E. Fischbach, D. E. Krause and R. D. Newman, Constraining the couplings of massive pseudoscalars using gravity and optical experiments, *Phys. Rev. D* **68**, 062002 (2003).
17. M. Bordag, G. L. Klimchitskaya, U. Mohideen and V. M. Mostepanenko, *Advances in the Casimir Effect* (Oxford University Press, Ox-

ford, 2015).
18. V. B. Bezerra, G. L. Klimchitskaya, V. M. Mostepanenko and C. Romero, Constraints on the parameters of an axion from measurements of the thermal Casimir-Polder force, *Phys. Rev. D* **89**, 035010 (2014).
19. V. B. Bezerra, G. L. Klimchitskaya, V. M. Mostepanenko and C. Romero, Stronger constraints on an axion from measuring the Casimir interaction by means of a dynamic atomic force microscope, *Phys. Rev. D* **89**, 075002 (2014).
20. V. B. Bezerra, G. L. Klimchitskaya, V. M. Mostepanenko and C. Romero, Constraining axion-nucleon coupling constants from measurements of effective Casimir pressure by means of micromechanical oscillator, *Eur. Phys. J. C* **74**, 2859 (2014).
21. V. B. Bezerra, G. L. Klimchitskaya, V. M. Mostepanenko and C. Romero, Constraints on axion-nucleon coupling constants from measuring the Casimir force between corrugated surfaces, *Phys. Rev. D* **90**, 055013 (2014).
22. G. L. Klimchitskaya and V. M. Mostepanenko, Constraints on the axion and corrections to Newtonian gravity from the Casimir effect, *Grav. Cosmol.* **21**, 1 (2015).
23. G. L. Klimchitskaya and V. M. Mostepanenko, Improved constraints on the coupling constants of axion-like particles to nucleons from recent Casimir-less experiment, *Eur. Phys. J. C* **75**, 164 (2015).
24. Y.-J. Chen, W. K. Tham, D. E. Krause, D. López, E. Fischbach and R. S. Decca, Isoelectronic measurements yield stronger limits on hypothetical Yukawa interactions in the 40–8000 nm range, arXiv:1410.7267v1.

Geometry of stationary black holes near the ring singularity

H. García-Compeán and V. S. Manko*

Departamento de Física, Centro de Investigación y de Estudios Avanzados del IPN, A.P. 14-740, 07000 México D.F., Mexico

**E-mail: vsmanko@fis.cinvestav.mx*

In this article based on the plenary talk of one of the authors, we review recent results on the geometry of the two-surface $r = 0$, $t =$ const in various stationary black-hole spacetimes which, contrary to the general belief, turns out to be different from that of a disk generically. The sources of the historically erroneous interpretation of that surface are pointed out, and a novel approach to performing analytical extensions of the known black-hole stationary solutions is outlined.

Keywords: Black holes; Kerr solution; ring singularity; analytic extensions.

1. Introduction

It is well known that the stationary black-hole solutions of the Einstein-Maxwell equations develop ring singularities, so that the knowledge of the geometry in the vicinity of the latter is very important not only from a physical viewpoint but also mathematically, for instance, for carrying out correctly the maximal analytic extensions of these solutions.

We first came to realize that the well-known interpretation given in the literature[1,2] to the two-surface $r = 0$, $t =$ const (henceforth $r = 0$ for simplicity) of the Kerr solution[3] really needs a serious inspection after our finding out[4] that the ring singularity in the Kerr spacetime with positive mass is different qualitatively (and in size) from the ring singularity determined by negative mass. This naturally led us to understanding that the known maximal analytic extension (MAE) of the Kerr solution, described for instance in Ref. 2, is unphysical and most probably non-analytic, as it represents a unification of two asymptotically flat regions in which the same singularity is seen as having positive or negative mass depending on the region from which the singularity is being observed. Apparently, the same kind of problems, resulting from a wrong interpretation of the surface

$r = 0$, must have the MAE of the Kerr-Newman solution,[5] as well as the MAE's of the Kerr-de Sitter (KdS) and Kerr-Newman-de Sitter (KNdS) solutions.[6,7]

Fortunately, in recent years, various aspects related to the geometry of the surface $r = 0$ and the MAE's of the black-hole solutions have been ultimately clarified,[8,9] clearly demonstrating that bad physics indeed implies bad mathematics, and in the next sections we shall discuss in some detail the present knowledge about the geometry of the above-mentioned surface, as well as our vision of why for many years a simple truth has been hidden from the eyes of the researchers.

2. The Surface $r = 0$ in Standard MAE's

It should be clearly stated from the very beginning that in all known MAE's of the aforementioned black-hole solutions the two-surface $r = 0$ is unambiguously interpreted as a *disk* encircled by the ring singularity that lies *entirely* in the equatorial plane. To emphasize the fact that the disk is a part of the equatorial plane, Israel for instance in his well-known paper[10] uses for the surface $r = 0$ of the KN solution the term "*equatorial disk*". Thus, it is assumed that by crossing such a disk starting, say, from the northern hemisphere of a black-hole spacetime, an observer would immediately re-emerge in the southern hemisphere of the same spacetime, with only one (known) reservation that this cannot be done in analytic manner.[11] Then it is clear that if one wants, within the disk interpretation of the surface $r = 0$, to extend a black-hole spacetime to negative values of the radial coordinate r, one must take another copy of that spacetime and glue it in a special manner, on the disk, with the initial copy. The use of the second copy with another spatial infinity becomes, therefore, inevitable!

For example, the Kerr metric

$$ds^2 = \Sigma\left(\frac{dr^2}{\Delta} + d\theta^2\right) + \frac{\sin^2\theta}{\Sigma}[adt - (r^2 + a^2)d\varphi]^2 - \frac{\Delta}{\Sigma}(dt - a\sin^2\theta d\varphi)^2,$$
$$\Delta = r^2 - 2Mr + a^2, \quad \Sigma = r^2 + a^2\cos^2\theta, \tag{1}$$

written in the Boyer-Lindquist coordinates, is invariant under the discrete transformation

$$r \to -r, \quad M \to -M, \tag{2}$$

which suggests the equivalence of the regions

$$r \leq 0,\ M > 0 \quad \text{and} \quad r \geq 0,\ M < 0. \tag{3}$$

Hence, the known extension of the Kerr spacetime to negative values of r consists in gluing the regions

$$r \geq 0,\ M > 0 \quad \text{and} \quad r \geq 0,\ M < 0 \tag{4}$$

on the disks $r = 0$ of these two regions, as it is described for instance in:[2,11] the upper side of one disk must be identified with the bottom side of the other disk, and vice versa. As a result, one gets a totally unphysical construction possessing two asymptotically flat regions that give birth to two ADM masses of opposite signs characterizing the same singularity.

At this point, it seems likely to clarify an important question of why the entire general relativistic community for nearly fifty years have been absolutely sure that the surface $r = 0$ in the black-hole spacetimes is a disk. Our own explanation for that is the following.

First, in 1966, a year before the publication of the paper[1] on the MAE of the Kerr solution, it was shown by O'Neill[12] that the Gaussian curvature of the surface $r = 0$ in Kerr's case is equal to zero, which was strongly pointing to the disk geometry of the latter surface, and it is plausible to suppose that Boyer and Lindquist were already aware of the O'Neill's result when they were writing their paper. However, they discarded (maybe as very improbable) another possibility for a two-surface to have a zero Gaussian curvature – the surface might be a cone as well. Apparently, the Boyer and Lindquist's "natural" (but in fact misleading) choice made for the Kerr's $r = 0$ geometry was later automatically extrapolated to the more general black-hole spacetimes; this follows at least from the fact that no calculation of the Gaussian curvature of the corresponding $r = 0$ surfaces has been even attempted until our recent work.

Second, and most important, the use of the quasi-Cartesian Kerr-Schild coordinates appears to be chiefly responsible for having induced an overall misleading idea about the disk geometry of the surface $r = 0$ in the Kerr and other cases. Indeed, suppose we have some stationary solution of the field equations defined by a metric written in the Boyer-Lindquist-like coordinates $\{r, \theta, \varphi, t\}$, whose surface $r = 0$ may in principle have an arbitrary geometry yet to be identified. Then, with the idea to "attain a better insight" of the geometrical properties of the solution, let us make a coordinate change, introducing the Cartesian-type coordinates $\{x, y, z, t'\}$ via the formulas

$$x + iy = (r + ia)e^{i\varphi'(\varphi, r)} \sin\theta, \quad z = r\cos\theta, \quad t' = t'(t, r), \tag{5}$$

where a is the usual rotation parameter. An immediate corrolary of (5) are the two equations,

$$x^2 + y^2 = (r^2 + a^2)\sin^2\theta \quad \text{and} \quad z^2 = r^2\cos^2\theta, \tag{6}$$

whence we first readily obtain the relation

$$\frac{x^2 + y^2}{r^2 + a^2} + \frac{z^2}{r^2} = 1, \tag{7}$$

which, when $r \neq 0$, shows that the surfaces $r =$ const are confocal ellipsoids, and then, in the limit $r = 0$, we yield the equations

$$x^2 + y^2 = a^2\sin^2\theta, \quad z = 0, \tag{8}$$

defining a disk of radius $|a|$ in the equatorial $z = 0$ plane.

Therefore, the coordinate transformation (5), which was first proposed in Kerr's celebrated paper[3] and nowadays better known under the name of the Kerr-Schild transformation, is undoubtedly a remarkable invention, permitting one to convert in a *magic* way any unknown surface $r = 0$ of arbitrary geometry into a disk (8), without even writing a concrete metric explicitly! It is not surprising that the above very simple and efficient method of presenting any surface $r = 0$ as a "disk" was employed not only for the analysis of the Kerr geometry by Boyer and Lindquist,[1] but also by Carter in the KN case,[13] as well as in the case of stationary black holes with a cosmological constant.[6]

In the next section we shall see how easily the disk interpretation of the surface $r = 0$ can be refuted already in the KN case.

3. The Surface $r = 0$ in the Kerr and KN Solutions

In order to prove that the surface $r = 0$ of the KN solution is not a disk, it is sufficient just to check that its Gaussian curvature is non-zero. Since the KN metric differs from (1) only in the form of Δ, namely,

$$\Delta = r^2 - 2Mr + a^2 + Q^2 \tag{9}$$

(the parameter Q representing a charge), the two-metric of the surface $r = 0$ in the KN case has the form

$$d\sigma^2 = a^2\cos^2\theta d\theta^2 + (a^2 - Q^2\tan^2\theta)\sin^2\theta d\varphi^2. \tag{10}$$

The Gaussian curvature K of (10) has been calculated in Ref. 8 and is defined by the following expression:

$$K = \frac{Q^2[3a^2 - (a^2 + Q^2)\sin^2\theta(2 + \sin^2\theta)]}{a^2\cos^4\theta(a^2\cos^2\theta - Q^2\sin^2\theta)^2}, \tag{11}$$

which clearly disproves the disk geometry of the surface $r = 0$ of the KN spacetime. As was shown in Ref. 8, equation (11) determines a rather sophisticated surface with the regions of positive and negative Gaussian curvature.

At the same time, in the negative mass case, the KN surface $r = 0$ can be written as[8]

$$\frac{\rho^2}{a^2 + Q^2} + \frac{z^2}{M^2} = 1, \tag{12}$$

using the usual Weyl-Papapetrou cylindrical coordinates (ρ, z) related to the Boyer-Lindquist coordinates via the formulas

$$r = M + \frac{1}{2}(r_+ + r_-), \quad \cos\theta = \frac{1}{2\kappa}(r_+ - r_-),$$
$$r_\pm = \sqrt{\rho^2 + (z \pm \kappa)^2}, \quad \kappa = \sqrt{M^2 - a^2 - Q^2}. \tag{13}$$

Equation (12) describes a spheroid with the poles $z = \mp M$, $\rho = 0$, the ring singularity $z = 0$, $\rho = \sqrt{a^2 + Q^2}$ being its equator. However, one should bear in mind that this is only a Euclidean picture of the surface $r = 0$, well demonstrating its non-disk geometry, while the precise properties of the surface are determined by its Gaussian curvature (11).

From (11) it also follows that in the Kerr case, when $Q = 0$, the Gaussian curvature of the corresponding $r = 0$ surface is equal to zero. The equation (12) nonetheless continues to describe a spheroid, and in this particular case the surface $r = 0$ must be interpreted as a *dicone* depicted in Fig. 1 (the reader is referred to our paper[8] for further details).

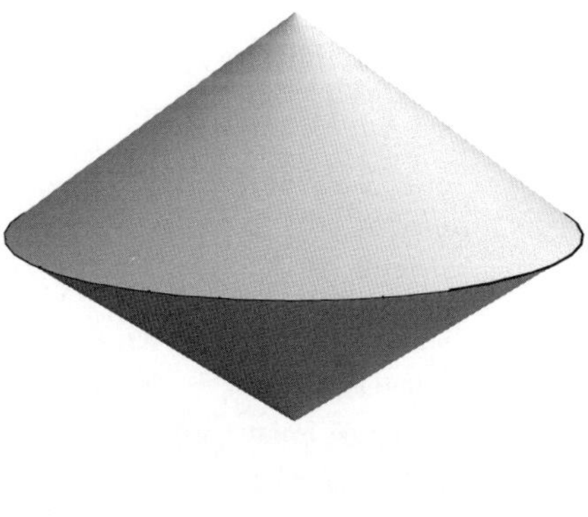

Fig. 1. The two-surface $r = 0$, $t = \text{const}$ of the Kerr solution – a dicone.

4. The Surface $r = 0$ in KdS and KNdS Solutions

Like in the case of the KN solution, the demonstration of the non-disk geometry of the surface $r = 0$ of the KNdS spacetime is straightforward and only requires the calculation of the Gaussian curvature for that surface. The $r = 0$ surface is defined by the two-metric

$$d\sigma^2 = \frac{a^2\cos^2\theta}{\Delta_\theta}d\theta^2 + (a^2\Xi - Q^2\tan^2\theta)\sin^2\theta d\varphi^2,$$
$$\Delta_\theta = 1 + \frac{\Lambda}{3}a^2\cos^2\theta, \quad \Xi = 1 + \frac{\Lambda}{3}a^2, \tag{14}$$

the parameter Λ standing for the cosmological constant.

The corresponding Gaussian curvature has the form[9]

$$K = \frac{\sec^4\theta}{3a^2(a^2\Xi\cos^2\theta - Q^2\sin^2\theta)^2}\{\Lambda a^6\Xi^2\cos^8\theta$$
$$+ a^2Q^2\Xi\cos^2\theta[9 + 3\sin^2\theta + \frac{\Lambda}{4}a^2\cos^2\theta(11 + \cos 4\theta)]$$
$$- Q^4\sin^2\theta[6 + 3\sin^2\theta + \frac{\Lambda}{8}a^2\cos^2\theta(15 + \cos 4\theta)]\}, \tag{15}$$

and in the limit $\Lambda = 0$ the above K reduces to the expression (11).

It is remarkable that in the particular case of vanishing electric charge ($Q = 0$) the formula (15) for the Gaussian curvature considerably simplifies, yielding the following constant value

$$K = \frac{\Lambda}{3}, \tag{16}$$

which characterizes the surface $r = 0$ of the KdS solution. In the paper[9] we have plotted two examples of such surfaces of constant Gaussian curvature in both $\Lambda > 0$ and $\Lambda < 0$ cases.

5. Towards the Construction of Correct MAE's

Obviously, the discovery of the non-disk geometry of the surface $r = 0$ discussed in the previous sections changes drastically the approach to the construction of the MAE's of stationary black-hole solutions. The main idea of the extension procedure in each case must be incorporating the region inside the closed surface $r = 0$ into the general manifold, which does not involve neither the gluing procedure nor the spatial negative infinity for r. To some extent, this would be similar to extending the usual 3-dimensional Euclidean space in which a sphere of radius r_0 is removed.

Acknowledgments

This work was partially supported by the CONACYT of Mexico.

References

1. R. H. Boyer and R. W. Lindquist, *J. Math. Phys.* **8**, 265 (1967).
2. S. W. Hawking and G. F. R. Ellis, *The Large Scale Structure of Space-Time* (Cambridge University Press, 1973).
3. R. P. Kerr, *Phys. Rev. Lett.* **11**, 237 (1963).
4. V. S. Manko, *Prog. Theor. Phys* **127**, 1057 (2012).
5. E. Newman, E. Couch, K. Chinnapared, A. Exton, A. Prakash and R. Torrence, *J. Math. Phys.* **6**, 918 (1965).
6. B. Carter, in *Black Holes* (Gordon and Breach, New York, 1973), p. 56.
7. G. W. Gibbons and S.W. Hawking, *Phys. Rev. D* **15**, 2738 (1977).
8. H. García-Compeán and V. S. Manko, *Prog. Theor. Exper. Phys.* **2015**, 043E02 (2015).
9. V. S. Manko and H. García-Compeán, *Phys. Rev. D* **90**, 047501 (2014).
10. W. Israel, *Phys. Rev. D* **2**, 641 (1970).
11. J. B. Griffiths and J. Podolský, *Exact Space-Times in Einstein's General Relativity* (Cambridge University Press, 2009), p. 200.
12. B. O'Neill, *Elementary Differential Geometry* (Academic Press, New York, 1966).
13. B. Carter, *Phys. Rev.* **174**, 1559 (1968).

Source of Kerr-Newman solution as supersymmetric bag model: 50 years of the problem

A. Burinskii*

Theor. Phys. Lab., NSI Russ. Acad. Sci.,
Moscow, 115191, Russia
**E-mail: bur@ibrae.ac.ru*

The ultra extreme Kerr-Newman (KN) solution($a = J/m >> m$) produces the gravitational and EM fields of the electron. It has a naked singular ring – a topological defect which may be regularized by a solitonic source forming the pseudo-vacuum bubble filled by Higgs condensate in a supersymmetric superconducting state. Structure and stability of this source is determined by Bogomolnyi equations as a BPS-saturated soliton. The Principal Null Congruences of the KN solution determine consistent embedding of the Dirac equation, which acquires the mass from the Higgs condensate inside the soliton, indicating that this soliton forms a bag model. Shape of this bag is unambiguously determined by BPS-bound. The bag turns out to be flexible and takes the form of a very thin disk, which is completed by a ring-string along its sharp boundary. The ring-string traveling waves generate extra deformations of the bag creating a circulating singular pole. Bag model of the KN source integrates the dressed and pointlike electron in a bag-string-quark system, which removes the conflict between gravity and the point-like electron of the Dirac theory.

Keywords: Kerr-Newman solution; Higgs field; dressed electron; soliton; bag model; strings; Dirac equation.

1. Introduction

This year debating of the source of the Kerr and Kerr-Newman (KN) solution marks 50 years. Reason of the complexity of this problem is the emergence of the Kerr singular ring instead of the pointlike singularity of the non-rotating black holes (BH). Radius of this ring, $a = J/m$, [a] is proportional to angular momentum J and inverse proportional to mass m of the BH. It leads to an unexpected behavior breaking the wide spread opinion that the effective range of the gravitational field r_g is proportional to mass and short, $r_g \sim m$. It is only true for the long distances from the source. The

[a]We use the dimensionless units $c = G = \hbar = 1$ and signature $(- + ++)$.

short distance Kerr-Newman field is extended as $r_g \sim a = J/m$, taking the form of a singular string, for which r_g is the more, than the smaller is mass of the BH. So, that for the elementary particles with very small masses, the effective range of gravity is tremendously expand, taking the form of the ring-string (gravitational waveguide[1,2]) surrounding the Compton zone of vacuum polarization. The spin/mass ratio of elementary particles turns out to be much more than their mass, $a = J/m >> m, \quad a/m \sim 10^{44}$, and the BH horizons disappear. In this ultra extreme case, the Kerr singular ring turns out to be naked, and the censorship principle requires some source, which should cover the naked singularity.

50 years ago, it was noted in the appendix to paper by Newman and Janis[3] that the simple interpretation of the KN source as a rotating singular ring does not go, because the Kerr singular ting turns out to be a branch line of space-time, and the observer, which passed through this ring, $r^+ \to r^- = -r^+$, turns out to be on a different space with different metric

$$g^+_{\mu\nu} = \eta_{\mu\nu} + 2Hk^+_\mu k^+_\nu \to g^-_{\mu\nu} = \eta_{\mu\nu} + 2Hk^+_\mu k^+_\nu, \tag{1}$$

and with the similar two-valuedness of the electromagnetic potential $A^+_\mu \to A^-_\mu$. The KN solution is *two-sheeted and contains the mirror "Aice" world* determined by two different Principal Null Directions of the Kerr congruence $k^\pm_\mu$ on the same auxiliary Minkowski background $\eta_{\mu\nu}$.

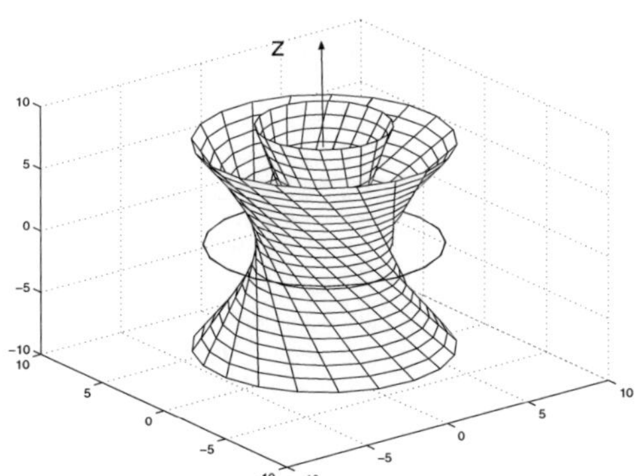

Fig. 1. Null directions of the Kerr congruence k^μ are focused on the Kerr singular ring, and passing through the Kerr singular ring forms the second sheet of space.

In addition, it was noted that the KN solution has the gyromagnetic ration $g = 2$ as that of the Dirac electron,[4] and therefore, it corresponds to the gravitational and em field of the electron. These features led to intensive study of the source of the KN solution. Taking into account the known conflict between gravity and quantum theory and, in particular, the unacceptance for gravity the known quantum statement on the point-like, structureless electron, because gravity requires a field model of the source leading to a stress-energy tensor. The problem of source of the KN solution is important because it could give a hint how to remove this conflict and answer the question: *what should be the structure of electron for compatibility with gravity?* We obtain that electron should have basic features of the known bag models:[5]

– to be based on the Higgs mechanism of symmetry breaking,
– to have consistent implementation of the Dirac equation,
– to be deformable, creating stringy structure,
– mass of the fermion should be generated by Higgs condensate.

We conclude that to ensure consistency with gravity the structure of electrons should be not far from the structure of hadrons.

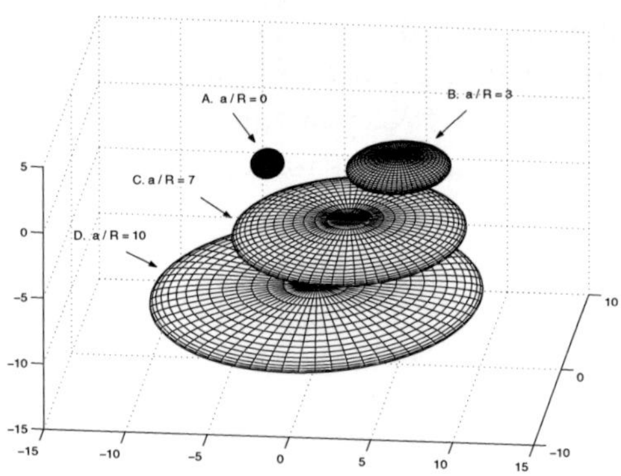

Fig. 2. (A): Spherical bad with zero rotation, $a/R = 0$, and the rotating disk-like bags for different ratios a/R, (B): $a/R = 3$;(C): $a/R = 7$, (D):$a/R = 10$.

2. Supersymmetric Domain Wall Phase Transition

Two lines were studied by development of this problem: the stringlike source[1,2,13] and the disk-like model of the KN sorce, started by Israel[7] and than modified by López to the oblate bubble model.[8] Both lines of development were recently joined into the bag-bubble model,[5] which, similar to the known MIT and SLAC bags, is based on the Higgs mechanism of symmetry breaking.

Structure of the bag is UNIQUELY determined by the requirements:

(R1) Quantum INTERIOR: no Gravity INSIDE the source!

(R2) Gravitating EXTERIOR: classical exact KN solution OUTSIDE!

(R3) A smooth transition between internal and external regions.

Indeed, SHAPE of the bubble is uniquely determined by the requirements **R1** and **R2** and by the KN metric

$$g_{\mu\nu}^{(KN)} = \eta_{\mu\nu} + 2H_{(KN)}k_\mu k_\nu, \tag{2}$$

where

$$H_{(KN)} = \frac{mr - e^2/2}{r^2 + a^2\cos^2\theta}. \tag{3}$$

Boundary of the source corresponds to *the surface of zero gravity potential*, $H_{(KN)} = 0$, which is $r = R = e^2/2m$, where r is the Kerr oblate spheroidal coordinate.[6] For parameters of an electron, thickness of the disk $R = e^2/2m$ is the classical radius of electron, while radius of te disk $\approx a = \hbar/2m$ is the Compton length. So, the ratio $R/a = 137^{-1}$ is the fine structure constant. The KN source takes the Compton region, thus, the bag-bubble describes a *dressed* electron.

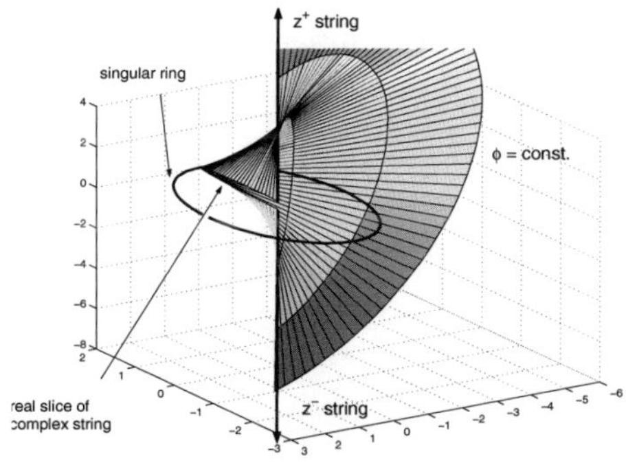

Fig. 3. Vector potential is dragged by Kerr singular ring, forming a closed Wilson loop.

The used in the known bag models non-linear self-interaction of the Higgs field, described by potential $V(r) = \lambda(|H|^2 - \eta^2),^2$ conflicts with external KN solution, and we shell use a supersymmetric model of phase transition, consisting from the triplet of the chiral fields $\Phi^{(i)} = \{H, Z, \Sigma\}$,

where H is taken as the Higgs field. The corresponding superpotential $W(\Phi^{(i)})$ (suggested by J. Morris[10]) provides two vacuum states[5]:
(I.VacIn) – supersymmetric internal state with $|H| = \eta$, and
(II.VacExt) – external vacuum state with $|H| = 0;\ Z = 0;\ \Sigma = \eta$, leading to unbroken gauge symmetry of the external KN solution.
The HLG Lagrangian leads to interaction of Higgs field $H(x) = |H|e^{i\chi(x)}$ with KN vector potential given by

$$\nabla_\nu \nabla^\nu A_\mu = I_\mu = \frac{1}{2}e|H|^2(\chi,_\mu + eA_\mu). \tag{4}$$

This equation leads to two important consequences:

(i) spin of the bag is quantized – closed flux of vector potential forms a quantum Wilson loop $\oint eA_\varphi d\varphi = -4\pi ma$, leading to , $J = ma = n\hbar/2,\ n = 1, 2, 3...$, see Fig. 2, and

(ii) the bag has oscillon structure – the Higgs condensate forms a coherent vacuum state oscillating with the frequency $\omega = 2m$.
The resulting bag represents a supersymmetric BPS-saturated soliton, which structure and stability are determined by the Bogomolnyi equations.[11]

3. Embedding the Dirac Equation

The Dirac quark-fermion gets the mass from the Higgs condensate and is trying to get energetically advantageous position with minimum mass. Thus, in the bag models the mass is variable function determined by vev of the Higgs condensate. In the KN bag, the Higgs condensate is enclosed inside the bad, see (**I.VacIn**), and therefore, outside the bag the Dirac field is massless, and splits into two massless equations for the Weyl spinors collinear to Kerr congruence, as it is required by the algebraically special structure of Kerr geometry. Congruences are determined by the **Kerr Theorem**[5,6,13] in terms of the projective twistor coordinates $T^a = \{Y,\ \ \zeta - Yv,\ \ u + Y\bar{\zeta}\}$, where $Y = \phi_1/\phi_0$, is equivalent to the Weyl spinor ϕ_α. Thus, the *TWISTOR/SPINOR correspondence* provides consistency of the Dirac field with KN gravity.

The two Weyl spinors ϕ^α and $\bar{\chi}_{\dot{\alpha}}$, generated by the Kerr theorem are not equivalent. They are collinear to different Kerr congruences, and only one of them is compatible with external KN solution.[5,12] Another one lives on the second (advanced) sheet of the Kerr geometry and is not consistent with the external KN solution. Non-equivalence of the "left" and "right" Weyl spinors corresponds to electroweak sector of the Standard Model.

4. Stringy Deformations of the KN Bag

The bags are to be soft and easily deformed to stringy flux-tubes, as in the known model of a meson built from the quark-antiquark pair. Similarly, the disk-like bag of the KN electron can be considered as if the originally spherical electron of the classical radius $r_e = e^2/2m$ is stretched under rotation, taking the form of thin disk of the Compton radius (Fig. 1). It forms a circular string on the sharp rim of the disk. The stringy traveling waves lead to additional deformations of the KN bag which break regularization and create a circulating singular pole, reproducing zitterbewegung of the pointlike Dirac electron.[11] It allows us to unite the dressed and point-like electron model in a single "bag-string-quark" system.

Acknowledgments

This work is supported by the RFBR grant 13-01-00602.

References

1. A. Ya. Burinskii, *Zh. Eksp. Teor. Fiz.* **66**, 406 (1974) [*Sov. Phys. JETP* **39** 193 (1974)].
2. D.D. Ivanenko and A.Ya. Burinskii, Gravitational strings in the models of elementary particles, *Izv. Vuz. Fiz.*, **5**, 135 (1975).
3. E.T. Newman and A.I. Janis, *J. Math. Phys,* **6**, 915 (1965).
4. B. Carter, *Phys. Rev.* **174**, 1559 (1968).
5. A. Burinskii, *JETP* **148 (2)** 228 (2015), `arXiv:1505.03439`.
6. G. C. Debney, R. P. Kerr and A. Schild, *J. Math. Phys.* **10**, 1842 (1969).
7. W. Israel, *Phys. Rev.* **D 2**, 641 (1970).
8. C. A. López, *Phys. Rev.* D **30**, 313 (1984).
9. A. Burinskii, *Journal of Physics: Conference Series* **361**, 012032 (2012).
10. J. R. Morris, *Phys. Rev.* **D 53**, 2078 (1996).
11. A. Burinskii, to appear in *JETP* **148** (2015).
12. A. Burinskii, *Grav. and Cosmol.* **21**, 28 (2015).
13. A. Burinskii, *Theor. Math. Phys.*, **177**(2), 1492 (2013).

Strong gravity and structure of topological solitons

Yu. P. Rybakov

Department of Theoretical Physics and Mechanics
Peoples' Friendship University of Russia
117198 Moscow, 6, Mikluho-Maklay str., Russia
E-mail: soliton4@mail.ru

The unification of Skyrme and Faddeev chiral models describing baryons and leptons respectively as topological solitons is suggested within the framework of 16-spinor field $\Psi = \Psi_1 \oplus \Psi_2$ nonlinear model containing two 8-semispinors Ψ_1 and Ψ_2. Using Brioschi identity for 8-spinors and special structure of the Higgs potential V implying the spontaneous symmetry breaking, it is possible to realize topological soliton-like excitations of two kinds due to the choice of S^2- or S^3- manifolds as phase spaces. The interactions with electromagnetic, Yang—Mills and gravitational fields are exhibited through the extention of derivatives via gauge invariance principle. Specific inclusion in the Higgs potential of the Kretschmann gravitational invariant $K = R_{\mu\nu\sigma\lambda}R^{\mu\nu\sigma\lambda}/48$ permits one to obtain the strong gravity behavior at small distances and guarantee the correspondence with Quantum Mechanics at large distances.

Keywords: 16-spinor; topological charge; solitons; Kretschmann invariant.

1. Introduction. Brioschi Identity and 16-Spinor Field Model

In the popular Skyrme model (SM)[1] baryons are considered as topological solitons (TS), the baryon number $\mathbb{B}$ being identified with the degree type topological charge: $\mathbb{B} = \deg(S^3 \to S^3)$ serving as generator of the homotopy group $\pi_3(S^3) = \mathbb{Z}$. Similarly, in the Faddeev model (FM)[2] leptons are considered as TS endowed with the Hopf invariant Q_H, the latter one playing the role of the lepton number $\mathbb{L} = Q_H$ serving as generator of the homotopy group $\pi_3(S^2) = \mathbb{Z}$. To unify these two approaches we introduce 16-spinor field $\Psi = \Psi_1 \oplus \Psi_2$, Ψ_i, $i = 1, 2$, being 8-spinors, and the two kinds of internal Pauli matrices: $\vec{\lambda} = I_4 \otimes \vec{\sigma} \otimes I_2$, $\vec{\Lambda} = I_8 \otimes \vec{\sigma}$, where $\vec{\sigma}$ stands for usual Pauli matrices and I_n- for $n-$ dimensional unit matrix. As was shown in Ref. 3, for each 8-spinor ψ the following Brioschi identity[4] holds:

$$j_\mu j^\mu - \tilde{j}_\mu \tilde{j}^\mu = s^2 + p^2 + \vec{v}^2 + \vec{a}^2, \tag{1}$$

where standard bilinear spinor quantities are introduced: $s = \bar{\psi}\psi$, $p = \imath\overline{\psi}\gamma_5\psi$, $\vec{v} = \overline{\psi}\vec{\lambda}\psi$, $\vec{a} = \imath\overline{\psi}\gamma_5\vec{\lambda}\psi$, $j_\mu = \overline{\psi}\gamma_\mu\psi$, $\tilde{j}_\mu = \overline{\psi}\gamma_\mu\gamma_5\psi$, $\overline{\psi} = \psi^+\gamma_0$, with γ_μ, $\mu = 0, 1, 2, 3$, being Dirac matrices. In view of the identity Eq.(1), one can use the special structure of the Higgs potential V implying the spontaneous symmetry breaking:

$$V = \frac{\sigma^2}{8}\left(j_\mu j^\mu - \varkappa_0^2\right)^2, \tag{2}$$

with σ and $\varkappa_0$ being parameters of the model and the natural boundary condition at space infinity being imposed:

$$\lim_{r\to\infty} j_\mu j^\mu = \varkappa_0^2, \quad r = |\vec{x}|. \tag{3}$$

As follows from Eq.(1) and Eq.(3), the two kinds of TS are possible[3] due to the choice of S^2- or S^3- manifolds as phase spaces, with the vacuum fixed point being determined by Eq.(3). In fact, S^2- manifold is determined by the $O(3)-$invariant $\vec{v}^2$ and corresponds to FM or lepton sector (LS). Similarly, S^3- manifold is determined by the chiral invariant $s^2 + \vec{a}^2$ and corresponds to SM or baryon sector (BS).

However, the unification of SM and FM supposes that the vacuum state Ψ_0 should be universal, that can be realized within the scope of 16-spinor field model, if one supposes that $\Lambda\Psi_0 = 0$, $\Lambda = (1 - \Lambda_3)/2$, and S^2- manifold is determined by the invariant $\vec{V}^2$, $\vec{V} = \overline{\Psi}\vec{\Lambda}\Psi$, S^3- manifold being unchanged. To choose LS, one remarks that condition $\vec{a} = \imath(\overline{\Psi}\gamma_5\vec{\lambda}\Psi) = 0$ implies $\mathbb{B} = 0$ and can be satisfied if one supposes the mirror symmetry (MS) of LS:

$$\Psi \to \gamma_0\Psi. \tag{4}$$

Introducing the 16-spinor of the form

$$\Psi = \oplus_{j=1}^2(\varphi_j \oplus \chi_j \oplus \xi_j \oplus \zeta_j), \tag{5}$$

where φ_j, χ_j, ξ_j, ζ_j stand for 2-spinors, one finds from invariance condition Eq.(4) that $\varphi_j = \chi_j$, $\xi_j = \zeta_j$ for the Weyl representation of $\gamma-$matrices, i. e. one gets an effective 8-spinor. In this case $\vec{V} = \overline{\Psi}\vec{\Lambda}\Psi \neq 0$ and in accordance with the Brioschi identity

$$\left(j_\mu j^\mu - \vec{V}^2\right)/16 = \left(|\varphi_2|^2 + |\zeta_2|^2\right)\left(|\varphi_1|^2 + |\zeta_1|^2\right) - |\varphi_1^+\varphi_2 + \zeta_1^+\zeta_2|^2 \geq 0$$

due to the Schwartz inequality.

Now let us choose BS, taking into account the charge independence of strong interactions, or MS of the isotopic space:

$$\Psi \to \gamma_0\gamma_5\gamma_2\lambda_2\Psi^*. \tag{6}$$

Inserting Eq.(5) into the invariance condition Eq.(6) one finds for BS $\xi_j = \imath\sigma_2\varphi*_j$, $\zeta_j = \imath\sigma_2\chi*_j$, i. e. again one gets an effective 8-spinor. In this case $p = \vec{v} = V_2 = 0$, i. e. $\mathbb{L} = 0$, but $s \neq 0$, $\vec{a} \neq 0$ and in accordance with the Brioschi identity

$$(s^2 + \vec{a}^2)/16 = |\varphi_1^+\chi_1 + \varphi_2^+\chi_2|^2 + |\varphi_1^T\sigma_2\chi_1 + \varphi_2^T\sigma_2\chi_2|^2 \leq (j_\mu j^\mu)/16.$$

Thus, the structure Eq.(2) of the Higgs potential is consistent with the topology constraints of LS or BS.

In view of these arguments, let us consider the following Lagrangian density[3]:

$$\mathcal{L}_{\text{spin}} = \frac{1}{2\lambda^2}\overline{\partial_\mu\Psi}\gamma^\nu j_\nu\partial^\mu\Psi + \frac{\epsilon^2}{4}f_{\mu\nu}f^{\mu\nu} - V, \tag{7}$$

where $f_{\mu\nu} = (\overline{\Psi}\gamma^\alpha\partial_{[\mu}\Psi)(\overline{\partial_{\nu]}\Psi}\gamma_\alpha\Psi)$ is the anti-symmetric tensor of Skyrme—Faddeev type and λ, ϵ are constant parameters. The first term in Eq.(5) is similar to the sigma-model one and includes the projector $P = \gamma^0\gamma^\nu j_\nu$ on the positive energy states.

2. Interaction with Physical Vector Fields

Due to the general gauge invariance principle, the group of phase transformations of Ψ with some charge generator Γ_e and left, right isotopic rotations with generators $P_{L,R}\lambda^a\Lambda/2$, $a = 1, 2, 3$, $P_{L,R} = (1 + \gamma_5)/2$, give rise to the interactions with the electromagnetic field A_μ and left, right Yang—Mills fields $A_\mu^{aL,R}$ respectively through the extention of the derivative $\partial_\mu\Psi \to D_\mu\Psi = \partial_\mu\Psi - \imath e_0\Gamma_e A_\mu\Psi + (A_\mu^L + A_\mu^R)\Psi$, where we adopt $\Gamma_e = P_3\Lambda$, $P_3 = (1 - \lambda_3)/2$, $A_\mu^{L,R} = P_{L,R}e_{1L,R}A_\mu^{aL,R}\lambda^a\Lambda/(2\imath)$. Here e_0, e_{1L}, e_{1R} stand for corresponding coupling constants. However, the Yang—Mills fields being responsible for strong interactions, they should not give any contribution to LS. In view of MS Eq.(4), this is possible only if the sum $A_\mu^{aL} + A_\mu^{aR}$ is proportional to γ_5, or the following constraint holds:

$$e_{1L}A_\mu^{aL} + e_{1R}A_\mu^{aR} = 0. \tag{8}$$

In view of Eq.(8) the extended derivative reads:

$$D_\mu\Psi = \partial_\mu\Psi - \imath e_0 P_3\Lambda A_\mu\Psi + g_0\gamma_5\mathbb{A}_\mu\Psi, \tag{9}$$

where $g_0 = e_{1L} = e_{1R}$, $\mathbb{A}_\mu = \lambda^a\Lambda A_\mu^{aL}/(2\imath)$, with the standard Yang—Mills Lagrangian taking the form:

$$\mathcal{L}_{\text{YM}} = \frac{1}{32\pi}\text{Sp}\left\{(\partial_\mu\mathbb{A}_\nu - \partial_\nu\mathbb{A}_\mu)(\partial^\mu\mathbb{A}^\nu - \partial^\mu\mathbb{A}^\nu) + g_0^2[\mathbb{A}_\mu, \mathbb{A}_\nu][\mathbb{A}^\mu, \mathbb{A}^\nu]\right\}. \tag{10}$$

It is worth-while to stress that our spinor model admits a new possibility to derive usually adopted hypothesis of the electric charge Q_e quantization, i. e. $Q_e = ne_0$, $n \in \mathbb{Z}$. To this end let us introduce the modified Maxwell Lagrangian density:

$$\mathcal{L}_{\rm em} = -\frac{1}{16\pi} F_{\mu\nu} F^{\mu\nu} f(u), \tag{11}$$

where $F_{\mu\nu} = \partial_\mu A_\nu - \partial_\nu A_\mu$ and $f(u)$ is some unknown function of the special Lorentz and dilatation invariant

$$u = (n_\mu A^\mu)^2 (-E_\nu E^\nu)^{-1/2}. \tag{12}$$

Here $E_\nu = n^\mu F_{\mu\nu}$, the unit time-like vector n^μ being identified with the normalized Dirac current: $n^\mu = j^\mu/j$, $j = (j_\nu j^\nu)^{1/2}$. In fact, for the static soliton with the charge q one finds the equation for the scalar potential $\Phi = A^0$:

$$\Phi \,\mathrm{div}\, [\vec{E}(2f - uf'(u))] = 2uf'(u)(\vec{E})^2, \quad \vec{E} = -\bigtriangledown \Phi. \tag{13}$$

At large distances $r \to \infty$ one gets $\Phi = (q/r)[1 + \mathcal{O}(1/r)]$, $u = q[1 + \mathcal{O}(1/r^2)]$ and the left hand side of Eq.(13) behaves as $\mathcal{O}(1/r^5)$, or $f'(q) = 0$. The latter equation being considered as the quantization condition for Q_e, one can suggest the simplest choice of the function $f(u) = 1 + \mu_0 \sin^2[\pi u/(2e_0)]$, where μ_0 stands for some small constant.

Finally, let us consider small excitations of our soliton configuration near the vacuum: $\Psi = \Psi_0 + \xi$ as $|\vec{x}| \to \infty$. Then the linearized equation for ξ after the substitution $\xi = k\Psi_0$ takes the form:

$$\partial_\mu \partial^\mu k + \frac{1}{2} M_0^2 (k^* + k) = 0, \quad M_0 \equiv 2\sigma\lambda\varkappa_0. \tag{14}$$

Thus, from Eq.(14) one derives the equations for real and imaginary parts of $k = k_1 + \imath k_2$:

$$\partial_\mu \partial^\mu k_1 + M_0^2 k_1 = 0, \quad \partial_\mu \partial^\mu k_2 = 0. \tag{15}$$

According to Eq.(15), our model admits two types of vacuum excitations: massive and massless ones. For massive soliton the excitation k_1 could be interpreted as the wave function if the parameter M_0 were replaced with the real mass M of the soliton. To satisfy this condition, we first introduce the interaction with the gravitational field via the new extended derivative $\bigtriangledown_\mu \Psi = (D_\mu - \Gamma_\mu)\Psi$, where Γ_μ stands for spinor connection. Then we generalize the Higgs potential Eq.(2) through replacing σ^2 with the new invariant:

$$\sigma^2 = -\frac{16 D^3}{\lambda^2 G^2 K^2 \varkappa_0^2}, \tag{16}$$

where G is the Newton gravitational constant, $D = \overline{\nabla_\mu \Psi \gamma^\alpha j_\alpha \nabla^\mu \Psi}$ and K is the Kretschmann invariant constructed with the help of the Riemannian curvature tensor: $K = R_{\mu\nu\sigma\lambda} R^{\mu\nu\sigma\lambda}/48$. The validity of this choice of the invariants in the Higgs potential becomes evident at large distances, since in view of the Schwarzschild metric $D = -r_g^2 \varkappa_0^2/(4r^4)$, $K = r_g^2/r^6$, where $r_g = GM$ and natural units $\hbar = c = 1$ are used.

3. Axially-Symmetric States and Strong Gravity

Let us consider static axially-symmetric states in geometric toroidal coordinates $x^1 = r$, $x^2 = \vartheta$, $x^3 = \phi$, $r \in [0, \infty]$, $\vartheta \in [-\pi, \pi]$, $\phi \in [0, 2\pi]$, where the scale parameter a is inserted in standard cylindrical coordinates: $z = r \sin\vartheta$, $\rho = a + r\cos\vartheta$. Therefore the squared interval takes the form:

$$ds^2 = e^{2\eta} dt^2 - e^{2\alpha} dr^2 - e^{2\beta} d\vartheta^2 - e^{2\gamma} d\phi^2, \tag{17}$$

where the metric coefficients η, α, β, γ depend only on r, ϑ. Let us now calculate the Kretschmann invariant for the metric Eq.(17), the result being:

$$K = \frac{1}{12}\left[\sum_{i=1}^{6} k_i^2 + 2e^{-2\alpha-2\beta}(k_7^2 + k_8^2)\right], \tag{18}$$

where the following quantities are introduced:

$$\begin{gathered}
k_1 = e^{-2\alpha}[\partial_1^2\eta + (\partial_1\eta)^2 - \partial_1\eta\partial_1\alpha] + e^{-2\beta}\partial_2\eta\partial_2\alpha; \\
k_2 = k_1(\alpha \Leftrightarrow \beta); \quad k_5 = k_1(\eta \Rightarrow \gamma); \quad k_6 = k_2(\eta \Rightarrow \gamma); \\
k_3 = e^{-2\alpha}\partial_1\eta\partial_1\gamma + e^{-2\beta}\partial_2\eta\partial_2\gamma; \\
k_4 = e^{-2\alpha}[\partial_1^2\beta + (\partial_1\beta)^2 - \partial_1\beta\partial_1\alpha] + (\alpha \Leftrightarrow \beta, 1 \Rightarrow 2); \\
k_7 = \partial_{12}\eta + \partial_1\eta\partial_2\eta - \partial_1\eta\partial_2\alpha - \partial_2\eta\partial_1\beta; \\
k_8 = k_7(\eta \Rightarrow \gamma).
\end{gathered}$$

Using invariant Eq.(18), one can find the structure of the metric at small distances $r \ll a$, where the higher derrivatives give the most important contribution to the gravitational equations, which can be obtained through minimizing K. To this end we first remark the symmetry of K with respect to transposition $\eta \Leftrightarrow \gamma$. The solution to the corresponding equations reads:

$$\alpha = 0, \quad \beta = \log r, \quad \eta = \gamma = \log(a + r\cos\vartheta), \tag{19}$$

with the Kretschmann invariant being

$$K = \frac{1}{12(a + r\cos\vartheta)^4}. \tag{20}$$

Thus we conclude that the strong gravity approximation Eq.(19) works well in the domain $r \ll a$, with the field configuration appearing as closed string of the large closure radius $a \gg r$.

4. Conclusions

The resulting Lagrangian of our 16-spinor model reads: $\mathcal{L} = \mathcal{L}_{\text{spin}} + \mathcal{L}_{\text{em}} + \mathcal{L}_{\text{Mie}} + \mathcal{L}_{\text{YM}} + \mathcal{L}_g$, with $\mathcal{L}_g$ being the Einstein gravitational Lagrangian and $\mathcal{L}_{\text{Mie}} = -(A_\mu A^\mu)^3 \nu_0/6$ standing for Mie correction term providing regular behavior of A_μ at small distances. It is worth-while to stress that the gravitational field plays an important role in our model, since the wave-particle duality principle of Quantum Mechanics has the gravitational origin and the vacuum excitation ξ can be considered as the wave function for the soliton center in the special stochastic representation.[5,6] On the other hand, at small distances the gravity becomes strong and provides the existence of topologically nontrivial field configurations with closed string structure.[2]

References

1. T. H. R. Skyrme, A unified field theory of mesons and baryons, *Nucl. Phys.* **31**, 556 (1962).
2. L. D. Faddeev, Gauge-invariant model of electromagnetic and weak interactions of leptons, *Rep. of Ac. Sci. USSR* **210**, 807 (1973).
3. Yu. P. Rybakov, Soliton configurations in generalized Mie electrodynamics, *Phys. of Atomic Nuclei* **74**, 1102 (2011).
4. E. Cartan, *The Theory of Spinors* (Hermann, Paris, 1966).
5. Yu. P. Rybakov, On the causal interpretation of Quantum Mechanics, *Found. of Physics* **4**, 149 (1974).
6. Yu. P. Rybakov, Solitons and Quantum Mechanics, *Dynamics of Complex Systems* **3**, 3 (2009).

100 years of general relativity: From equations to symmetry principles

V. N. Pervushin*[1], A. B. Arbuzov[1], A. Yu. Cherny[1], D.J. Cirilo-Lombardo[1,3], R. G. Nazmitdinov[1], Nguyen Suan Han[1], A. E. Pavlov[1] and A. F. Zakharov[1,2]

[1] *Bogoliubov Laboratory of Theoretical Physics, JINR, Dubna, Moscow region, Russia*

[2] *Institute of Theoretical and Experimental Physics, Moscow, 117218, Russia*

[3] *National Institute of Plasma Physics, (INFIP-CONICET), FCEyN, Universidad de Buenos Aires, Buenos Aires, 1428, Argentina*
* *E-mail: pervush@theor.jinr.ru*
www.theor.jinr.ru

The Supernovae data, the small value of the Higgs particle mass, and the matter content of the Universe are explained via joint nonlinear realization of the affine and conformal symmetry groups (in spirit of the effective Chiral Lagrangians). This nonlinear realization is used to classify observable physical processes in the Universe, including its origin from the quantum vacuum. All dimension parameters of the theory (including masses and field condensates) arise via quantum anomalies in the beginning of the Universe as consequences of the vacuum postulate. The obtained results are in agreement with last observational and experimental data. These results testify to a super-analog of the joint nonlinear realization as the super-unification of the GR and SM that can be based on the quaternionic supertwistor construction.

Keywords: QFT; quantum gravity; conformal symmetry breaking.

1. "Foundations of Physics"

In 1915 Albert Einstein introduced the field equations of gravitation.[1] The equations express the Riemannian space metric components through the energy-momentum ones. The mathematical basis of these equations was their (co)invariance with respect to general coordinate transformations. These solutions contain certain initial data. Einstein supposed that the general coordinate transformations have the same physical status as the initial data transformations under the Galileo group or their relativistic generalization known as the Poincaré symmetry group in the Special Relativity.

According to David Hilbert, "Foundations of Physics" are the Einstein intervals and the least action principle which give the the second order equations of motion. Then the Einstein equations can be derived from the Hilbert action by variation of the metric components. Along this line, Emmy Nöther proved two theorems. In accordance with these Nöther theorems, one should distinguish between the Poincaré initial data transformations (as the object of the first theorem) and the Einstein general coordinate transformations (i.e. diffeomorphisms as the object of the second theorem). This is a consequence of the Hilbert "Foundations", which is absent in the Einstein's approach.

The difference between the Einstein diffeomorphisms and transformations of reference frames was found in the approach to the General Relativity (GR) developed by Vladimir Fock.[2] Fock actually introduced a gauge-invariant orthogonal basis in order to include fermions in GR. In this way, Fock replaced the metrical formalism in GR by the diffeo-invariant simplex components in the tangent Minkowskian space-time.[2] By using the Fock modification of the Einstein's approach developed in 1915, one can introduce the Poincaré symmetry group of initial data transformations in the tangent space-time.

In the framework of the Hilbert "Foundations", the next steps forward in the conception of the initial data transformations were made by Dirac and by Arnowitt, Deser and Misner,[3,4] namely, they formulated so called Dirac-ADM Hamiltonian approach to the constrained gravitation systems described by the Hilbert action. In this case, the Einstein diffeomorphism group transformations is reduced to the time reparameterizations. The reparameterization time invariance means that there are two invariant evolution parameters, the geometrical and dynamical ones. The first parameter is the geometrical interval, and the second can be associated with zero mode of one of the dynamical fields in the Wheeler–DeWitt (WDW) field superspace of events.[5,6] Then the energy of events in the WDW field superspace space appears as a solution of the energy constraint with respect to the canonical momentum of this time-like field.

Misner[7] supposed that the dynamical evolution time can be identified with the cosmological scale factor logarithm that can be considered as zeroth mode of the spatial metric determinant logarithm. The scalar field zeroth mode is defined by means of averaging over finite spatial volume in accordance with the cosmological principle suggested by Einstein in 1917.[8] The Misner definition of the dynamical evolution time by means of the cosmological principle restricts spatial diffeomorphisms by uniform transformations that hold spatial metric determinant.

Following Dirac,[9] one can call the spatial metric determinant logarithm the *dilaton* field. Dirac and Deser[10] replaced the Einstein intervals by the dilaton free conformal intervals and identified these intervals with the observational distances. In this case, the dilaton field describes the both redshift in cosmology and relativistic effects, such as the Mercury anomaly or double Eddington angle of the photon deviation in the Sun mass gravitational field. If the Hubble parameter and all masses are equal to zero, then the Einstein theory becomes the conformal invariant. Therefore, one can say that the dilaton field is the Goldstone mode accompanying the spontaneous conformal symmetry breaking.

The Goldstone mode approach to the Einstein GR was developed in paper.[11] The authors obtained the Hilbert action of 1915 as joint nonlinear realization of the affine symmetry group of all linear transformations [A(4)] and conformal symmetry transformations [C=SO(4, 2)] and supposed that the dilaton and all metric components are the Goldstone modes. This Goldstone mode approach collects all previous modification discussed before including Hilbert's action, Fock's simplex, and Dirac's dilaton.

The papers Refs. 12, 13 adduce theoretical and observational arguments that quantum theory requires diffeo-invariant initial data of the independent degrees of freedom and needs the separation between the initial data symmetry group and the subgroup of kinemetric uniform diffeomorphisms. We shall call this quantum theory the unitary representations of the affine symmetry group. It is just the object of the present paper.

2. Unitary Representations of the Affine Symmetry Group

The separation between the initial data symmetry group and the subgroup of kinemetric uniform diffeomorphisms can be fulfilled by using all the modifications discussed before

$$ds^2 = \underbrace{g_{\mu\nu}dx^\mu dx^\nu}_{\text{Einstein(1915)}} = e^{-2D}\underbrace{\widetilde{g}_{\mu\nu}dx^\mu dx^\nu}_{\text{Dirac(1973)}} = e^{-2D}\underbrace{\omega_{(\alpha)}\otimes\omega_{(\beta)}}_{\text{Fock(1929)}}\underbrace{\eta^{(\alpha)(\beta)}}_{\text{tangent}}. \tag{1}$$

The Fock simplex components in the 3+1 foliation take the form

$$\omega_{(0)} = e^{-2D}Ndx^0, \quad \omega_{(b)} = \mathbf{e}_{(b)j}dx^j + N_{(b)}dx^0 \tag{2}$$

where $|\mathbf{e}|=1$, D is dilaton, $d\eta = a^2N_0dx^0 = a^2d\tau$ is diffeo-invariant time interval, and $N_0^{-1} \equiv \langle N^{-1}\rangle$ is the global lapse function. This choice of variables allows us to split the Einstein's general coordinate transformations

$$x^\mu \to \widetilde{x}^\mu = \widetilde{x}^\mu(x^\mu),\ g_{\mu\nu} \to \widetilde{g}_{\mu\nu} = g_{\alpha\beta}(\widetilde{x})\frac{d\widetilde{x}^\mu}{dx^\alpha}\frac{d\widetilde{x}^\nu}{dx^\beta}$$

onto the kinemetric uniformal diffeomorphisms and the initial data transformations. The latter include of the dilaton shift $D(x) \to D(x) + \text{Constant}$ and the Lorentz rotations of Fock's simplex components. This separation enables us to restate Einstein's cosmological principle of the volume average[8] in diffeo-invariant form $\langle D\rangle \equiv V_0^{-1}\int_{V_0}\omega_{(1)}\wedge\omega_{(2)}\wedge\omega_{(3)}D$. In particular, the average of the dilaton can be identify with cosmological scale factor logarithm $\langle D\rangle = -\ln a = \ln(1+z)$. This quantity is known in the observational astrophysics as luminosity.

From the "quantum" point of view, the "Foundation of Physics" are the unitary irreducible representations of the initial data symmetry group. By analogy with the unitary irreducible representation of the Poincaré group[14] in quantum field theory, we obtain the general operator solution of the Wheeler–DeWitt equation for the universe as a sum of two exponents ordered with respect to the luminosity evolution parameter $\langle D\rangle$

$$\hat{\Psi}_{\langle D\rangle_I,\langle D\rangle_0} = \hat{A}^+_{\langle D\rangle_I}\hat{\mathbb{U}}^0_I\frac{1}{\sqrt{2\mathbf{E}_{0\mathrm{U}}}} + \hat{A}^-_{\langle D\rangle_I}\frac{1}{\sqrt{2\mathbf{E}_{0\mathrm{U}}}}\hat{\mathbb{U}}^{I\dagger}_0,$$

$$\hat{\mathbb{U}}^0_I = T_{\langle D\rangle}\exp\left\{-\imath\int\limits_{\langle D\rangle_I}^{\langle D\rangle_0} d\langle D\rangle\mathbf{E}_{\mathrm{U}}\right\}. \tag{3}$$

It is an unitary operator of the universe evolution in the space of events $[\langle D\rangle|F]$ with the dilaton energy $\mathbf{E}_{\mathrm{U}}$ (as the energy constraint solution) and the evolution parameter $[\langle D\rangle]$. The WDW functional describes the creation of the universe at the time of the Planck epoch $\langle D\rangle_I = \ln(1+z_I)$, where the current values of the Planck mass and the Hubble parameter coincide. The new fact is that this conformal Planck epoch coincides with the electro-weak one

$$\frac{M^*_{\mathrm{Pl}}}{(1+z_I)^2} \approx \frac{M_{\mathrm{Higgs}}}{(1+z_I)} \approx T_{\mathrm{CMB}} \approx H_{0,\mathrm{SNe\ Ia\ data}}(1+z_I)^2. \tag{4}$$

where T_{CMB} is the cosmic microwave background temperature. This coincidence points out that both the theories have the same nature of the spontaneous conformal symmetry breaking.

The reasons and sources of the spontaneous conformal symmetry breaking are the vacuum postulate and corresponding quantum anomalies. In other words, we suppose that both the Hubble parameter and masses appear at the Planck epoch as consequences of quantum anomalies and their primordial parameters including the spatial volume dimension.[12,13]

In both the General Relativity and the Standard Model quantum anomalies yield the following experimental and observational facts. The arrow of time arises due to the vacuum postulate at the Planck scale $(1+z_I) = \left[\frac{M^*_{\rm Pl}}{H_0}\right]^{1/4} \approx 10^{15}$. SNe Ia data mean dominance of the vacuum Light energy with the rigid state cosmic evolution and primordial Hubble parameter $H_I = H_0(1+z_I)^2$. The vacuum elementary field condensates yield[15] $M_{\rm Higgs} = H_0(1+z_I)^3 \simeq 125$ GeV. The intensive vacuum creation of electro-weak bosons and their decay product give the present day number of the CMB photons $N_\gamma \approx \alpha_W^2 \cdot (1+z_I)^6 \simeq 10^{87}$ and baryon asymmetry $N_b \approx \alpha_{K_0-\bar{K}_0} N_\gamma$ that arises via triangle anomaly of W-bosons and the $K_0 - \bar{K}_0$ transition coupling $\alpha_{K_0-\bar{K}_0} \approx 10^{-9}$.[12,13,16]

These results and the GR/SM hierarchy (4) suggest that the historical chain of the initial data symmetry group can be extended to the (4+4*) affine symmetry group

$$\underbrace{\text{Galileo}}_{\text{Mechanics}} \to \underbrace{\text{Poincaré}}_{\text{SR}} \to \underbrace{A(4)}_{\text{GR}} \to \underbrace{A(4+4^*)}_{\text{GR}\oplus\text{SM}}, \tag{5}$$

where all SM fields should be the Goldstone particles in the corresponding nonlinear realization of the group $A(4+4^*)$.[17] It can be presented as the boson sector of the superspace that follows from a super-conformal group $SU(2;2|4+4)$[18] in the field of quaternions[19] $\mathbb{H}$.

3. The Quaternionic Supertwistor Basis of the Unification of GR and SM

It is interesting to perform a super-analog of the Borisov-Ogievetsky nonlinear realization[11] in order to obtain the corresponding number of Goldstone fields for both the general relativity and the minimal standard model of the electro-weak and strong interactions (10 tensors, 48 vectors, 96 fermions, and 4 scalars).

We have shown in Ref. 19 that this super-nonlinear realization as the last unification of the GR and SM can be based on the quaternionic superextension of the supertwistor construction given in Ref. 18. According to this quaternionic supertwistor construction, the number of fermionic fields can be extended beyond the pure supertwistorial one with coherent super-quaternionic states spanning the super-Hilbert spaces.

3.1. *Twistor theory*

In the twistor theory our starting point is a complex space $\mathbb{C}M \sim \mathbb{C}_{2,4}(T)$. By using the correspondence with null twistors we can successfully separate from it the real Minkowski space M invariant with respect to the conformal group. The complex space CM in its compactified form, also contains S^4. For this space, the twistor correspondence can be introduced as follows. Reality of the space S^4 does not follow from the nullification of some kind on twistors, as in the Minkowski case. It follows from the invariance under an antilinear mapping $\rho : \mathbb{C}^4 \to \mathbb{C}^4, \rho^2 = -1$. Then, is clear that this mapping represents the multiplication by the standard quaternionic unit due $\mathbb{C}^4 \sim \mathbb{H}$. Because the the space S^4 is not invariant under $SU(2,2)$ we have to take the covering group of the complexified $SU(2,2)$ group. The direct possibility is, for example, $SL(4,\mathbb{C})$ which covers simultaneously $SO(6,\mathbb{C})$ and S^4 being consequently invariant with respect to $SL(2,\mathbb{H})$: its real form which covers $SO(5,1)$.

When we pass to supertwistor space there exists similar mapping in analogy with the ρ in the non-susy case but only for N even. Then, if we assume $N = 2(p+q)$ the quaternionic structure can be introduced. This can be realized taking into account that the quaternions can be written as $\mathbb{H} = \mathbb{C} \otimes \mathbb{C}$ e.g.: the first quaternionic unit in the field of $\mathbb{H}$ is identified with the imaginary unit in $\mathbb{C}$. Consequently, any $q \in \mathbb{H}$ can be written as

$$q = a + b\widehat{i}_2, a, b \in \mathbb{C}\left(\widehat{i}_1\right),$$

where $\mathbb{C}\left(\widehat{i}_1\right)$ is the complex space with the first quaternionic generator as imaginary unit.

To make explicit connection between the quaternionic structures in supertwistor spaces, we only need to introduce 2×2 matrices by each element of the standard supertwistors operators. For instance, is clear that under conjugation any supertwistor(standard notation, internal fermion indices dropped) (ω, π, θ) goes to $\left(\overline{\omega} = \epsilon\omega^*, \overline{\pi} = \epsilon\pi^*, \overline{\theta} = \epsilon\theta^*\right)$ where[19]:

$$\epsilon = \begin{pmatrix} & 1 \\ -1 & \end{pmatrix}.$$

3.2. *Generalized quaternion-valuated supertwistors*

We introduce the generalized quaternion-valuated supertwistors

$$T_S = \begin{pmatrix} \overline{\xi}_A \\ \eta_M \end{pmatrix}, \qquad \widetilde{T}^R = \left(\xi^A, -\overline{\eta}^M\right)$$

such that

$$\overline{\xi}_A, \eta_M \in \mathbb{H}_n,$$

where

$$\left[\widetilde{T}^R, T_S\right\} = \delta^R_S.$$

The superspace Z takes the general form (we follow notation from[18])

$$Z^{\dagger\ A}_{B} = \overbrace{\begin{pmatrix} X_m^{\ a} & \theta_m^{\ i}, \\ \chi_l^{\ a} & \lambda_l^{\ i} \end{pmatrix}}^{2n+2q} \Bigg\} 2m+2p.$$

It represents the space of parameters of quaternionic supercoherent states, depending on the structure of the supercoset space. The above quaternionic supermatrix acts over the following quaternionic supervectors

$$\xi^A = \left(a^c, -\xi^i\right); \ \overline{\xi}_A = \begin{pmatrix} a_c^\dagger \\ \xi_i^\dagger \end{pmatrix} \Big\} 2n+2q,$$

$$\overline{\eta}^M = \overbrace{\left(b^{\dagger m}, \eta^{\dagger l}\right)}^{2m+2p}; \ \eta_M = \begin{pmatrix} b_m \\ \eta_l \end{pmatrix},$$

where we have defined

$$a_c^\dagger = \frac{1}{\sqrt{2}}\left(\lambda_\alpha + \overline{\mu}^{\dot\alpha}\right),$$

$$b_m = -\frac{1}{\sqrt{2}}\left(\lambda_\alpha - \overline{\mu}^{\dot\alpha}\right).$$

The explicit superfield coherent state reads as

$$\begin{aligned} \left|\Phi^{A\ldots}_{B\ldots}(Z)\right\rangle &= e^{\overline{\eta}^M Z^{\dagger\ A}_{M}\overline{\xi}_A}\left|\Phi^{A\ldots}_{B\ldots}\right\rangle_0 \\ &= \exp\left[b^{\dagger m}\left(X_m^{\ a}a_a^\dagger + \theta_m^{\ i}\xi_i^\dagger\right) + \eta^{\dagger l}\left(\chi_l^{\ a}a_a^\dagger + \lambda_l^{\ i}\xi_i^\dagger\right)\right]\left|\Phi^{A\ldots}_{B\ldots}\right\rangle_0. \end{aligned}$$

The Grassmann character of the matrix coefficients, restricts the number of terms in the superfield coherent state

$$\begin{aligned} \Phi^{A\ldots}_{B\ldots}(Z) = &\sum_{n_1=0}^{2p}\frac{1}{n_1!}\left(b^{\dagger m}\theta_m^{\ i}\xi_i^\dagger\right)^{n_1}\sum_{n_2=0}^{2q}\frac{1}{n_2!}\left(\eta^{\dagger l}\chi_l^{\ a}a_a^\dagger\right)^{n_2} \\ &\times\sum_{n_3=0}^{\min(p,q)}\frac{1}{n_3!}\left(\eta^{\dagger l}\lambda_l^{\ i}\xi_i^\dagger\right)^{n_3} f_{B\ldots}^{\ A\ldots}(x), \end{aligned}$$

where

$$f_{B\ldots}^{A\ldots}(x) = e^{b^{\dagger m} X_m^{\ a} a_a^{\dagger}} \left|\Phi_{B\ldots}^{A\ldots}\right\rangle_0 .$$

i) The basis in the simplest cases is the superalgebra $U(1,1 \mid 1,\mathbb{H})$ that contains as subgroups $U(1,1;\mathbb{H}) \sim SO(1,4)$ and $U(1,\mathbb{H}) \sim SO(2)$. It is determined by infinitesimal transformations preserving the scalar product $\overline{q}_a q_a - \overline{q}_b q_b + \overline{\eta} e_2 \eta$ where η is a standard Grassmann quaternion and e_2 is the B_1 part of the supermetric.

ii) For example, a more complicated case is in the field of $\mathbb{H}$ with $p = 2$ and $q = 2$, (that is the quaternionic analog of the $SU(2,2 \mid 8)$ with $N = 8 = p + q \rightarrow p = 4, q = 4$). In this case we have the following scalar quaternionic superfield:

$$\Phi(Z) = f(x) + \overline{\theta}_m^{\ i} \overline{\chi}_k^{\ a} f(x)_{(ia)}^{\{mk\}} + \overline{\theta}_m^{\ i} \overline{\theta}_n^{\ j} \overline{\chi}_k^{\ a} \overline{\chi}_l^{\ b} f(x)_{(ijab)}^{\{mnkl\}}$$

which is a supermultiplet with helicities ranging up to $|s| = 2$ and the multiplicities of the quaternionic $N = 4$ (complex $N = 8$) Maxwell supermultiplet. There is one more condition to be fulfilled by the quaternionic wave function:

$$\frac{1}{2}\widetilde{T}T \left|\Phi_{B\ldots}^{A\ldots}(Z)\right\rangle = 0$$

because, in this particular case, $\widetilde{T}T$ is a $U(1,1 \mid p+q,\mathbb{H})$ invariant quantity, this condition transforms into:

$$\left[\widehat{s} - \frac{1}{2}(F_\xi + F_\eta) + \frac{1}{4}(p+q)\right] \left|\Phi_{B\ldots}^{A\ldots}(Z)\right\rangle = 0,$$

where $\widehat{s}$ is the $U(1,1,\mathbb{H})$- invariant helicity operator and F_ξ, F_η are the quaternion-valuated fermion number operators.

In paper Ref. 20 we shall discuss the interesting cosets from which we are able to perform the nonlinear realization in order to obtain the super-analog of the Borisov-Ogievetsky one.[11] This super-analog yields the unification of the general relativity and the standard model, where all fields are Goldstone modes. All dimension parameters of the theory (including masses and field condensates) arise via quantum anomalies in the beginning of the Universe as consequences of the vacuum postulate.[15]

Acknowledgments

The authors would like to thank Profs. V.N. Melnikov and A.A. Zheltukhin for useful discussions. D.J.C-L. is grateful to the JINR Directorate for hospitality and to CONICET for financial support.

References

1. A. Einstein, Sitzungsber. Preuss. Akad. Wiss. Berlin, 844 (1915).
2. V. Fock, Zs. f. Fiz. **57**, 261 (1929).
3. P. A. M. Dirac, Proc. Roy. Soc. London. **A 246**, 326 (1958).
4. R. Arnowitt, S. Deser, C. W. Misner, *The dynamics of general relativity.* In Gravitation: an Introduction to Current Research. Witten, L. (ed.). Wiley, New York (1963).
5. B. S. De Witt, Phys. Rev. **160**, 1113 (1967).
6. J. A. Wheeler, in: *Battelle Rencontres: 1967 Lectures in Mathematics and Physics*, edited by C. M. DeWitt, and J. A. Wheeler (Benjamin, New York, 1968), p. 242.
7. C. W. Misner, Phys. Rev. **186**, 1319 (1969).
8. A. Einstein, Sitzungsber. Preuss. Akad. Wiss. Berlin, 142 (1917).
9. P. A. M. Dirac, Proc. Roy. Soc. Lond. A **333**, 403 (1973).
10. S. Deser, Ann. Phys. **59**, 248 (1970).
11. A. B. Borisov, V. I. Ogievetsky, Theor. Math. Phys. **21**, 1179 (1975).
12. A. B. Arbuzov, et al. Phys. Lett. **B 691**, 230 (2010).
13. V. N. Pervushin, et al. Gen. Relativ. Gravit. **44**, 2745 (2012).
14. E. P. Wigner, Ann. Math. **40**, 149 (1939).
15. A. B. Arbuzov, *et. al.* arXiv:1411.5124 [hep-ph];
    V. N. Pervushin, *et. al.* PoS Baldin ISHEPP XXII (2015) 136.
16. A. B. Arbuzov, B. M. Barbashov, A. Borowiec, V. N. Pervushin, S. A. Shuvalov, A. F. Zakharov, Grav. Cosmol. **15**, 199 (2009).
17. Yu. S. Vladimirov and A. N. Gubanov, G & C, **5**, 277 (1999).
18. L. B. Litov, V. N. Pervushin, Phys. Lett. **B 147**, 76 (1984).
19. D. J. Cirilo-Lombardo and V. N. Pervushin, arXiv:1509.03552 [hep-ph].
20. D. J. Cirilo-Lombardo and V.N. Pervushin, work in progress.

Description of dark energy and dark matter by vector fields

Boris E. Meierovich

P.L.Kapitza Institute for Physical Problems,
Moscow, 119334, Russia
E-mail: meierovich@mail.ru
http://kapitza.ras.ru/

A simple Lagrangian (with squared covariant divergence of a vector field as a kinetic term) turned out an adequate tool for macroscopic description of dark sector. The zero-mass field acts as the dark energy. Its energy-momentum tensor is a simple additive to the cosmological constant. Space-like and time-like massive vector fields describe two different forms of dark matter. The space-like field is attractive. It is responsible for the observed plateau in galaxy rotation curves. The time-like massive field displays repulsive elasticity. In balance with dark energy and ordinary matter it provides a four-parametric diversity of regular solutions of the Einstein equations describing different possible cosmological and oscillating non-singular scenarios of evolution of the Universe. In particular, the singular "big bang" turns into a regular inflation-like transition from contraction to expansion with accelerated expansion at late times. The fine-tuned Friedman-Robertson-Walker singular solution is a particular limiting case at the boundary of existence of regular oscillating solutions (in the absence of vector fields). The simplicity of the general covariant expression for the energy-momentum tensor allows analyzing the main properties of the dark sector analytically, avoiding unnecessary model assumptions.

Keywords: Dark sector; vector fields; regular cosmology; galaxy rotation curves.

1. Vector Fields Describing Dark Sector

Two most intriguing long standing problems in astrophysics (plateau in galaxy rotation curves and accelerated expansion of the Universe) strictly pointed to the existence of "hidden sector", containing "dark energy" and "dark matter", whose interaction with the ordinary matter (baryons and leptons) is observed only via gravitation. The accelerated expansion of the Universe indicates the existence of a hidden mechanism of repulsion, while a plateau of galaxy rotation curves is the result of some additional attraction. Nevertheless, my approach to the dark sector problems,[1] based on the analysis of vector fields in general relativity, provided an appropriate universal tool for theoretical description of both these phenomena.

The space-like massive vector field is attractive. It is responsible for the observed plateau in galaxy rotation curves. The time-like massive vector field displays repulsive elasticity. In the scale of the whole Universe it is the source of accelerated expansion. Inclusion of repulsive dark matter into consideration allows the existence of nonsingular solutions describing various possible regular scenarios of evolution of the Universe. My talk contained a discussion of physical nature of manifestations of dark sector. Analytical derivations are presented briefly by final results.

In regular solutions of the Einstein equations all invariants of the Riemann curvature tensor are finite. Hence, the invariants of the Ricci tensor are finite too. By virtue of Einstein equations the requirement of regularity automatically excludes a possibility to achieve an infinite value for all invariants of the energy-momentum tensor T_{IK}. Necessary restrictions, if any, on the existing parameters arise as a consequence of the condition of regularity.

In general relativity, the standard approach, starting from a general form of the Lagrangian of a vector field, is capable to describe not only the already known particles. Starting from a general form of the Lagrangian of a vector field, one derives field equations and energy-momentum tensor. Then, excluding the terms associated with the ordinary matter, one has a chance to separate a Lagrangian describing the dark sector. The separation is necessary, especially if the ordinary matter is considered as a continuous medium with the macroscopic energy-momentum tensor

$$T_{\mathrm{om}\ IK} = (\varepsilon + p)\, u_I u_K - p g_{IK}.$$

Otherwise, the ordinary matter would be taken into account twice: as a medium with the energy-momentum tensor $T_{\mathrm{om}\ IK}$, and as quantum particles described by a vector field.

It turns out that the simplest Lagrangian of a vector field

$$L_{\mathrm{dark}} = a \left(\phi^M_{;M}\right)^2 - V\left(\phi^L \phi_L\right), \qquad (1)$$

allows describing the main observed manifestations of dark sector completely within the frames of minimal general relativity. In case of ordinary particles the term with covariant divergence is set to zero ($a = 0$). For this reason L_{dark} with $a \neq 0$ gets separated from a Lagrangian of ordinary matter. The massless field corresponds to the dark energy, the massive space-like field ($\phi^L \phi_L < 0$) is responsible for a plateau in galaxy rotation curves, and the massive time-like vector field ($\phi^L \phi_L > 0$) displays a repulsive elasticity. Repulsive dark matter and attractive ordinary matter, acting together, provide a variety of possible regular scenarios of evolution of the Universe.

In the case of weak vector fields the second and higher derivatives of the potential $V(\phi^L \phi_L)$ can be neglected. Field equations and the energy-momentum tensor of a massive field are

$$a \frac{\partial \phi^M_{;M}}{\partial x^I} = -V_0' \phi_I, \tag{2}$$

$$T_{\text{dark } IK} = g_{IK} \left[a(\phi^M_{;M})^2 + V(0) \right] + V_0' \left(2\phi_I \phi_K - g_{IK} \phi^M \phi_M \right). \tag{3}$$

Here $V_0' = dV/d\left(\phi_L \phi^L\right)$ at $\phi_L \phi^L \to 0$. The energy-momentum tensor $T_{\text{dark } IK}$ of a zero-mass ($V_0' = 0$) vector field reduces to $T_{(0)IK} = g_{IK} \left[a(\phi_0')^2 + V(0) \right]$, where $\phi_0' \equiv \phi^M_{;M} = const$ in accordance with (2). It acts as a simple addition to the cosmological constant in the Einstein equations, changing Λ to $\widetilde{\Lambda} = \Lambda - \varkappa \left(a \phi_0'^2 + V_0 \right)$. $\varkappa$ is the gravitational constant.

2. Galaxy Rotation Curves

The velocity V of a star, orbiting around the center of a galaxy and satisfying the balance between the centrifugal and centripetal accelerations, should decrease with radius r of its orbit as $V(r) \sim 1/r^{1/2}$ at $r \to \infty$. However, numerous observed dependencies $V(r)$, named galaxy rotation curves, practically remain constant at far periphery of a galaxy.

Most galaxies are concentrated around super heavy objects, be it a black hole, or a neutron star. So, applying general relativity to the galaxy rotation problem, it is reasonable to consider a static centrally symmetric metric - the same way as for a Schwarzschild solution. Omitting details (one can see the derivation in[1]), I present here the analytical formula for the velocity $V(r)$ of a star, rotating around a black hole far outside the Schwarzschild radius r_{Sch}:

$$V\left(r\right) = \sqrt{V_{\text{pl}}^2 \left(1 - \frac{\sin 2mr}{2mr} \right) + c^2 \frac{r_{Sch}}{2r}}, \qquad r \gg r_{Sch}. \tag{4}$$

Here $V_{\text{pl}} = c\sqrt{\frac{\varkappa |a|}{2} \frac{\phi_0'}{m}}$ is the plateau velocity at $r \to \infty$, $m = \sqrt{|V_0'/a|}$ is the mass of a space-like vector field, $\phi_0' \equiv \phi^M_{;M}(0)$, and c is the speed of light. Without dark matter ($\phi_0' = 0$) (4) would give the Newton's $V(r) \sim r^{-1/2}$ at $r \to \infty$. In the presence of dark matter ($\phi_0' \neq 0$) the velocity of rotation $V(r)$ tends to V_{pl} at $r \to \infty$ with damping oscillations.

Though the rotation curves of galaxies differ from one another, the deviation from the Newton's $r^{-1/2}$ on the periphery of a galaxy is their common feature. In order to compare with observations, it looks natural to choose the galaxies having stars outside the main disc. Fitting the rotation curves of two such galaxies by the universal function $\sqrt{1-\sin 2x/2x}$ is shown in Figure 1.

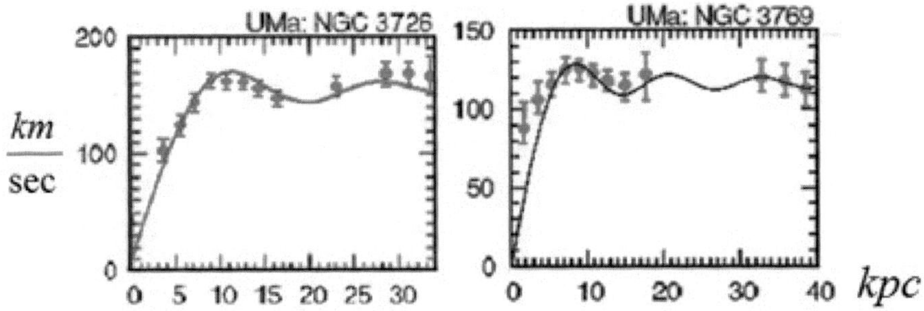

Fig. 1. Fitting the rotation curves of two galaxies in the Ursa Major cluster by the universal curve $\sqrt{1-\sin 2x/2x}$.

Damping oscillations of a rotation curve at far periphery of a galaxy is actually a "signature of dark matter". I strongly recommend this observational test. It confirms the existence of dark matter, along with its adequate description by a longitudinal non-gauge vector field.

From the physical point of view, a more strong attraction to the center at the periphery of a galaxy (than predicted by the Newton's theory) is a consequence of a finite velocity of propagation of interactions.

As a matter of fact, appearance of a plateau in a rotation curve can be interpreted as a manifestation of gravitational waves in the galactic scale. Meanwhile, huge efforts and funds are being spent in vain to detect gravitational waves on the Earth, just to prove their existence.

3. Evolution of the Universe

Discovery of accelerated expansion of the Universe shows that the source of acceleration continues to exist for a long time after the "big bang".

If one includes repulsive dark sector into consideration, then a singularity ceases to be the inevitable property of evolution of the Universe. Additional parametric freedom allows forgetting the fine-tuning problem.

The uniform and isotropic expansion is characterized by a single metric function $F(x^0)$, and the rate of expansion is $dF/dx^0 \equiv F'$. Longitudinal massive vector field ϕ_I in this case is time-like, its only nonzero component is ϕ_0. In contrast to a space-like field, a massive time-like field demonstrates elastic repulsion.

The energy-momentum tensor of a massless field (dark energy) acts in the Einstein equations as a part of the cosmological constant $\widetilde{\Lambda}$. The metric function $F(x^0) = \pm\sqrt{-\widetilde{\Lambda}/3}(x^0 - x_0^0)$ is a regular solution of the Einstein equations, provided that $\widetilde{\Lambda} < 0$. This solution belongs to de Sitter (1917). It describes either expansion of the Universe at a constant rate $F' = H$ (green horizontal line $F'/H = +1$ in Figure 2), or contraction (blue horizontal line $F'/H = -1$ in Figure 2). H is the Hubble constant. As long as the physical nature of vacuum is not known, the "geometrical"

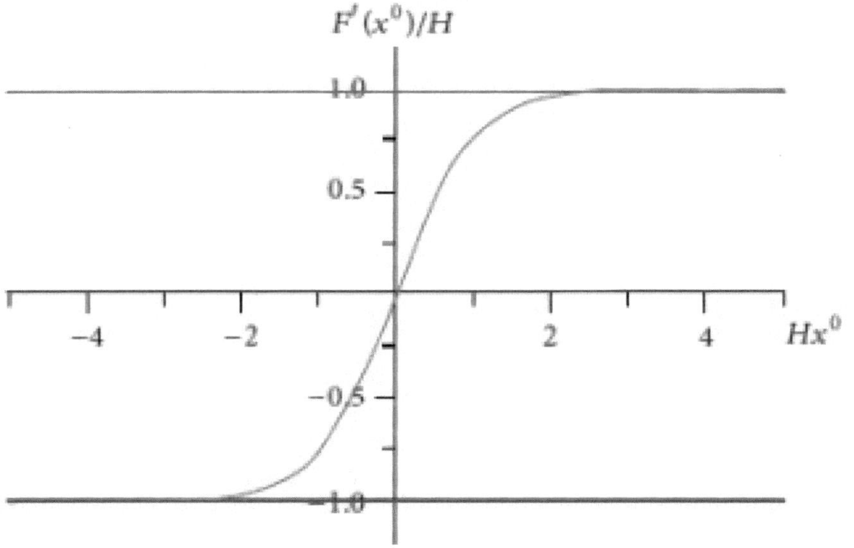

Fig. 2. Rate of evolution of the Universe. Upper green horizontal line is expansion. Lower blue horizontal line is compression. Red curve is a transition from compression to expansion.

origin of the cosmological constant $\widetilde{\Lambda}$ and the "material" contribution to $\widetilde{\Lambda}$ by a zero-mass vector field can not be separated from each other. The combined action of the massless field and/or the cosmological constant is described by the single parameter Hubble constant H. Without a massive field $F''=0$. The dark energy by itself can be responsible only for either contraction, or expansion at a constant rate. In particular, a zero-mass longitudinal vector field alone can not explain the observed switch from deceleration to acceleration at about a half of the age of the Universe.[2]

With account of a massive time-like field $F''=|a|\varkappa m^2\phi_0^2$ is positive, it is repulsion. The massive time-like vector field makes the rate of evolution $F'(x^0)$ a monotonically growing function from $-H$ in the past to $+H$ in future (red curve in Figure 2). If we set the origin $x^0=0$ at the moment when $F'=0$, then the Universe contracts at $x^0<0$, and expands at $x^0>0$. $x^0=0$ is the moment of maximum compression.

In the case of a small mass, $m \ll H$ (in dimensional units $mc^2 \ll \hbar H$) a symmetric compression-to-expansion transition is described by the analytical solution

$$F'\left(x^0\right)=H\tanh\left(3Hx^0\right), \qquad \phi_0\left(x^0\right)=\sqrt{\frac{|\widetilde{\Lambda}/a|}{\varkappa m^2}\frac{1}{\cosh\left(3Hx^0\right)}}, \qquad m \ll H.$$

Without ordinary matter there is only repulsion of a massive vector field (dark matter). With account of attraction of the ordinary matter the rate of evolution is

$$F'\left(x^0\right)=H\frac{\sinh\left(3Hx^0\right)}{\cosh\left(3Hx^0\right)-\left(1+\frac{2}{\Omega}e^{3F_0}\right)^{-1}}, \qquad \widetilde{\Lambda}<0.$$

As usual, parameter $\Omega=-\frac{\varkappa\varepsilon_0}{\widetilde{\Lambda}}=\frac{\varkappa\varepsilon_0}{3H^2}$ denotes the ratio of today's energy density of the ordinary matter to the density of kinetic energy of expansion at a constant rate H. Regular solutions are free from any fine tuning. Moreover, the existing parametric freedom leads to a wide variety of possible regular scenarios of evolution.

With ordinary matter taken into account, there are two kinds of regular solutions: cosmological, and oscillating. Cosmological solutions (when $\widetilde{\Lambda}<0$) describe a transition from contraction to expansion. The parameter $F_0=F(0)<0$ determines the degree of maximum compression at the turning point $F'(0)=0$. The peak value of the rate of expansion grows exponentially with increasing negative value of F_0, while the width of the transition decreases exponentially. It resembles inflation, except that there is no singularity. The regular contraction-to-expansion transition is often referred to as "nonsingular bounce".

If the ordinary matter is taken into account, there appears a possibility for regular oscillating solutions (with $\widetilde{\Lambda}>0$). The necessary condition of regular solutions with $\widetilde{\Lambda} > 0$ is the existence of an extremum moment $(F'(0) = 0)$ when the energy density of the ordinary matter exceeds the kinetic energy of expansion: $\Omega e^{-3F_0} = \varkappa\varepsilon(0)/\widetilde{\Lambda} > 1$, $\widetilde{\Lambda} > 0$. In this case the rate of evolution

$$F'\left(x^0\right) = H\frac{\sin\left(3Hx^0\right)}{\left(1 - \frac{2}{\Omega}e^{3F_0}\right)^{-1} - \cos\left(3Hx^0\right)}$$

and the scale factor

$$R\left(x^0\right) = e^{F_0}\left[\left(1 - \frac{1}{2}\Omega e^{-3F_0}\right)\cos\left(3Hx^0\right) + \frac{1}{2}\Omega e^{-3F_0}\right]^{1/3}$$

are periodic functions with no singularity. In the case $\widetilde{\Lambda} > 0$ the origin $x^0 = 0$ is a moment when the scale factor $R(x^0) = e^{F(x^0)}$ reaches a maximum. The points of minimum (where $cos(3Hx^0) = -1$) are

$$x_n^0 = (\pi/3H)(1 + 2n), \qquad n = 0, \pm 1, \pm 2, \ldots.$$

Without a massive field ($\phi_0 = 0$) the solutions with positive $\widetilde{\Lambda}$ are fine-tuned ($\Omega\exp(-3F_0) = 1$) and have a periodic singularity at $x^0 = x_n^0$. In the vicinity of each singular point x_n^0, as well as at $H \to 0$, the Hubble constant H drops out, and the scale factor reduces to the one of the Friedman-Robertson-Walker cosmology:

$$R\left(x^0\right) = (3\varkappa\varepsilon_0/4)^{1/3}\left|x^0 - x_n^0\right|^{2/3}, \quad \left|x^0/x_n^0 - 1\right| \ll 1.$$

Dark matter, described by a longitudinal time-like vector field $\phi_0 \neq 0$, removes a singularity and rules out the problem of fine tuning.

Conclusion: As simple a Lagrangian as possible (1) turns out an appropriate tool for phenomenological description of dark sector by vector fields. The dark substance is described by covariant vector field equations (2) and the energy-momentum tensor (3). It no longer needs inventing every time a new model for dark matter to understand various astrophysical phenomena.

References

1. B. E. Meierovich. "Macroscopic Theory of Dark Sector". *Journal of Gravity*, **2014**, *Article ID 586958*, 23 pages (2014).
2. N. Suzuki, D. Rubin, and C. Lidman, "The Hubble Space Telescope Cluster Supernova Survey: V. Improving the Dark Energy Constraints above $z > 1$ and building an early-type-hosted supernova sample". *The Astrophysical Journal*, **746**, *no. 1*, (2011).

Precession of stars in the Galactic center and dark matter annihilation limits

V. I. Dokuchaev[1,2] and Yu. N. Eroshenko[1,*]

[1] *Institute for Nuclear Research of the Russian Academy of Sciences, 60th October Anniversary Prospect 7a, Moscow, 117312, Russia*

[2] *National Research Nuclear University MEPhI, Kashirskoe sh. 31, Moscow, 115409, Russia*

[*] *E-mail: eroshenko@inr.ac.ru*

The group of the fast moving S0 stars is gravitationally bound with the supermassive black hole SgrA* in the Galactic center. Probably the stars' orbits deviate from the precise ellipses due to the gravitational perturbations from dim stars, compact stellar remnants or dark matter. The observational data provide only upper limits on these additional masses. Our calculation of the the Newtonian precession provides some new constraints. If the invisible mass in the Galactic center is composed by dark matter particles, it can be the source of the gamma-radiation due to annihilation. The annihilation limits on the invisible mass can be compared with the dynamical limits from the precession angle. In the nearest future the observations can provide the weighing of the invisible matter in the Galactic center.

Keywords: Galactic center; precession; black hole; dark matter; annihilation.

1. Introduction

Several high velocities stars are observed at the center of Galaxy around supermassive black hole SgrA*. According to the parameters of their orbits[1] and from the effect of quasiperiodic oscillations[2] the mass of SgrA* was found to be $M_{\rm BH} \simeq 4 \times 10^6 M_\odot$.

Besides SgrA*, there are additional masses such as gas, stars and their remnants, and dark matter at the center of Galaxy. According to the fitting of the stars' orbits,[1] the mass within the orbit of S0-2 star is no more than $3 - 4$ % of $M_{\rm BH}$. This additional mass result in the deviation of the gravitational potential from the potential of a point mass $U = -GM_{\rm BH}/r$. As a result, the orbits of stars would be unclosed and precess. Probably, this effect will be measured in the next one or two years. The nonrelativistic precession can significantly exceed the relativistic one. Therefore, the existence of fast S0 stars provides a unique possibility of measuring the mass distribution at the center of the Galaxy.

We consider the method for studying the distribution of the matter at the center of the Galaxy by measuring the precession angle of orbits of S0 stars. In particular cases, numerical calculations of S0 stars precession were performed.[3–5] We obtained general analytical formulas for the precession angle with a power-law density profile of the additional matter.

The independent method for determining the distribution of the dark matter is the search for annihilation signals. This possibility of the common constraints on annihilation and on the dynamics of stars was proposed in Ref. 4. We calculated[6] the mass of the dark matter necessary for the explanation of the excess of gamma radiation detected recently by the Fermi-LAT space gamma telescope from the center of the Galaxy.[7,8]

2. Precession

The precession angle of a stellar orbit in one turn due to a small correction δU to the Coulomb potential[9] (Sect. 15, Problem 3) is

$$\delta\phi = \frac{\partial}{\partial L}\left(\frac{2m}{L}\int\limits_0^\pi r^2(\phi)\delta U d\phi\right). \tag{1}$$

Here, the integration is performed over the unperturbed elliptic orbit. Note that, in the case of relativistic precession, the orbit would rotate in the direction of the rotation of the star, but Newtonian precession (1) occurs in the opposite direction, i.e., $\delta\phi < 0$. We consider the power-law density profile of additional matter responsible for δU:

$$\rho(r) = \rho_h\left(\frac{r}{r_h}\right)^{-\beta}, \tag{2}$$

where ρ_h, r_h, and β are parameters.

The calculation of the precession angle of the orbit by Eqs. (1) and (2) gives the expression

$$\delta\phi = -\frac{4\pi^2\rho_h r_h^\beta p^{3-\beta}}{(1-e)^{4-\beta}M_{\rm BH}}\,{}_2F_1\left(4-\beta, \frac{3}{2}; 3; -\frac{2e}{1-e}\right), \tag{3}$$

where e is the orbit's eccentricity, ${}_2F_1(a,b;c;z)$ is the hypergeometric function. The method of osculating elements[10] gives the same result. The function $\delta\phi$ given by Eq. (3) is shown at Fig. 1, where $\xi = \Delta M/M_{\rm BH}$ is the mass fraction of additional matter within the orbit of the S0 star.

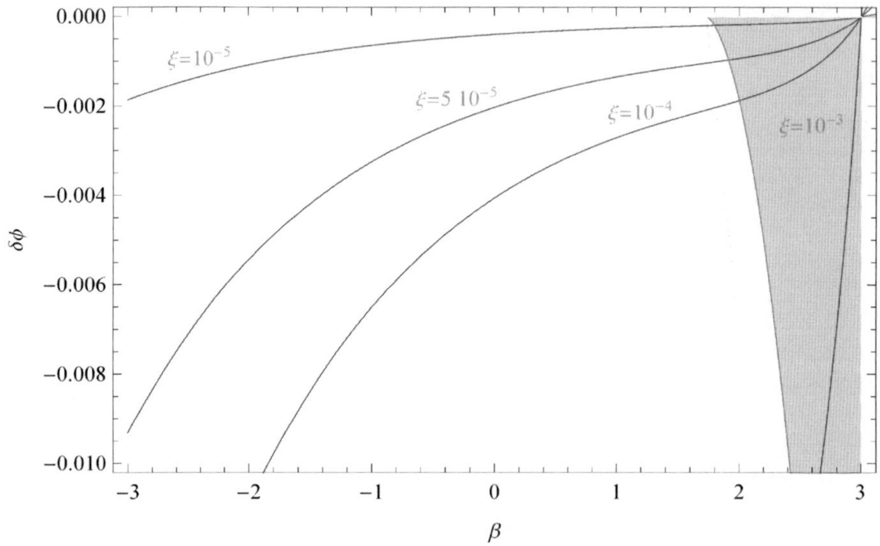

Fig. 1. Angle of precession $\delta\phi$ from (1) versus the exponent β in (2) for different mass fractions ξ. The indicated region can be excluded by the annihilation.

The angle (3) is in qualitative agreement with the numerical calculations of precession in Refs. 3–5, but coinsides with Iorio[11] only at $e \ll 1$.

3. Annihilation

The observed by the Fermi-LAT telescope excess of gamma-radiation from the center of the Galaxy was fitted[7,8] with the generalized Navarro–Frenk–White profile $\rho_{\rm H}(r) = \rho_0(r/d)^{-\gamma}(1 + r/d)^{\gamma-3}$, where $d = 20$ kpc, and the best fit is obtained with $\gamma \approx 1.26$.

The central region of the dark matter density peak is called the cusp. But the density profile should be significantly modified by the gravity of the black hole. The formation of an additional density peak (spike) with $\propto r^{-\beta}$ was considered in Ref. 12. For the adiabatically formed spike $\beta = 2.36$, and the spike would be a very bright gamma-source.[13] The signal from the spike is a factor of ~ 35 stronger than the Fermi-LAT excess.[8] So, the spike should be formed nonadiabatically or destroyed.

Following Ref. 8, we write the density of the dark matter in the spike in the form of Eq. (2), and the radius of the maximum density is limited by the annihilation effect. It was shown in Ref. 8 that contradiction with the bright point source is absent if $\beta = 1.8$. The $\langle\sigma v\rangle = const$ was assumed above. However, in a number of models of the dark matter $\langle\sigma v\rangle$ can depend on v.[14] Owing to high Kepler velocities $v(r) \approx (GM_{\rm BH}/r)^{1/2}$ near the black hole, the annihilation signal from the center can be reduced. We consider the cross section in the form[15,16]

$$\langle\sigma v\rangle = \langle\sigma v\rangle_0 \left(\frac{v_0}{v}\right)^\eta, \tag{4}$$

where $\langle\sigma v\rangle_0 = const$ and $v_0 = const$. The $\langle\sigma v\rangle = const$ case and the Sommerfeld effect correspond to the particular cases $\eta = 0$ and $\eta = 1$.

We assume that the integral annihilation signals from the cusp and spike in model (4) are the same as in Ref. 8. The results of the calculations with cross-section (4) are the following. The adiabatic density profile $\beta = 2.36$ is reached at $\eta = 3.13$. The mass of the dark matter $M_{\rm DM}$ within the orbit of the S0-2 star in particular cases $\eta = 0$, 1, 3.13, and 3.5 is $45M_\odot$, $144M_\odot$, $1.8\times10^3M_\odot$, and $2.8\times10^3M_\odot$, respectively. These values correspond to the values $\xi = 1.2\times10^{-5}$, 3.6×10^{-5}, 4.4×10^{-4}, and 6.9×10^{-4} and $\beta = 1.8$, 1.9, 2.36, and 2.4, respectively. The real prospect of the measurement of the additional mass of the dark matter from the precession of S0 stars appears at $\eta > 3$. Note however that for $\eta \geq 2$ the "freezes-out" abundance of dark matter particles in the early universe is suppressed.[15]

4. Conclusion

The currently existing observation accuracy is still insufficient for the measurement of the precession angle of fast S0 stars and the distributed invisible mass. However, there is a high probability of reaching this in the near future. The annihilation signals from the center of the Galaxy can provide additional possibilities for the calculation of the distributed mass or for the determination of bounds on it.[6]

This work was supported by the Russian Academy of Sciences program OFN-17, and by the Leading Scientific Schools grant NSh-3110.2014.2.

References

1. S. Gillessen, F. Eisenhauer, S. Trippe, et al., Monitoring Stellar Orbits Around the Massive Black Hole in the Galactic Center, *Astrophys. J.* **692**, 1075 (2009).

2. V. I. Dokuchaev, Spin and mass of the nearest supermassive black hole, *Gen. Relativ. Gravit.* **46**, 1832 (2014).
3. G. F. Rubilar and A. Eckart, Periastron shifts of stellar orbits near the Galactic Center, *Astron. Astrophys.* **374**, 95 (2001).
4. A. F. Zakharov, A. A. Nucita, F. De Paolis et al., Apoastron shift constraints on dark matter distribution at the Galactic Center, *Phys. Rev. D* **76**, 062001 (2007).
5. F. de Paolis, G. Ingrosso, A. A. Nucita et al., Estimating the parameters of the Sgr A* black hole, *Gen. Rel. Gravit.* **43**, 977 (2011).
6. V. I. Dokuchaev, Yu. N. Eroshenko, Weighing of the Dark Matter at the Center of the Galaxy, *Pisma v ZhETF* **101**, 875 (2015).
7. T. Daylan, D. P. Finkbeiner, D. Hooper, et al., The Characterization of the Gamma-Ray Signal from the Central Milky Way: A Compelling Case for Annihilating Dark Matter, arXiv:1402.6703 [astro-ph.HE].
8. B. D. Fields, S. L. Shapiro, J. Shelton, Galactic Center Gamma-Ray Excess from Dark Matter Annihilation: Is There A Black Hole Spike?, *Phys. Rev. Lett.* **113**, 151302 (2014).
9. L. D. Landau and E. M. Lifshitz, *Mechanics* (Pergamon Press, New York, 1988).
10. P. E. El'yasberg, *Introduction to the Theory of Flight of Artificial Earth Satellites* (Israeli Program for the Translation of Scientific Publications, Jerusalem, 1967), Chapter XI, Eq. (11.39).
11. L. Iorio, Exact Expressions for the Pericenter Precession Caused by Some Dark Matter Distributions and Constraints on Them from Orbital Motions in the Solar System, in the Double Pulsar and in the Galactic Center, *Galaxies* **1**, 6 (2013).
12. P. Gondolo, J. Silk, Dark Matter Annihilation at the Galactic Center, *Phys. Rev. Lett.* **83**, 1719 (1999).
13. V. S. Berezinsky, A. V. Gurevich, K. P. Zybin, Distribution of dark matter in the galaxy and the lower limits for the masses of supersymmetric particles, *Phys. Lett.* **B 294**, 221 (1992).
14. M. Lattanzi, J. Silk, Can the WIMP annihilation boost factor be boosted by the Sommerfeld enhancement?, *Phys. Rev. D* **79**, 083523 (2009).
15. R. J. Scherrer, M. S. Turner, On the relic, cosmic abundance of stable, weakly interacting massive particles, *Phys. Rev. D* **33**, 1585 (1986).
16. K. M. Belotsky, A. A. Kirillov, M. Yu. Khlopov, Gamma-ray evidence for dark matter clumps, *Gravitation and Cosmology* **20**, 47 (2014).

Detection of slow gravity field variations with GW-interferometers

K. V. Rudenko

Physical Department, Moscow State University,
Moscow, 119991, Russia
E-mail: kroky@ilc.edu.ru
www.phys.msu.ru

P. S. Solenov

Bauman Moscow State Technical University,
Moscow, 105005, Russia
E-mail: solenovps@gmail.com
www.bmstu.ru/en/

V. N. Rudenko

Sternberg Astronomical Institute of Moscow State University,
Moscow, 119991, Russia
E-mail: valentin.rudenko@gmail.com
www.sai.msu.su

Detection of very slow gravitational perturbations with large scale gravitational wave interferometers at the free spectral range (FSR) frequency was considered in [1]. The mechanism of penetration of such perturbations at the main interferometer output has a parametrical nature. For a reliable readout of such signals the two component resonance optical pump was introduced in [2]. The alternative method of using an additional input modulator at the FSR frequency was proposed in the paper [3] with a numerical simulation of signal detection using the FINESSE code [4]. In this paper we present our results of similar computer simulation with discussion of possible application for the measuring some weak gravitational effects.

Keywords: Weak gravitational effects; GW interferometers; FINESSE.

Attention to an ability of the large scale gravitational wave interferometer to register gravitational signals at very low frequencies was attracted by the paper [1]. A mechanism of penetration of quasi static gravitational perturbations at the main interferometer dark fringe output was analysed in [2].

It was shown that it has a parametrical nature associated with slow variations of the optical transfer function of the interferometer. It is resulted in the correspondent modulation of the harmonic component at the photon circulation frequency in the interferometer arms. For a reliable readout of this signal the two components resonance optical pump was introduced in the paper [2] with a correspondent numerical calculation of a value of the effect. A technically more convenient method through the additional modulation of interferometer input beam at the FSR frequency was considered in [3] with the numerical simulation of the interferometer operation using the program FINESSE code [4].

This code belongs to the fast interferometer simulation program. For a given optical setup, it computes the light field amplitudes at every point of the interferometer optical configuration in the frame of a steady state approach. Description of the interferometer is given as a set of linear equations that may be solved numerically. FINESSE allows carry out the analysis using the plane-wave approximation with the Hermits-Gaussian modes. It provides a computing of characteristics of multi mirrors optical systems (like telescopes etc.) taking into account the required mode matching and correction of mirrors angular positions.

In this note we present our result of independent simulation of GW-interferometer type of LIGO and VIRGO [5] with numerical reconstruction of its transfer function for the FSR channel and short final discussion of a possible application for measuring some weak global gravitational effects. In the process of simulation we follow the conventional interferometer structure typical for the LIGO and VIRGO setups (before the modernization to their advance version which is under commissioning at present) having only one specific node: - the input optical modulator at FSR frequency (which is close to 37.473 KHz for LIGO), see also [3]. Such element provides an illumination of interferometer arm modes neighbour the central one at the laser pump frequency. It provides the appearance of the FSR harmonic in the main output of the setup after synchronous photo detection (demodulation) subsequently at the radio side band and FSR frequencies [2,3]. Results of our simulation of the interferometer work is presented at the figures 2-5 below.

The optical scheme of the model which was taken under simulation is presented at the Fig 1. In the main features it corresponds to the LIGO and VIRGO construction demonstrating the complex multi mirror Michelson type interferometer complicated by FP resonators in its arm and a recycling mirror at the laser power input (see [5]). As it was mentioned the

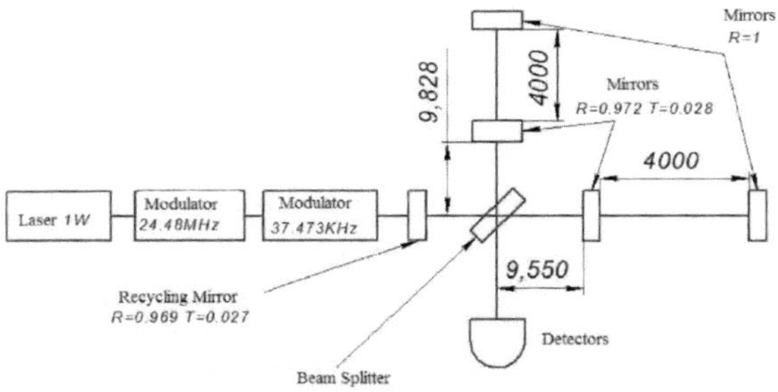

Fig. 1. Optical scheme of the interferometer.

only difference consists in the additional input electro-optical modulator at the FSR frequency. For the signal simulation on can produce vibration of the end mirror at the given frequency and amplitude. In such a manner it is possible to reconstruct the transfer function of the interferometer presented at the Fig.2 Constancy of the transfer function at low frequencies

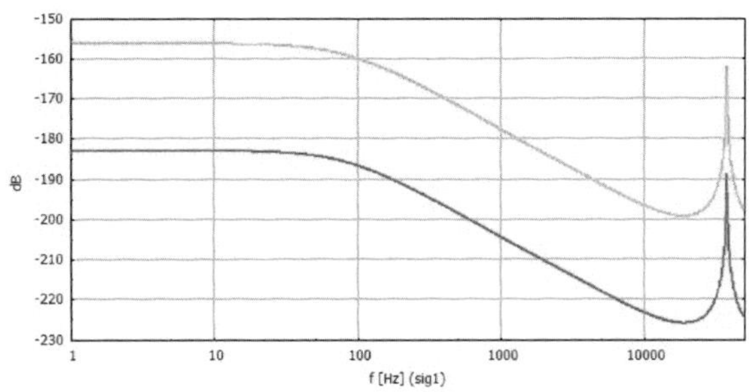

Fig. 2. Transfer function (green - at the carrier frequency; red - on the FSR sideband frequency).

corresponds to the reception bandwidth of the LIGO GW detector. The falling down in the region above the 100 Hz reflects the inertia of the instrument for high frequency signals (the signal oscillation period becomes close to the photon round trip time in the arms). The reconstruction of the transfer function at the FSR frequency (peak at 37 KHz) takes place. It occurs due to the resonating neighbour mode light into at the FP arms. It was shown analytically [2] that simultaneous illumination of both (symmetrical) modes around the central one cancels the transfer coefficient in the first order. So in the simulation only the one neighbour mode (upper) was implemented. During the interferometer operation arm's mirrors are kept their position against of low frequency gravitational disturbances due to feed back control circuits but with a limited accuracy. For displacements below 10^{-12} meter the feedback control signal became too small to keep mirrors and they start to move producing the modulation of the output power in the signal frequency band of GW interferometer (20 Hz - 2 KHz) as well as at FSR frequency. At the Fig. 3 the value of the output signal is presented versus the mirrors displacement amplitude. (Note, that there is a strong dark spot on the carrier and a much weaker one on the sideband). The graph shows that the interferometer at the carrier frequency and the

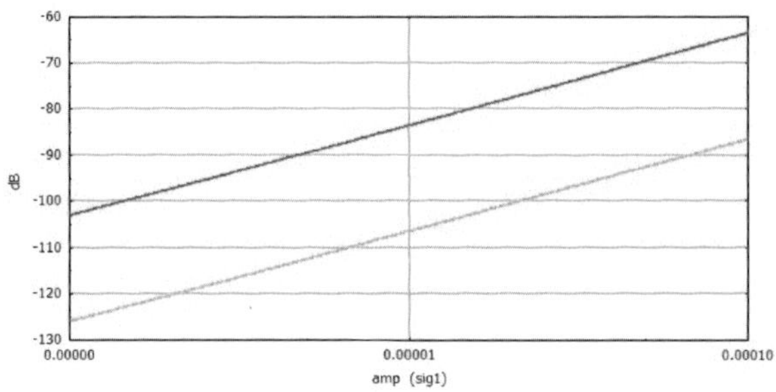

Fig. 3. Response of carrier frequency (red) and sideband (green) to the mirrors sinusoidal displacement

neighbour sideband frequency responds equally proportional to the external (say gravitational wave) perturbations. It provides a potential opportunity of detecting low-frequency gravitational signals at FSR harmonics.

To consider the systems response in vicinity of FSR frequency (37.473 kHz), we varied the modulation frequency of the second modulator at the Fig. 1. It resulted in the unusual peaks on the graph reflected in a the complex structure of the equivalent FP cavity between recycling mirror and the residual parts of the interferometer.

As seen from Fig.4, the positive peak occurs when the modulation frequency is just equal to 37,473 KHz, which confirms our assumption about maximum intensity at FSR frequency.

Finesse software allows to calculate the signal to noise ratio (S/N) at the shot noise background via the Schottky formula. We introduced a variable displacement (signal) at the 1 Hz frequency applied to the "end" mirrors

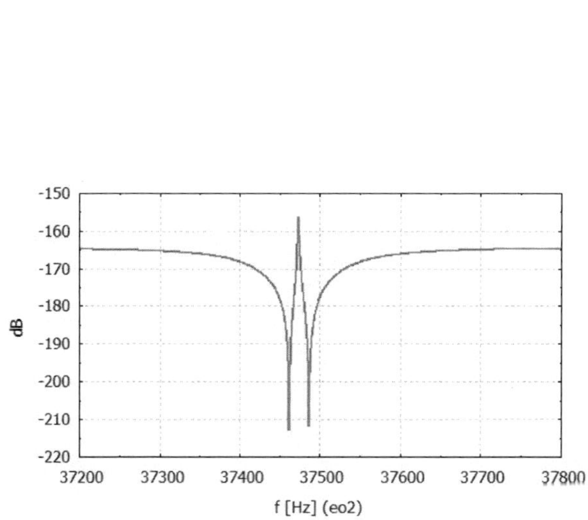

Fig. 4. Signal intensity versus the modulation frequency variations.

and calculated the S/N at the output, see graph at Fig. 5. For sideband

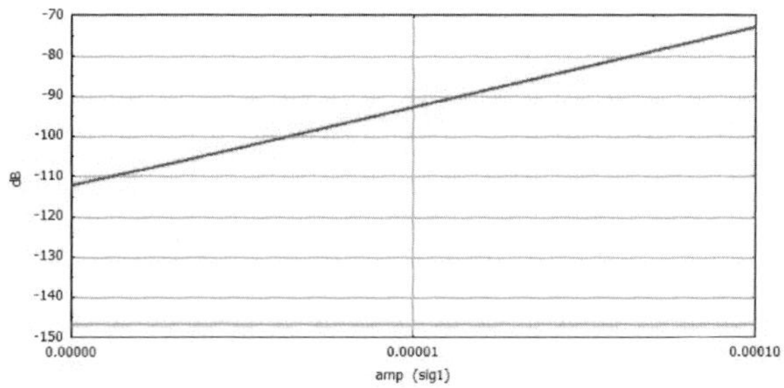

Fig. 5. S/N for sideband (green) and for the sideband demodulation by 1 Hz (red).

the condition of the "dark spot (fringe)" was fulfilled, so the output signal was absent: there was the constant level of output (green curve)). But for the carrier (S/N) increases proportional to the excitation amplitude (red curve). The numerical experiment gave the value of the short noise relative amplitude spectral density 10^{-22} Hz$^{-1/2}$. This value can be also estimated according to the simple formula for minimal registered arm's deformation: $h_{min} \geq (1/Q\mu)^{-1}(h\nu/\eta P)^{1/2}(\triangle f)^{1/2} \approx 5 \cdot 10^{-22}(\triangle f)^{1/2}$ which is in agreement with results of our simulation (the following parameters where used: $P_0 = 1\ Wt$, modulation index $\mu = 0.4$, $Q = 2.5 \cdot 10^{12}$, quantum efficiency $\nu = 0.7$, $h\nu \approx 2 \cdot 10^{12}\ erg$).

Summary: Results of our numerical simulation confirmed analytical conclusion concerning the possibility to pick up low frequency gravitational perturbations at the main interferometer output by filtering the FSR frequency. We believe that this method will have some advantage in respect of the slow signal filtering at the flicker noise background. Between such signal could be slow global geodynamics gravitational perturbation like very weak tidal harmonics, earth inner core oscillation [2] etc. or low frequency gravitational waves influence [3]. But for this a possible suppression of seismic perturbation has to be addressed specially by numerical modelling of the seismic background.

Acknowledgment

Authors express their gratitude to prof. A. Melissinos (University of Rochester, USA) for his permanent attention to this work and many fruitful discussions.

References

1. A. Melissinos,*Proc. MG-12* (World Scientific, Singapore, 2012, Part C, p. 1718).
2. A. V. Gusev, V. N. Rudenko, I. S. Yudin. *JETP* **119**, 687 (2014).
3. A. Melissinos (2 Oct.2014), `arXiv:1410.0854v1[astro-ph.IM]`.
4. A. Freise and K. Strain, *Interferometer Techniques for Gravitational-Wave Detection* (Living Review, 2010). `http://relativity.livingreviews.org/Articles/lrr-2010-1/`.
5. R. X. Adhikari, *Rev. Mod. Phys.* **86**, 121 (2014) (DOI: 10.1103/RevModPhys.86.121.

KAGRA
-Large-scale Cryogenic Gravitational wave Telescope-

Takayuki Tomaru* on the behalf of KAGRA

High Energy Accelerator Research Organization,
1-1 Oho, Tsukuba Ibaraki, 305-0801, Japan
**E-mail: tomaru@post.kek.jp*

KAGRA is a second generation gravitational wave (GW) telescope under constructing in Japan. Features in KAGRA are underground site and cryogenic mirror systems. It is well studied that underground site of Mt. Ikenoyama in Kamioka has over two orders of magnitude smaller seismic vibration than that of typical urban area, and it is very effective to reduce seismic noise. The cryogenic mirror system is an unique technology developed in Japan to reduce mirror and suspension thermal noises in GW interferometer. These two unique features allow KAGRA to aim to detect GWs from neutron-star binary coalescence from 280 Mpc away in best direction. The details of these features and their status, including the status of the constructions site of KAGRA are reported in this paper.

Keywords: gravitational wave; KAGRA, cryogenic; seismic noise; thermal noise.

1. Introduction

It is expected that new avenues in the field of GW astronomy will be opened by second generation interferometric gravitational wave (GW) telescopes. In the case of ground base interferometric GW telescopes, main observation band is around between several dozens Hz and several kHz. The GW sources with such high frequencies are limited. Compact and heavy astronomical objects such as binary neutron-star coalescence, black-hole binary coalescence, supernova, pulsar are promising targets of observation. Especially, it is expected that the GWs radiated from black-hole just after merger of binary neutron-stars has peculiar waveform and one can investigate strong gravity field. However, such astronomical events occur in very rare rate, and sensitivity to be able to search 200 - 300 Mpc range is required to detect several GW events every year. Now five second generation GW telescopes in the world are under constructing to realize the sensitivity. Advanced LIGO (aLIGO, two telescopes)[1] in USA and advanced VIRGO (aVIRGO)[2] in Italy are 3 - 4 km scale large interferometers upgraded from 1st generation detectors. GEO-HF[3] in Germany is also upgraded interferometer with 600 m length, but it chose different strategy from others; high frequency GW search. KAGRA is a new GW telescope with 3 km arm length under constructing in Japan. KAGRA has two unique features; underground site and cryogenic mirror system.

2. KAGRA project

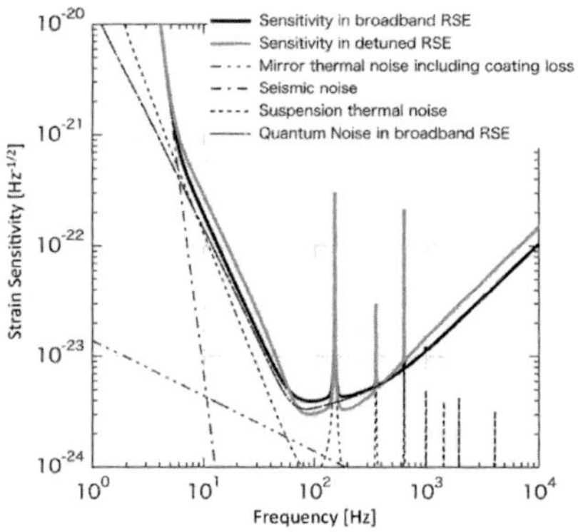

Fig. 1. Design sensitivity of KAGRA with major noises.

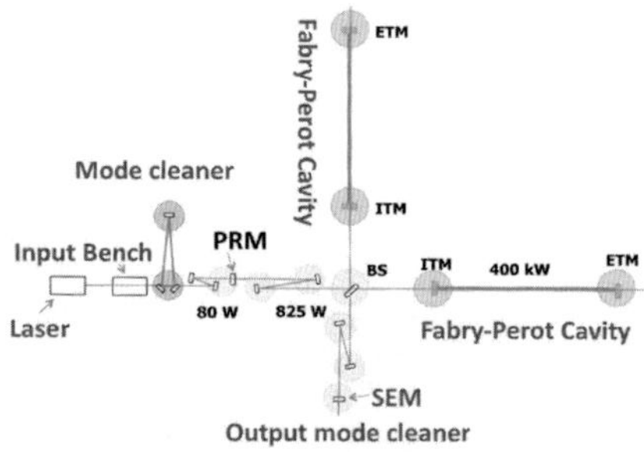

Fig. 2. Layout of KAGRA. This configuration is called as Dual Recycled Fabry-Perot Michelson Interferometer.

KAGRA can detect GW signals from neutron-star binary coalescence and the range of the detector is 280 Mpc in best direction and is 173 Mpc in whole sky average. Figure 1 shows design sensitivity of KAGRA. The best sensitivity is orders of $10^{-24}/\sqrt{\mathrm{Hz}}$ at 100 Hz. KAGRA has two operation modes; broadband resonant sideband extraction (RSE) and detuned RSE[5]. A RSE mirror exists at the dark port in interferometer shown in Fig. 2 and acts as GW signal amplifier. By tuning the position of RSE mirror, two modes are switched. In both the modes, quantum noise limited sensitivity, which consists of photon shot noise and radiation pressure noise[4], are realized. The Fabry-Perot cavities in arms and the power recycling mirror (PRM) helps in increasing the laser power inside the arm cavities, and they also help in reducing photon shot noise. The arm cavities store 400 kW of laser power.

The sensitivity of the KAGRA detector will be limited by quantum noise, however, there are important noise sources such as seismic noise and thermal noise from the mirrors and suspensions, which must be reduced. In KAGRA, these two noise sources are reduced by, (a) having an underground site and (b) employing cryogenic mirror suspension system.

3. Underground Site

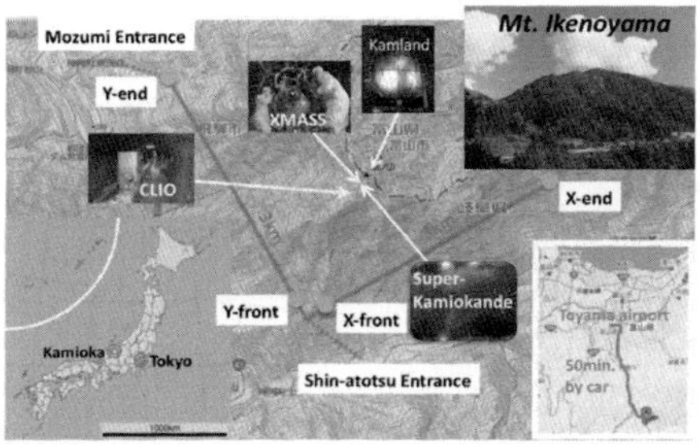

Fig. 3. Underground site of KAGRA.

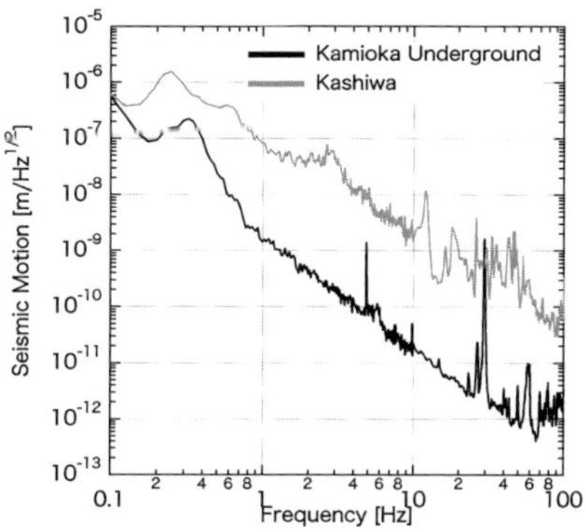

Fig. 4. Spectrum of seismic vibration at University of Tokyo in Kashiwa and Kamioka underground. Data for Kamioka was measured at about 1000 m underground of Mt. Ikenoyama from ground surface.

The seismic noise is a noise induced by mirror vibration due to steady ground motion. A lot of seismic vibration attenuators for mirror are developed, however, still the seismic noise dominated the sensitivity below several dozens Hz. We decided to construct KAGRA in an underground site with very small seismic motion. Figure 3 shows underground site of KAGRA in Kamioka in Japan. Figure 4 is a comparison of seismic vibration in Mt. Ikenoyama in Kamioka with typical urban area in Kashiwa. Apparently, seismic motion in Kamioka underground is two or three orders of magnitude smaller than that in Kashiwa, and is also about one order of magnitude smaller than that in aLIGO site. Hence the underground site at Kamioka is ideal for the construction of a GW interferometer

Figure 5 shows vibration attenuation strategy for main mirror in the KAGRA tunnel. A 14 m tall pendulum is installed to have low frequency seismic attenuation. Usually, this is not easy since one also needs tall tower to suspend tall pendulum and structural resonance of the tower contami-

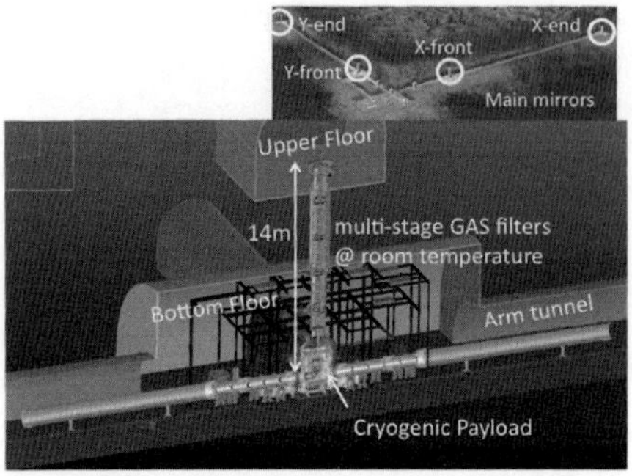

Fig. 5. Vibration Attenuation for main mirror in the KAGRA tunnel.

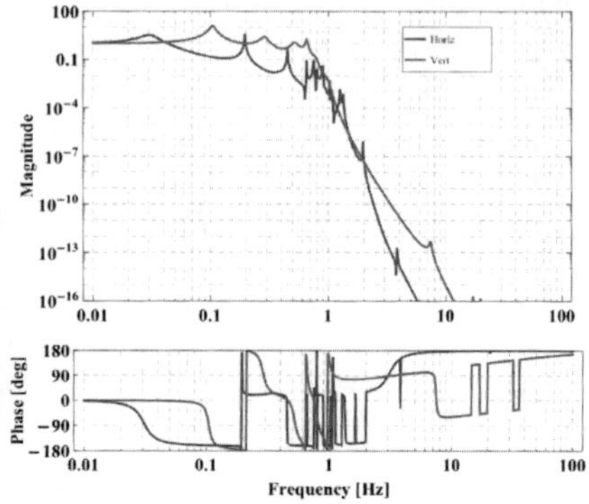

Fig. 6. Estimated transfer function of vibration attenuator system for a main mirror.

nates seismic attenuator performance. But it is possible to excavate vertical tunnel in the KAGRA site, so a tower-free tall vibration attenuator is realized. This tall pendulum has a structure of five stage pendulum at room temperature and cryogenic mirror suspension. Each stage of room temperature pendulum has Geometric Ati-Spring (GAS) filter[6] to attenuate vertical seismic motion. Figure 6 is an estimated transfer function of this vibration attenuator system[7]. And cryogenic mirror suspension is another feature in KAGRA, which is described in next section.

4. Cryogenic Mirror System

The best sensitivity range of around 100 Hz in an interferometer can be limited by mirror and suspension thermal noises. Thermal noise is a kind of Brownian motion of mirror and suspension wires, and they cause mirror vibration. To realize quantum noise limited sensitivity, it is important to suppress the thermal noises. The thermal noise amplitude is proportional to the square of the temperature and mechanical loss. Where mechanical loss means acoustic dissipation in material. So low temperature and use of materials with small mechanical loss are critical to reduce thermal noises. The aLIGO and the aVIRGO focused to use mirror and suspension materials with very small mechanical loss; fused silica at room temperature. On the other hand, the KAGRA chose another story, development of cryogenic sapphire mirror and suspension. Cryogenic sapphire realizes not only low temperature but also very small mechanical loss. It is well studied that mechanical loss of sapphire at cryogenic temperature decreases about one or two orders of magnitude from that at room temperature[8,9]. So by the effect of both low temperature and small mechanical loss, we can expect about one order of magnitude smaller amplitude of thermal noise than that at room temperature.

Fig. 7. A drawing of cryogenic sapphire mirror system with cryogenic suspension.

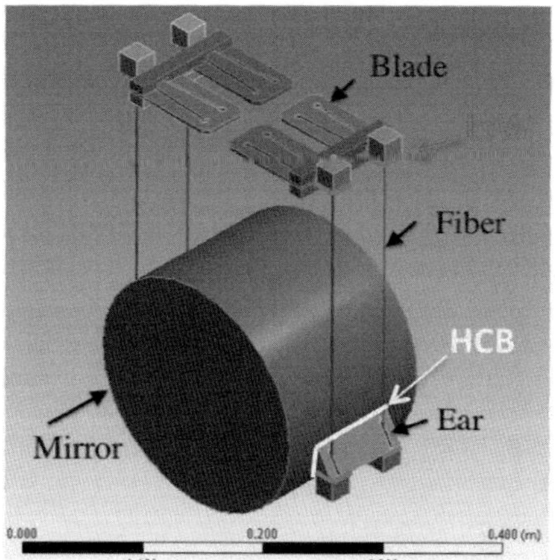

Fig. 8. Semi-monolithic sapphire suspension.

Figure 7 shows a drawing of cryogenic mirror and suspension system. This is a four-stage pendulum. The top part is a GAS filter. The GAS filter attenuates vibration through heat links to cool the cryogenic pendulum. The second part from the top is a tilting control stage named Marionette. The third is a intermediate mass and the last part is a mirror and its recoil mass. Figure 8 shows CAD model of mirror and suspensions made of sapphire to obtain ultra low mechanical loss. The suspensions system consists of sapphire mirror, sapphire ears to be hanged by sapphire fibers with sapphire blocks and sapphire blade as vertical spring. These parts have to be assembled with keeping small mechanical loss condition. Two type of contact technologies, Indium Welding[10] and Hydro-Catalysis Bonding[11], are developed. Studies have shown that using very thin layer of bonds is essential to obtain low mechanical loss in the suspensions, and these two contacts satisfies the requirement[10].

And extremely quiet cooling technologies have also been developed[12]. A small vibration pulse-tube cryocooler unit was developed and it has been observed that this system has same vibration level with Kamioka seismic motion. Soft heat links with large thermal conductivity were also developed. We plan to use 6N pure aluminium stranded cable and it is effective to reduce vibration conduction through cooling path.

The fundamental technologies of cryogenic mirror has been demonstrated in CLIO experiment, which is a prototype cryogenic interferometer in Kamioka[13].

5. Construction Status of KAGRA

Fig. 9. Assembled cryostat at X-front location.

Fig. 10. Vacuum tubes in an arm tunnel.

Excavation of KAGRA tunnel was done at March 2014. Soon after, installations of major devises were started. The first installed device inside tunnel was the cryostat for cryogenic mirror at Y-end location. The size of cryostat is 4.3 m in height and 2.6 m in diameter, and this is comparable size with the KAGRA tunnel. After installing two cryostats to both ends, vacuum tubes were installed into arm tunnels. A single vacuum tube size is 800 mm in diameter and 12 m in length, and 482 tubes and 16 tanks are installed in total. About 90 % vacuum vessels were connected in this stage, and no vacuum leakage over 10^{-10} Pa m^3/s were found. Figure 9 and 10 show pictures of installed cryostat and vacuum tubes.

We plan to have brief test of KAGRA by using simple Michelson interferometer at the end of 2015. At between 2016 and 2017, high performance vibration attenuators, cryogenic mirror systems and a high power laser are installed, and a science run will be done by the end of 2017.

6. Conclusion

KAGRA is an unique 2nd generation GW telescope, which will employ advanced features like underground site and cryogenic mirror systems. KAGR will detect GW signals from 280 Mpc in the best detector orientation. Installation of vacuum vessels have been almost completed and optical components and vibration attenuators are being installed rapidly. The construction of the KAGRA detector is expected to be completed by the end of 2017, following which the science runs will begin.

Acknowledgments

This project was supported by MEXT, JSPS Leading-edge Research Infrastructure Program, JSPS Grant-in-Aid for Specially Promoted Research 26000005, MEXT Grant-in-Aid for Scientific Research on Innovative Areas 24103005, JSPS Core-to-Core Program, A. Advanced Research Networks, and the joint research program of the Institute for Cosmic Ray Research, University of Tokyo."

References

1. G. M, Harry (for the LIGO Scientific Collaboration), Class. Quantum Grav. **27**, 084006 (2010).
2. The Virgo Collaboration, Advanced Virgo Baseline Design: VIR-0027A-09, https://pub3.ego-gw.it/itf/tds/.
3. B. Willke et al., "The GEO-HF project", Class. Quantum Grav. **23**, S207-S214 (2006).
4. P. R. Saulson, "Fundamentals of Interferometric Gravitational Wave Detectors", World Scientific Pub Co Inc. (1994).
5. J. Mizuno et al., Phys. Lett. A **175** 273 (1993).
6. G. Cella et al., Nucl. Instrum. Methods A, **540**, 502 (2005).
7. E. Hirose et al, Class. Quantum Grav. **31** 224004 (2014).
8. T. Uchiyama et al., Phys. Lett. A **261** 5 (1999).
9. T. Uchiyama et al., Phys. Lett. A **273** 310 (2000).
10. Dan Chen, Ph. D thesis, University of Tokyo (2014).
11. A. Anna-Maria et al., Adv. Opt. Techn. **3** 293 (2014).
12. T. Tomaru et al., Cryocoolers **13**, 695 (2005); T. Tomaru et al., Int. J. Mod. Phys. (Proc. ECRS) **20**, 7063 (2005); Y. Ikushima et al., Cryogenics **48**, 406 (2008).
13. T. Uchiyama et al., PRL **108**, 141101 (2011).

Evolution of density perturbations in classical General Relativity and in modified gravity

E. V. Arbuzova[*,1,2] and A.D. Dolgov[**,1,3]

[1]*Physics Department, Novosibirsk State University, Novosibirsk, 630090, Russia*
[2]*Department of Higher Mathematics, Dubna University, Dubna, 141980, Russia*
[3]*Dipartimento di Fisica e Scienze della Terra, Università degli Studi di Ferrara Polo Scientifico e Tecnologico - Edificio C, Via Saragat 1, Ferrara, 44122, Italy*
E-mail: [*]*arbuzova@uni-dubna.ru,* [**]*dolgov@fe.infn.it*

Gravitational instability in classical Jeans theory, General Relativity, and modified gravity is considered. A simple suggestion to eliminate the problem with zero order approximation in the Jeans theory is proposed. It is shown that density perturbations in the background with an increasing density rise faster than in the standard case of time independent background. Evolution of density and metric perturbations in the background of high frequency curvature oscillations in F(R) gravity is studied. In addition to the usual Jeans-like instability new effects of amplification of perturbations, associated with parametric resonance and antifriction phenomena, are found.

Keywords: Jeans theory; general relativity; modified gravity.

1. Introduction

The instability of self-gravitating systems was first investigated by Jeans[1] in non-relativistic Newtonian gravity. It was extended to General Relativity (GR) by Lifshitz[2] and nowadays it is widely used in cosmology to study the rise of density perturbations in the expanding universe.[3–5] The comparison between theoretical calculations and astronomical data is a powerful tool for testing the Standard Model of cosmology.

In our paper 6 we study the Jeans instability in classical and modified gravity in a background with rising energy density, assuming that the background metric changes slowly as a function of space and time compared to the typical frequency and the Compton wavelength of the scalaron field. In work 7 we investigate gravitational instability in a quickly oscillating curvature background. The time evolution of first-order perturbations is governed by a fourth-order differential equation instead of the usual second order one, therefore new types of unstable solutions are induced.

There appear not only the usual Jeans solution with a slightly reduced length scale, but also parametric resonance amplification of density perturbations, and an amplification of the perturbations due to the "antifriction" behaviour of the coefficients of odd derivatives in the equation.

2. Jeans Instability in Newtonian Theory with Space and Time Dependent Background

We consider a spherically symmetric cloud of particles with initially vanishing pressure and velocities, and study the classical non-relativistic Jeans problem in Newtonian gravity. The essential equations are the well known Poisson, Euler, and continuity equations:

$$\Delta\Phi = 4\pi G\rho, \tag{1}$$

$$\partial_t(\rho\mathbf{v}) + \rho(\mathbf{v}\nabla)\mathbf{v} + \nabla P + \rho\nabla\Phi = 0, \tag{2}$$

$$\partial_t\rho + \nabla(\rho\mathbf{v}) = 0. \tag{3}$$

We consider the time dependent problem taking as initial value a homogeneous distribution $\rho = const$ inside a sphere with radius r_m, while outside this sphere $\rho = 0$. The initial values of particle velocities and pressure are taken to be zero and the potential Φ is supposed to be a solution of the Poisson equation (1):

$$\Phi_0(r > r_m) = -MG/r, \quad \Phi_0(r < r_m) = 2\pi G\rho_0 r^2/3 + C_0, \tag{4}$$

where $C_0 = -2\pi G\rho_0 r_m^2$ to make the potential continuous and the total mass of the gravitating sphere is $M = 4\pi\rho_0 r_m^3/3$.

In what follows we will be interested in the internal solution for $r < r_m$. Now we can find how the background quantities ρ, v, and P evolve with time at small t. In particular we find from the continuity equation (3):

$$\rho_b(t,r) = \rho_0 + \rho_1 = \rho_0\left(1 + \frac{2\pi}{3}G\rho_0 t^2\right). \tag{5}$$

The evolution of perturbations over this time-dependent background can be found as usually writing $\rho = \rho_b(r,t) + \delta\rho$, $\Phi = \Phi_b(r,t) + \delta\Phi$, $v = v_1(r,t) + \delta v$, and $\delta P = c_s^2\delta\rho$, where c_s is the speed of sound. All δ-quantities are infinitesimal, so we find the linear system:

$$\Delta(\delta\Phi) = 4\pi G\delta\rho, \tag{6}$$

$$\partial_t\delta\mathbf{v} + \nabla\delta\Phi + \delta\rho/\rho_0\nabla\Phi_b + \nabla\delta P/\rho_0 = 0, \tag{7}$$

$$\partial_t\delta\rho + \rho_0\nabla(\delta\mathbf{v}) = 0. \tag{8}$$

Making the Fourier transform of Eqs. (6) - (8) and neglecting the r-dependent term $(\delta\rho/\rho_0)\nabla\Phi_b$, we obtain the eigenvalue equation:

$$k^2(\gamma^2 - c_s^2 k^2 + 4\pi G\rho_0) = 0\,. \tag{9}$$

For small k we find the usual exponential Jeans instability:

$$\frac{\delta\rho_{J1}}{\rho_0} \sim \exp\left[t\left(4\pi G\rho_0 - c_s^2 k^2\right)^{1/2}\right]. \tag{10}$$

We can estimate the impact of the rising background energy density on the rise of perturbations making an adiabatic approximation, that is replacing the exponent in Eq. (10) with the integral:

$$\frac{\delta\rho_{J2}}{\rho_0} \sim \exp\left\{\int_0^t dt\left[4\pi G\rho_b(t,r) - k^2 c_s^2\right]^{1/2}\right\}. \tag{11}$$

where $\rho_b(t,r)$ is given by Eq. (5).

Estimating the above integral for small k we find that the enhancement factor $\delta\rho_{J2}/\delta\rho_{J1}$ is equal to 1.027 after a time $t = t_{grav}$, where $t_{grav} = 1/\sqrt{4\pi G\rho_0}$, and grows with time. We can interpret this fact as an indication that the rise of fluctuations is indeed faster then in the usual Jeans scenario.

3. Density Perturbations in General Relativity

The treatment of the Jeans instability in General Relativity starts from the Einstein equations:

$$G_{\mu\nu} \equiv R_{\mu\nu} - \frac{1}{2}\,g_{\mu\nu}R = 8\pi G\,T_{\mu\nu} \equiv \tilde{T}_{\mu\nu}\,, \tag{12}$$

which automatically include the equations of motion of matter.

As previously, we consider a spherically symmetric cloud of matter with initially constant energy density inside the limit radius $r = r_m$. We choose Schwarzschild-like isotropic coordinates in which the metric takes the form:

$$ds^2 = Adt^2 - B\,\delta_{ij}\,dx^i dx^j\,, \tag{13}$$

where the functions A and B may depend upon r and t.

The energy-momentum tensor is taken in the perfect fluid form without dissipative corrections and has the following components:

$$T_{tt} = (\rho + P)U_t^2 - PA \approx \rho A\,, \tag{14}$$

$$T_{jt} = (\rho + P)U_t U_j \approx -(\rho + P)Bv_j\,, \tag{15}$$

$$T_{ij} = (\rho + P)U_i U_j - Pg_{ij} \approx PB\delta_{ij}\,. \tag{16}$$

We assume that the background metric changes slowly as a function of space and time and study small fluctuations around the background quantities: $\rho = \rho_b + \delta\rho$, $\delta P = c_s^2 \delta\rho$, $\mathbf{v} = \delta\mathbf{v}$, $A = A_b + \delta A$, $B = B_b + \delta B$.

The corresponding linear equations for infinitesimal perturbations are:

$$-\Delta \delta B = \delta \tilde{\rho}\,, \tag{17}$$

$$\partial_i \partial_j (\delta A + \delta B) = 0\,. \tag{18}$$

$$\delta\dot{\rho} + \rho_b\, \partial_j \delta v^j + \frac{3}{2}\rho_b\, \delta\dot{B} = 0\,, \tag{19}$$

$$\rho\, \delta\dot{v}_j + \partial_j \delta P + \frac{1}{2}\rho_b\, \partial_j \delta A = 0\,. \tag{20}$$

We look for the solution in the form $\sim \exp[-i\gamma t + i\mathbf{k}\cdot\mathbf{x}]$ and obtain the following expressions for the frequency eigenvalues:

$$\gamma^2 = \frac{c_s^2 k^2 - \tilde{\rho}_b/2}{1 + 3\tilde{\rho}_b/(2k^2)}\,, \tag{21}$$

where $\tilde{\rho}_b = 8\pi G \rho_b$. This result almost coincides with the Newtonian one (9). An extra term in the denominator is induced by the volume variation, and it is small for $k \sim k_J = \sqrt{4\pi G \rho_0}/c_s$.

4. Gravitational Instability in Oscillating Background

In modified gravity a nonlinear function of curvature $F(R)$ is added to the standard Einstein-Hilbert Lagrangian, so the modified Einstein equations have the form:

$$(1 + F_{,R})\, R_{\mu\nu} - \frac{1}{2}\,(R + F)\, g_{\mu\nu} + (g_{\mu\nu} D_\alpha D^\alpha - D_\mu D_\nu)\, F_{,R} = \tilde{T}_{\mu\nu}\,, \tag{22}$$

where $F_{,R} = dF/dR$. Some examples of $F(R)$-theories which realistically describe the cosmological acceleration can be found in Refs. 8–11.

In these models, the function $F(R)$ takes very different values for $|R| \ll |R_c|$, $|R| \sim |R_c|$, and $|R| \gg |R_c|$, where R_c is the present average cosmological curvature. We consider the case $|R| \gg |R_c|$, which is realized in astronomical systems with the energy density grossly exceeding the cosmological one. In many models presented in the literature the following conditions are also fulfilled in this regime: $|F| \ll |R|$ and $|F_{,R}| \ll 1$.

Therefore, equation (22) can be approximated as:

$$G_{\mu\nu} + \frac{1}{3\omega^2}(D_\mu D_\nu - g_{\mu\nu} D^2)R = \tilde{T}_{\mu\nu}\,, \tag{23}$$

where $G_{\mu\nu} = R_{\mu\nu} - g_{\mu\nu} R/2$ is the Einstein tensor and $\omega^{-2} = -3F''_{RR}$. Once written in this form, the equation is largely independent of the specific model considered except of course of the value of ω.

As usual the metric and the curvature tensors are expanded around their background values at first order in infinitesimal perturbations, i.e.

$$A = A_b + \delta A\,, \quad B = B_b + \delta B\,, \quad R = R_b + \delta R\,, \tag{24}$$

where A_b and R_b are quickly oscillating functions of time with possibly large amplitude ("spikes") which were found in our previous works[12,13].

Using technique described in papers 6,7 we derive the 4th order equation for the single function δB. We introduce the dimensionless time and define:

$$\tau = \omega t\,, \quad \delta B \equiv z\,, \quad R_b = -\tilde{\rho}_b y\,, \quad a = \frac{\tilde{\rho}_b}{k^2}\,, \quad b = \frac{k^2}{\omega^2}\,, \quad c = c_s^2\,. \tag{25}$$

With these quantities equation takes a form convenient for the analysis:

$$z'''' + \alpha y' z''' + (\Omega^2 + 2\alpha y'')z'' + \alpha(y''' + bcy')z' + \mu z = 0\,, \tag{26}$$

where

$$\alpha = \frac{a}{2}\left(1 + \frac{2b}{3}\right), \quad \Omega^2 = 1 - \frac{a}{2}\left(1 + \frac{8b}{3}\right) + b(1+c)\,, \tag{27a}$$

$$\mu = b\left[c(1+b) - \frac{a}{2}\left(1 + \frac{4b}{3}\right) - 2ac\left(1 + \frac{2b}{3}\right)\right]. \tag{27b}$$

Let us compare our equation (26) with the usual Jeans equation

$$\delta\ddot{\rho} + \left(c_s^2 k^2 - \frac{\tilde{\rho}_b}{2}\right)\delta\rho = 0\,. \tag{28}$$

As is well known, the sign of the term $c_s^2 k^2 - \tilde{\rho}_b/2$ determines the stability of solutions, which are sound waves for $c_s^2 k^2 > \tilde{\rho}_b/2$ and unstable modes in the opposite regime. These unstable modes appear at the Jeans scale

$$k^2 \leq (k_J^{\rm GR})^2 \equiv \frac{\tilde{\rho}_b}{2c_s^2}\,. \tag{29}$$

In modified gravity the condition of stability is determined by the sign of μ (27b). In the limit of small amplitude of curvature oscillations or very low frequency, we can neglect $y(\tau)$ in Eq. (26) so that it is reduced to a simple equation with constant coefficients which is solved by the substitution $z = \exp(\gamma\tau)$. The eigenvalues γ are thus determined by the expression:

$$\gamma^2 = -\frac{\Omega^2}{2} \pm \sqrt{\frac{\Omega^4}{4} - \mu}. \tag{30}$$

If $\mu < 0$, then one of the roots $\gamma^2 > 0$, so one of the eigenvalues is positive. It corresponds to the usual exponential Jeans instability, though the values of the Jeans wave vector in modified gravity and in GR are different:

$$(k_J^{\rm MG})^2 = (k_J^{\rm GR})^2 \left[1 + \frac{(k_J^{\rm GR})^2}{3\omega^2}\right]. \tag{31}$$

Eq. (31) recovers the GR result in the limit $\omega \to \infty$ and shows that in modified gravity the Jeans wavenumber is larger than in GR. This corresponds to a reduced minimum length scale associated to structure formation.

If $y(\tau)$ is non-negligible, quite interesting new effects can show up. The function $y(\tau)$ entering the 4th order dimensionless equation (26) is an oscillating function of "time" τ. It can induce an analogue of the parametric resonance instability resulting in a very fast rise of perturbations at a certain set of frequencies. Another new effect can be called "antifriction". It appears at sufficiently large amplitudes of oscillations of $y(\tau)$ such that the coefficients in front of the odd derivative terms become periodically negative. This phenomenon leads to an explosive rise of z in a wide range of frequencies.

Both effects do not exist in the standard General Relativity and, if discovered, would be a proof of modified gravity. On the contrary, the non-observation of these effects would allow to put stringent restrictions on the parameters of $F(R)$-theories.

We consider two possible forms of periodic function $y(\tau)$:

(1) purely harmonic ones:

$$y_{harm}(\tau) = y_{eq}(\tau) + y_0 \cos(\Omega_1 \tau + \theta), \tag{32}$$

where $\Omega_1 = \omega_1/\omega$ is the dimensionless frequency, y_0 is the amplitude of oscillations and θ is a constant phase, and y_{eq} is the equilibrium point of the potential around which the curvature oscillates.

(2) spiky solutions found in Ref. 12, which we approximate as

$$y_{sp}(\tau) = \frac{y_0\, d^2}{d^2 + \sin^2(\Omega_2 \tau + \theta)}, \tag{33}$$

where $d \ll 1$, so that we have narrow peaks with a large separation between them.

We solved Eq. (26) numerically for different values of y_0 and $\Omega_{1,2}$. The parametric resonance excitation for harmonic oscillations is observed at the expected frequency $\Omega_1 = 2\Omega$, see Fig. 1 (left panel). Here and in what follows the parameters of the medium and the wave number of the fluctuations were taken according to $a = b = 0.01$ and $c = c_s^2 = 0.02$. Fig. 1 (right panel) clearly shows parametric resonance in the case of spike-like oscillations at $\Omega_2 = \Omega/2 \simeq 0.5013$ which corresponds to the second mode of the resonance, see Eq. (27a). The parametric resonance is rather sensitive to variations of y_0 and of course of $\Omega_{1,2}$. In fact, the exponential growth is slower as we depart from the resonance frequency.

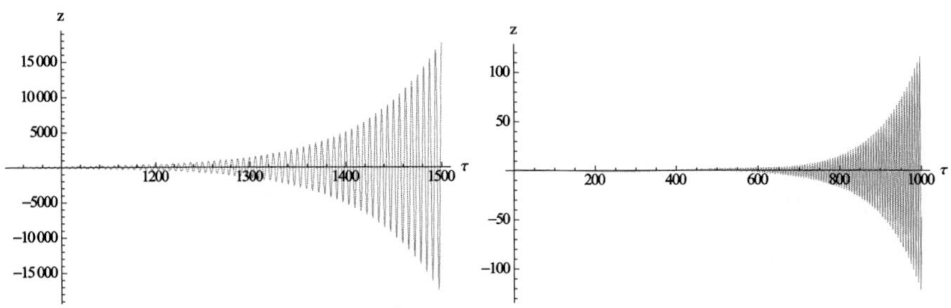

Fig. 1. *Left panel*: Parametric resonance excitation of $z(\tau)$ for harmonic curvature oscillations with $y_0 = 5$ and $\Omega_1/\Omega = 2$. *Right panel*: Parametric resonance excitation of $z(\tau)$ for spike-like curvature oscillations with $y_0 = 30$ and $\Omega_2/\Omega = 0.5$.

The antifriction amplification is observed at the frequencies away from the resonance values, if y_0 exceeds a threshold value, y_{th}. The farther away the frequency is from the resonance, the larger is the threshold. For example for harmonic oscillations with $\Omega_1/\Omega = 3.2$ the threshold value is $y_{th} = 169$, while for $\Omega_1/\Omega = 4.4$ the threshold is $y_{th} = 267$. The evolution of $z(\tau)$ in the first case is depicted in Fig. 2 (left panel).

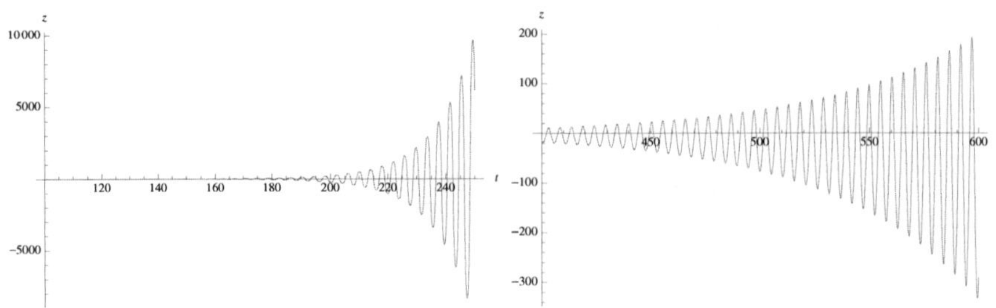

Fig. 2. *Left panel*: Antifriction effect in evolution of $z(\tau)$ for harmonic curvature oscillations with $y_0 = 169$, $\Omega_1/\Omega = 3.2$. *Right panel*: Antifriction effect in evolution of $z(\tau)$ for spike-type curvature oscillations with $y_0 = 400$, $\Omega_2/\Omega = 0.6$.

In the right panel of Fig. 2 we present the evolution of $z(\tau)$ for spike-like oscillations with out-of-resonance frequencies $\Omega_2/\Omega = 0.6$. Similarly to the case of harmonic $y(\tau)$, the farther away the frequency is from the resonant one, approximately equal to 0.5, the larger the threshold value of y_0 necessary for generating an unstable solution.

Acknowledgments

EA and AD acknowledge the support of the grant of the Russian Federation government 11.G34.31.0047.

References

1. J.H. Jeans, Philosophical Transactions of the Royal Society A 199, 1 (1902).
2. E.M. Lifshitz, ZhETF, **16**, 587 (1946).
3. Ya.B. Zeldovich, I.D. Novikov Relativistic Astrophysics. Vol. 2. The Structure And Evolution Of The Universe Chicago, Usa: Chicago Univ., 1983; S. Weinberg, Cosmology, Oxford University Press, 2008.
4. V. Mukhanov, Physical Foundations of Cosmology, Cambridge University Press, 2005.
5. D.S. Gorbunov, V.A. Rubakov, Introduction to the theory of the early universe: Cosmological perturbations and inflationary theory, Hackensack, USA: World Scientific, 2011.
6. E.V. Arbuzova, A.D. Dolgov and L. Reverberi, Phys. Lett. B **739**, 279 (2014).
7. E.V. Arbuzova, A.D. Dolgov and L. Reverberi, arXiv:1507.02152.
8. A.A. Starobinsky, JETP Lett. **86**, 157 (2007).
9. W. Hu, I. Sawicki, Phys. Rev. D **76**, 064004 (2007).
10. A. Appleby, R. Battye, Phys. Lett. B **654**, 7 (2007).
11. S.A. Appleby, R.A. Battye, A.A. Starobinsky, *JCAP* **1006**, 005 (2010).
12. E.V. Arbuzova, A.D. Dolgov, L. Reverberi, Eur. Phys. J. C **72**, 2247 (2012); Phys. Rev. **D 88** 024035 (2013).
13. E.V. Arbuzova, A.D. Dolgov, L. Reverberi, Astropart. Phys. **54**, 44 (2014).

Early astrophysical objects and cosmological antimatter

A. D. Dolgov

Departemnt of Physics University of Ferrara, via Saragat, 1Ferrara 40100, Italy
Department of Physics Novosibirsk State University Ul. Pirogova, 2, Novosibirsk, 630090, Russia,
E-mail: dolgov@fe.infn.it

Astronomical data on $z \sim 10$ universe, which are in odds with the accepted mechanisms of star and black hole formation are reviewed. A model resolving the contradiction leads to abundant creation of cosmological antimatter.

Keywords: Early universe; supermassive black holes; quasars.

1. Introduction

On top of the standard ΛCDM cosmology, there are a few clouds in the sky which could change the standard lore. The astronomical data, accumulated during recent years, indicate that the high redshift universe is densely populated by supermassive black holes (quasars), supernovae, gamma bursters, and very bright galaxies. Such an early creation and evolution implies that the usually accepted mechanisms should be essentially modified.

The traces of this early created objects are also observed in the present day universe through central black holes (especially in small galaxies), stars "older tan the universe", black hole mass distribution in the present day universe and some other poorly explained data. For more detail see Ref. 1.

2. Observed High-z Objects with No Time for Their Creation

There is a large "zoo" of astronomical objects formed in surprisingly short times. Several galaxies have been observed at high redshifts, e.g. a galaxy at $z \approx 9.6$ which was created when the universe was about 0.5 Gyr old.[2]

Moreover a galaxy at $z \approx 11$ has been observed[3] which was formed earlier than the universe age was 0.41 Gyr (or even shorter with larger H).

An observation of not so young but extremely luminous galaxy was recently reported[7]: $L = 3 \cdot 10^{14} L_{\odot}$; age: ~ 1.3 Gyr. For creation of such galaxy the galactic seeds, e.g. embryonic black holes, must be bigger than thought possible. The BH was already billions of $M_\odot$, when our universe was only a tenth of its present age of 13.8 billion years.

Another and even more striking example of early formed objects are quasars observed at high-z e.g. quasar with maximum $z = 7.085$,[4] which was formed at $t < 0.75$ Gyr. Its luminosity and mass are respectively $L = 6.3 \cdot 10^{13} L_\odot$ and $M = 2 \cdot 10^9 M_\odot$. The quasars are supposed to be supermassive black holes (BH) and their formation in such short time looks problematic by conventional mechanisms.

According to[5] rapid emergence of high-z galaxies so soon after big bang may actually be in conflict with current understanding of how they came to be. This problem is very reminiscent of the better known (and probably related) premature appearance of supermassive black holes at $z \sim 6$. It is difficult to understand how $10^9 M_\odot$ black holes appeared so quickly after the big bang without invoking non-standard accretion physics and the formation of massive seeds, both of which are not seen in the local Universe.

Very recently another monster was discovered[6]: "An ultraluminous quasar with a twelve billion solar mass black hole at redshift 6.30". There is already a serious problem with formation mechanism of lighter and less luminous quasars which is now multifold deepened with this new "creature". About 40 quasars with $z > 6$ are known, each quasar containing BH with $M \sim 10^9 M_\odot$. Such black holes, when the Universe was less than one billion years old, present substantial challenges to theories of the formation and growth of black holes and the coevolution of black holes and galaxies. The new one is an order of magnitude heavier, $M \approx 10^{10} M_\odot$.

The universe at $z \sim 10$ is quite dusty.[8] The medium around the observed early quasars contains considerable amount of "metals" (elements heavier than He). According to the standard picture, only elements up to ^{4}He and traces of Li, Be, B were formed by BBN, while heavier elements were created by stellar nucleosynthesis and dispersed in the interstellar space by supernova explosions. If so, prior to or simultaneously with the QSO formation a rapid star formation should take place. These stars could produce plenty of supernovae which enriched interstellar space by metals.

Observations of high redshift gamma ray bursters (GBR) also indicate a high abundance of supernova at large redshifts. The highest redshift of the observed GBR is $z = 9.4$ and there are a few more GBRs with smaller but still high redshifts. The necessary star formation rate for explanation of these early GBRs is at odds with the canonical star formation theory.

3. Strange Phenomena in the Milky Way

Old stars in the Milky Way. With an increased accuracy of the nuclear chronology the ages of several stars have been recently determined to be much older than expected. Below we quote a few recent results.

Employing thorium and uranium abundances in comparison with each other and with several stable elements the age of metal-poor, halo star BD+17° 3248 was estimated as[9] 13.8 ± 4 Gyr. For comparison the age of inner halo of the Galaxy is 11.4 ± 0.7 Gyr.

The age of a star in the galactic halo, HE 1523-0901, was estimated to be about 13.2 Gyr.[11] First time many different chronometers, such as the U/Th, U/Ir, Th/Eu and Th/Os ratios to have been employed.

The metal deficient *high velocity* subgiant in the solar neighborhood HD 140283 has the age 14.46 ± 0.31 Gyr.[12] The central value exceeds the universe age by two standard deviations, if $H = 67.3$ and $t_U = 13.8$, while for $H = 74$, and thus $t_U = 12.5$ the excess is by nine standard deviations. So the star really looks older than the universe.

A possible explanation of this discrepancy could be an unusual initial chemical content of such stars. Normally a pre-stellar cloud consists of 25% of He^4 and 75% of hydrogen. However, in the suggested model BBN could create much heavier elements. The early supernovae could also enrich the interstallar gas with heavy elements, so the initial chemical content of some stars would be much different from the traditional one. Such stars could evolve to their present state considerably faster than the usual ones.

The mass distribution of BHs observed in the Milky way demonstrates some features not understood by the conventional theory. It is found that their masses are concentrated in the narrow range $(7.8 \pm 1.2)M_\odot$.[13] This result agrees with Ref. 14, where a peak around $8M_\odot$, a paucity of sources with masses below $5M_\odot$, and a sharp drop-off above $10M_\odot$ are observed. This is not explained if these BHs were formed by stellar collapse. However the log-normal mass distribution (2) of the primordial black holes naturally explains such a form of the mass spectrum of the stellar mass black holes in the Galaxy. There is an indication to a two-peak mass distribution of the PBHs and compact stars, which is not explained up to now.[15] Quoting this work: "sample of black hole masses provides strong evidence of a gap between the maximum neutron star mass and the lower bound on black hole masses". These results fit the distribution (2) assuming that the neutron stars were mostly created by the conventional mechanism.

4. Supermassive BHs in (Near) Contemporary Universe

As observed every large galaxy and some smaller ones contain a central supermassive BH with masses larger than $10^9 M_\odot$ in giant elliptical and compact lenticular galaxies and $\sim 10^6 M_\odot$ in spiral galaxies like Milky Way. Their masses are about 0.1% of the mass of the stellar bulge of galaxy[16] but some galaxies may have huge black holes: e.g. NGC 1277 has the central BH of $1.7 \times 10^{10} M_\odot$, or 60% of its bulge mass.[17] This fact creates serious problems for the standard scenario of formation of central supermassive BHs by accretion of matter in the central part of a galaxy. An inverted picture looks more plausible, when first a supermassive black hole was formed and attracted matter serving as seed for subsequent galaxy formation.

More examples can be found in Ref. 18. As the authors say, although supermassive black holes correlate well with their host galaxies, there is an emerging view that outliers exist. Henize 2-10, NGC 4889, and NGC1277 are examples of super massive black holes at least an order of magnitude more massive than their host galaxy suggests.

A discovery of an ultra-compact dwarf galaxy older than 10 Gyr, enriched with metals, and probably with a massive BH in its center also seems to be at odds with the standard model.[19] The dynamical mass of the galaxy is $2 \times 10^8 M_\odot$ and the radius is $R \sim 24$ pc, so its density is very high. The Chandra data reveal a variable central X-ray source with $L_X \sim 10^{38}$ erg/s that could be an active galactic nucleus associated with a massive black hole or a low-mass X-ray binary. The object has the solar metallicity, with elevated [Mg/Fe] and strongly enhanced [N/Fe].

Recently an observation of a quasar quartet embedded in giant nebula was reported[20] at redshift $z \approx 2$. It reveals rare massive structure in distant universe. The chance of finding a quadruple quasar is $\sim 10^{-7}$. It implies that the most massive structures in the distant universe have a tremendous supply ($\sim 10^{11} M_\odot$) of cool dense ($n \approx 1/\mathrm{cm}^3$) gas, in conflict with current cosmological simulations.

5. Possible Explanation

The mechanism which can explain all these puzzles was suggested in Ref. 21. It is based on the Affleck and Dine scenario of baryogenesis.[22] A minor modification of AD-scenario can lead to very early formation of compact stellar-type objects such as e.g. quasars, PBH, or compact stars, formed in the regions with a huge baryon asymmetry, and naturally to a comparable amount of anti-objects, plus sub-dominant observed homogeneous baryonic background. The amount of antimatter may be comparable or even larger than of *known* baryons, but such "compact" (anti)baryonic objects would not contradict any existing observations.

The necessary generalization of the AD-baryogenesis was suggested and discussed in Ref. 21. A general renormalizable coupling of the Affleck-Dine field χ to inflaton, Φ, has been introduced so the total potential of χ included the interaction with the inflaton, Φ, gets the form:

$$U = g|\chi|^2(\Phi - \Phi_1)^2 + \lambda|\chi|^4| \ln\left(\frac{|\chi|^2}{\sigma^2}\right) + \lambda_1\left(\chi^4 + h.c.\right) + (m^2\chi^2 + h.c.). \tag{1}$$

When Φ is far from Φ_1, the effective mass squared of χ is positive and large, so the gate to the flat directions is closed. It is open only during relatively short period, when $\Phi \approx \Phi_1$, but still at inflationary epoch. Because of that cosmologically small but possibly astronomically large bubbles with very high β could be created, occupying a small fraction of the universe volume, while the rest of the universe has the normal baryon asymmetry $\beta \approx 6 \cdot 10^{-10}$, created in the regions with small initial χ, occupying the bulk of space. In the simplest version of the model the high-β objects would have positive and negative baryon asymmetry with practically equal probability. The mass distribution of high baryon density bubbles has the form:

$$\frac{dN}{dM} = C_M \exp\left[-\gamma \ln^2(M/M_0)\right] \tag{2}$$

where C_M, γ, and M_0 are constant model dependent parameters.

There are two periods when the density contrast between the regions with high baryonic density and those with the normal low one arose:
1. After formation of the domains with high value of χ the initial density contrast between the high-B bubbles and the rest of the universe would be zero, because quarks, at this stage were massless. Later a density contrast would evolve due to higher amplitude of non-relativistic field χ inside.
2. The second period of perturbation rise took place after the QCD phase transition at $T \sim 100$ MeV when quarks made non-relativistic protons. During this stage compact bubbles with high baryonic mass density would be formed. The parameters allow for formation of PBHs, or dense stellar like objects. The PBH masses could range from a few $M_\odot$ up to $10^{6-7} M_\odot$.

6. Anti-byproduct

In grand unified SUSY models the baryonic number of χ is naturally non-conserved. Initially χ is naturally away from the origin and when inflation is over, it starts to evolve down to the equilibrium point, $\chi = 0$, according to the equation of Newtonian mechanics for point-like particle:

$$\ddot{\chi} + 3H\dot{\chi} + U'(\chi) = 0. \tag{3}$$

Baryonic number of χ, $B_\chi = \dot{\theta}|\chi|^2$, is analogous to the mechanical angular momentum in the two-dimensional complex plane $[Re\,\chi,\ Im\,\chi]$. The decays of χ transferred the baryonic charge accumulated in the rotational motion of χ into that of quarks in B-conserving processes. The rotation might be directed both clock- or anticlock-wise and hence baryonic and antibaryonic bubbles could be formed.

According to our estimates the compact anti-objects mostly survived in the early universe, (especially if they are PBHs). The model predicts an existence of early dense stars form in the hot universe. Such "stars" may evolve quickly and, in particular, make early SNs, enrich the universe with heavy (anti)nuclei and reionize the universe. Phenomenology and bounds on compact antimatter objects and disperse anti-clouds was studied in the papers[23,24]. The early formed stars and black holes are discussed in Ref. 25.

7. Conclusion

The observational data at high redshifts, which became available in the recent years show that the universe is populated by the early formed bright galaxies, quasars, gamma-bursters, and contains a lot of metals and dust. All that is not expected in the standard theory.

An existence of supermassive BHs is a huge problem. The standard scenario says that such BHs have been formed through accretion of matter to initial solar mass BHa. However, it demands much more time than all universe age. This is especially troubling for the supermassive BHs (quasars) created at the dawn of the universe at high-z. The inverted picture when primordial BHs were first created and then served as seeds for the galaxy formation looks much more probable. BHs found in small galaxies, having masses about half of the total galaxy mass surely cannot be created by the conventional mechanism.

The model,[21] very nicely fits all the new observations. A natural, though not obligatory, outcome of this model is a prediction of cosmic antimatter, mostly in the form of compact high velocity stars in the Galaxy. The observational bounds on the abundance of such stars are quite vague, so we may have a lot of antimatter practically at hand. The early formed stars would behave as CDM. and so should have much larger velocity than the normal stars. Hence they would populate the galactic halo. The mysterious MACHOs[26] can be just them.

An observation of antistars in the Galaxy would present a very strong support of the suggested scenario. Since antimatter was created in clouds with the unusually high baryon-to-photon ratio β, the BBN in these clouds would produce anomalous abundances of light elements, so there is a good chance that this region is made of antimatter.

Acknowledgement

This work was supported by the grant of the Russian Federation government 11.G34.31.0047.

References

1. A.D. Dolgov, Early formed astrophysical objects and cosmological antimatter, arXive 1508.07398 and references therein.
2. W. Zheng,, *et al*, A highly magnified candidate for a young galaxy seen when the Universe was 500 Myrs old" arXiv:1204.2305.
3. D. Coe, *et al* CLASH: Three Strongly Lensed Images of a Candidate $z \sim 11$ Galaxy, *Astrophys. J.* **762**, 32 (2013).
4. D.J. Mortlock, *et al*, A luminous quasar at a redshift of z = 7.085, *Nature* **474**, 616 (2011).
5. F. Melia, The Premature Formation of High Redshift Galaxies", arXiv: 1403.0908.
6. Xue-BingWu, *et al*, An ultraluminous quasar with a twelve-billion-solar-mass black hole at redshift 6.30, *Nature* **518**, 512 (2015).
7. Chao-Wei Tsai, *et al*, The Most Luminous Galaxies Discovered by WISE, arXiv:1410.1751.
8. L. Mattsson, The sudden appearance of dust in the early Universe, arXiv:1505.04758.
9. J.J. Cowan, *et al*, The chemical composition and age of the metal-poor halo star BD+17° 3248, *Astrophys. J.* **572**, 861 (2002).
10. J. Kalirai, The Age of the Milky Way Inner Halo, *Nature*, **486**, 90 (2012).
11. A. Frebe, *et al* Discovery of HE 1523-0901: A Strongly r-Process Enhanced Metal-Poor Star with Detected Uranium, *Astrophys. J.* **660**, L117 (2007).
12. H. E. Bond, *et al* HD 140283: A Star in the Solar Neighborhood that Formed Shortly After the Big Bang, *Astrophys. J. Lett.* **765**. L12 (2013).

13. F. Ozel, *et al* The Black Hole Mass Distribution in the Galaxy, arXiv:1006.2834.
14. L. Kreidberg, C.D. Bailyn, W.M. Farr, *et al*, Mass Measurements of Black Holes in X-Ray Transients: Is There a Mass Gap? arXiv:1205.1805.
15. W.M. Farr, *et al*, The Mass Distribution of Stellar-Mass Black Holes, arXiv:1011.1459.
16. N. Häring, H.-W. Rix, On the Black Hole Mass - Bulge Mass Relation, *Astrophys. J. Lett.* **604**, L89 (2004);
E. Sani, *et al* The Spitzer/IRAC view of black hole - bulge scaling relations, *Mon. Not. .R. Astron. Soc.* **413**, 1479 (2011).
17. R. C. E. van den Bosch, *et al*, An Over-Massive Black Hole in the Compact Lenticular Galaxy NGC1277, *Nature,* **491**, 729 (2012).
18. F. Khan, K. Holley-Bockelmann, P. Berczik, Ultramassive Black Hole Coalescence, arXiv:1405.6425.
19. J. Strader, *et al*, The Densest Galaxy, *Astrophys. J. Lett.* **775**, L6 (2013).
20. J.F. Hennawi *et al*, Quasar quartet embedded in giant nebula reveals rare massive structure in distant universe *Science*, **348**, 779 (2015).
21. A. Dolgov, J. Silk, Baryon isocurvature fluctuations at small scales and baryonic dark matter, *Phys. Rev.* **D 47**, 4244 (1993);
A. D. Dolgov, M. Kawasaki, N. Kevlishvili, Inhomogeneous baryogenesis, cosmic antimatter, and dark matter, *Nuclear Physics* **B 807**, 229 (2009).
22. I. Affleck, M. Dine, A New Mechanism for Baryogenesis, *Nuclear Physics*, **B 249**, 361 (1985).
23. C. Bambi, A. D. Dolgov, Antimatter in the Milky Way, *Nuclear Physics,* **B 784**, 132 (2007).
24. S.I Blinnikov, A.D. Dolgov, K.A. Postnov, Antimatter and antistars in the universe and in the Galaxy, *Phys. Rev.* **D92**, 2, 023516 (2015).
25. A.D. Dolgov, S.I. Biinnikov Stars and Black Holes from the very Early Universe *Phys. Rev.* **D89**, 2, 021301 (2014).
26. C. Alcock, *et al* The MACHO Project: Microlensing Results from 5.7 Years of LMC Observations, *Astrophys. J.* **542**, 261 (2000);
P. Tisserand, *et al* Limits on the Macho Content of the Galactic Halo from the EROS-2 Survey of the Magellanic Clouds, *Astron. Astrophys*, **469**, 387 (2007);
A. Riffeser, S. Seitz, R. Bender, The M31 microlensing event WeCAPP-GL1/Point-AGAPE-S3: evidence for a MACHO component in the dark halo of M31? *Astrophys. J.* **684**, 1093 (2008).

Is there an ordinary supermassive black hole at the Galactic Center?

A. F. Zakharov*

Institute of Theoretical and Experimental Physics, Moscow, 117218, Russia
**E-mail: zakharov@itep.ru*

Bogoliubov Laboratory of Theoretical Physics, JINR,
Dubna, Moscow region, Russia

National Research Nuclear University MEPhI
(Moscow Engineering Physics Institute), 115409, Moscow, Russia

North Carolina Central University, Durham, NC 27707, USA

Institute for Computer Aided Design of RAS, 123056, Moscow, Russia

Now there are two basic observational techniques to investigate a gravitational potential at the Galactic Center, namely, a) monitoring the orbits of bright stars near the Galactic Center to reconstruct a gravitational potential; b) measuring a size and a shape of shadows around black hole giving an alternative possibility to evaluate black hole parameters in mm-band with VLBI-technique. At the moment one can use a small relativistic correction approach for stellar orbit analysis (however, in the future the approximation will not be not precise enough due to enormous progress of observational facilities) while now for smallest structure analysis in VLBI observations one really needs a strong gravitational field approximation. We discuss results of observations, their conventional interpretations, tensions between observations and models and possible hints for a new physics from the observational data and tensions between observations and interpretations. We will discuss an opportunity to use a Schwarzschild metric for data interpretation or we have to use more exotic models such as Reissner – Nordström or Schwarzschild – de-Sitter metrics for better fits.

Keywords: General Relativity; alternative theories of gravity; black holes; Sgr A*; the Galactic Center.

1. Introduction

There were only three famous tests of GR at the beginning, namely, deflection of light, the Mercury anomaly and gravitational redshifts,[1] but there is a number of phenomena where predictions of GR have been checked or they will be checked in the future.[2]

Supermassive black holes have been found in center of 85 galaxies,[3] however, usually astronomers do not use GR approaches for such claims about black hole existence. Around 40 quasars with redshifts $z > 6$ have been found and each quasar has a supermassive black hole with a mass around one billion solar masses $10^9 M_\odot$ and recently[4] found the ultraluminous quasar SDSS, J010013.02+1280225.8, at redshift $z \sim 6.30$, and the quasar has the black hole with a mass around $1.2 \times 10^{10} M_\odot$. Remarkably, that the initial optical spectroscopy of the quasar was carried out with the Chinese Lijiang 2.4-m telescope (it means that discoveries may be done with relatively modest facilities). Later, spectroscopic observations were conducted for the object with the 6.5-m Multiple Mirror Telescope(MMT) and the twin 8.4-m mirror Large Binocular Telescope (LBT) in the USA and initial estimates of black hole mass and redshift have been confirmed.

To evaluate a gravitational potential at the Galactic Center two teams of astronomers observe trajectories of bright stars in the IR band for several years. One group led by A. Ghez (UCLA, USA) uses the twin 10-meter optical/infrared telescopes on Mauna Kea (Hawaii), and according to the Keck Strategic Mission, the first important goal is *high angular resolution astrophysics*[a] and practically it gives an opportunity to be a world leader in the field, see results of the observations of bright stars. Another group led by R. Genzel (ESO, MPE) uses four 8.2-meter VLT telescopes at Paranal (Chile). The European group got very important results[5] which are consistent results presented by US team. The important case of G2 gas cloud is a very useful tracer of the gravitational potential at the Galactic Center[5] (later, the object has been called Dusty S-cluster Object (DSO/G2) since further detailed observations were not completely consistent with the gas cloud model). It occurs that very likely this may be another example of a close peribothron[b] passage of a dust-enshrouded star.[7–9] The analysis showed that the DSO/G2 is rather a young star than a coreless gas and dust cloud.[10] However, a detailed analysis showed that a gas cloud model is consistent with observations.[11]

The ESO and MPE formed a team to construct the GRAVITY, the Very Large Telescope Interferometer for precise astrometry and interferometric

[a]http://www.keckobservatory.org/

[b]The word "peribothron" has been introduced by[6] following W.R. Stoeger's suggestion. Greek bothros means pit or hole. So, the peribothron means the point of least distance of an object orbiting a black hole.

imaging. The interferometer will provide precise astrometry of order 10 micro-arcseconds.[12,13] The GRAVITY equipment should be shipped to the VLT observatory in Chile in 2015 and commissioning is planned to start in October 2015.[13]

2. Shadows for the Black Hole at the Galactic Center

Several years ago a formation of images for supermassive black holes has been simulated.[14,15] The authors used a toy model for their analysis and they concluded that a strong gravitational field is bent trajectories of photons emitted by accreting particles and an observer can see a dark spot (shadow) around a black hole position. For the black hole at the Galactic Center a size of shadow is around 50 μas. Based on results of simulations, the authors concluded that the shadow may detectable at mm and sub-mm wavelengths, however, scattering may be very significant at cm wavelength, so there are very small chances to observe the shadows at the cm band.[14,15] Importantly, that the results are rather general in spite of their specific model.[14,15]

Based on ideas introduced earlier,[16,17] different types of shadow shapes for Kerr black holes and different position angles of a distant observer have been considered.[18] Moreover, it was shown that for an equatorial plane position of a distant observer, maximal impact parameter $|\beta_{\max}|$ in z-direction (which coincides with a black hole rotation direction) is $\sqrt{27}$, while the corresponding impact parameter in the perpendicular direction for the $\beta_{\max}$ is $\alpha_{\max} = 2a$,[18] if we consider the function $\beta(\alpha)$ for critical impact parameters separating a capture and scattering of photons.

Theories with extra dimensions admit astrophysical objects (supermassive black holes in particular) which are rather different from standard ones. There were proposed tests which may help to discover signatures of extra dimensions in supermassive black holes since the gravitational field may be different from the standard one in the GR approach. So, gravitational lensing features are different for alternative gravity theories with extra dimensions and general relativity. Some time ago, an opportunity that the black hole at the Galactic Center is described by the tidal Reissner– Nordström metric which may be admitted by the Randall–Sundrum II braneworld scenario has been discussed.[19] Bin-Nun suggested an opportunity of evaluating the black hole metric analyzing (retro-)lensing of bright stars around the black hole in the Galactic Center. A minimal size of a spot for the black hole at the Galactic Center has been evaluated.[20] According to a theoretical consideration and simulations a minimal size of spot practically has to coincide with the shadow size.[14,15]

We derive an analytic expression for the black hole shadow size as a function of charge for the tidal Reissner– Nordström metric. We conclude that observational data concerning shadow size measurements are not consistent with significant negative charges, in particular, the significant negative charge $Q/(4M^2) = -1.6$ (discussed in Ref. 19) is practically ruled out with a very high probability (the charge is roughly speaking is beyond 9σ confidence level, but a negative charge is beyond 3σ confidence level).[23] We could evaluate a shadow size for the black hole at the Galactic Center assuming that the black hole mass is about $4 \times 10^6 M_\odot$ and a distance toward the Galactic Center is about 8 *kpc*. In this case a diameter of shadow is about 52 μas for the Schwarzschild metric and about 40 μas for the extreme Reissner–Nordström metric.

A size of the smallest spot near the black hole at the Galactic Center such as 37^{+16}_{-10} microarcseconds at a wavelength of 1.3 mm with 3σ confidence level has been evaluated.[20] Theoretical analysis and observations show that the size of shadow can not be smaller than a minimal spot size at the wavelength.[14,15,18,21] Roughly speaking, it means that a small positive q is consistent with observations but a significant negative q is not. For $q = -6.4$ (as it was suggested in Ref. 19) we have a shadow size 84.38 μas. It means that the shadow size is beyond of shadow size with a probability corresponding to a deviation about 9σ from an expected shadow size. Therefore, a probability to have so significant tidal charge for the black hole at the Galactic Center is negligible. So, we could claim that the tidal charge is ruled out with observations and corresponding theoretical analysis.[22,23]

A positive Λ-term could also reduce a shadow size since we have for shadow size in the Kottler (Schwarzschild – de-Sitter) metric[24]

$$\xi^2_{\rm cr} = \frac{27}{1 - 9\Lambda M^2}. \qquad (1)$$

3. Constraints on Black Hole Parameters and Gravity Theories from Trajectories of Bright Stars at the Galactic Center

As it was mentioned earlier an enormous progress in monitoring bright stars near the Galactic Center has been reached. The astrometric limit for bright stellar sources near the Galactic Center with 10 meter telescopes is today $\delta\theta_{10} \sim 1$ mas and the Next Generation Large Telescope (NGLT) will be able to improve this number at least down to $\delta\theta_{30} \sim 0.5$ mas or even to

$\delta\theta_{30} \sim 0.1$ mas in the K-band (see also perspectives for observations with GRAVITY facilities and TMT[c] or E-ELT.[d] Therefore, it will be possible to measure the proper motion for about ~ 100 stars with astrometric errors several times smaller than errors in current observations.

GR predicts that orbits about a massive central body suffer peribothron shifts yielding rosette shapes. However, the classical perturbing effects of other objects on inner orbits give an opposite shift[25–28] (the effect is rather general and it weakly depends on a choice of extended mass distribution between peribothron and apobothron). Since the peribothron advance depends strongly on the compactness of the central body, the detection of such an effect may give information about the nature of the central body itself. This would apply for stars orbiting close to the GC, where there is a "dark object", the black hole hypothesis being the most natural explanation of the observational data. A cluster of stars in the vicinity of the GC (at a distance < 1 arcsec) has been monitored by ESO and Keck teams for several years.

Based on results of simulations, one could conclude that now even around 5% of the BH mass (or around $2 \times 10^5 M_\odot$) distributed in bulk between apobothron and peribothron could lead to measurable shift for S2 star orbit,[25,26] so one could constrain total extended mass distribution for the stellar cluster and dark matter concentration near the Galactic Center. These conclusions are weakly depend on assumptions about specific extended mass distributions of stellar cluster and dark matter concentrations. Therefore, in the future precise astrometrical measurements of bright star trajectories could be treated as a tool for indirect searches of dark matter or these measurements could rule out the hypothesis about the significant concentration of dark matter near the Galactic Center (astroparticle physicists assumed that γ-flux from the Galactic Center region is caused by neutralino annihilation).

Similarly to the Mercury anomaly, dark matter and dark energy problems may be explained by a change of the fundamental gravity law, for instance, one could change Einstein – Hilbert Lagrangian R with a function $f(R)$ and we have $f(R) = R$ for the standard GR.[29] For instance, in the framework of $f(R) = R^n$ (we have GR limit for $n = 1$), one could explain acceleration of Universe without dark energy and rotation curves for spiral galaxies without dark matter, however, we have $n \approx 1$ from Solar sys-

---

[c]http://www.tmt.org/sites/default/files/documents/application/pdf/srd%20ccr20_final.pdf.

[d]http://www.eso.org/sci/facilities/eelt/science/doc/eelt_sciencecase.pdf.

tem data.[30] One could arrive at the same conclusion from S2 orbit data[31] (see also, Ref. 32, where an extended distribution near the Galactic Center has been taken into account).

In paper[33] parameters of Yukawa gravity from the S2 star trajectory have been constrained and it was shown that the Yukawa potential may be a weak gravitational field limit for a wide class of alternative gravitational theories

$$\Phi(r) = -\frac{GM}{(1+\delta)r}\left[1+\delta e^{-\left(\frac{r}{\Lambda}\right)}\right], \tag{2}$$

where Λ is a parameter with a length dimension and δ is a dimensionless constant.

It was obtained the most probable Λ in the case of S2 star, is around 5000 - 7000 AU and that from the current observations it is very hard to obtain the reliable constraints on the universal constant δ and it was found also the same universal constant $\delta = 1/3$ which was successfully applied to clusters of galaxies and rotation curves of spiral galaxies also gives a good agreement in the case of observations of S2 star orbit.[33]

4. Conclusions

One can conclude that there are tensions between a size of the smallest spot at the Galactic Center and an expected shadow size, therefore one should use Reissner – Nordström or/and the Kottler (Schwarzschild – de-Sitter) metrics or we have to conclude that there are systematic errors. Concerning the best fits for trajectories S2 like stars with alternative theories of gravity, one concludes that R^n is practically has to be ruled out with the observational data, there are hints for the Yukawa potential from an analysis of these data, because the Yukawa potential provides a slightly better fit in comparison the Newtonian fit.

Acknowledgments

The work was supported in part by the NSF (HRD-0833184) and NASA (NNX09AV07A) at NASA CADRE and NSF CREST (NCCU, Durham, NC, USA) and RFBR 14-02-00754a at ICAD of RAS (Moscow).

References

1. P.G. Bergmann, *Introduction to the theory of relativity*, (New York, Prentice-Hall, 1942).

2. C. Will, *Living Reviews in Relativity*, **17**, 4 (2014).
3. J. Kormendy, L. C. Ho, *ARAA* **51**, 511 (2013).
4. X.-B. Wu, F. Wang, X.-H. Fan *et al.*, *Nature* **518**, 512 (2015).
5. S. Gillessen, *et al.*, *Nature* **481**, 51 (2012).
6. J. Frank, M.J. Rees, *MNRAS* **176**, 633 (1976).
7. K. Phifer, T. Do, L. Meyer, A.M. Ghez, *et al.*, *ApJ*, **773**, L13 (2013).
8. M. Zajaček, V. Karas, A. Eckart, *A & A*, **565**, A17 (2014).
9. M. Zajaček, A.M. Eckart, F. Peissker, G.D. Karssen, V. Karas, arXiv:1507.00237v1[astro-ph.GA] (2015).
10. M. Valencia-S., A. Eckart, M. Zajaček, *et al.*, *ApJ*, **800**, 125 (2015).
11. O. Pfuhl, S. Gillessen, F. Eisenhauer, *et al.*, *ApJ* **798**, 111 (2015).
12. F. Eisenhauer *et al.*, *GRAVITY: Observing the Universe in Motion*, *The Messenger* **143**, 16 (2011).
13. N. Blind et al. 2015, arXiv:1503.07303 [astro-ph.IM] (2015).
14. H. Falcke, F. Melia, E. Agol, *ApJ* **528**, L13 (2000).
15. F. Melia & H. Falcke, *ARAA* **39**, 309 (2001).
16. S. Chandrasekhar, *Mathematical Theory of Black Holes*, (Clarendon Press, Oxford, 1983).
17. D. Holz and J.A. Wheeler, *ApJ* **578**, 330 (2002).
18. A.F. Zakharov, *et al.*, *New Astron.* **10**, 479 (2005).
19. A.Y. Bin-Nun, *Phys. Rev.* D **81**, 123011 (2010).
20. S.S. Doeleman *et al.*, *Nature* **455**, 78 (2008).
21. A.F. Zakharov, A.A. Nucita, F. De Paolis, G. Ingrosso, *A & A* **442**, 795 (2005).
22. A.F. Zakharov, *et al.*, *New Astron. Rev.* **56**, 64 (2012).
23. A.F. Zakharov, *Phys. Rev.* D **90**, 062007 (2014).
24. A.F. Zakharov, arXiv:1407.2591[astro-ph.GA] (2014).
25. A.A. Nucita, *et al.*, *PASP* **119**, 349 (2007).
26. A.F. Zakharov, *et al.*, *Phys. Rev.* D **76**, 062001 (2007).
27. A.F. Zakharov, *et al.*, *Phys. At. Nucl.* **73**, 1870 (2010).
28. A.F. Zakharov, *et al.*, *Space Sci. Rev.* **48**, 301.
29. S. Capozziello, *Int. J. Mod. Phys.* D, **11**, 483 (2002).
30. A.F. Zakharov, *et al.*, *Phys. Rev.* D **74**, 107101 (2006).
31. D. Borka, *et al.*, *Phys. Rev.* D **85**, 124004 (2012).
32. A.F. Zakharov, *et al.*, *Adv. Space Res.*, **54**, 1108 (2014).
33. D. Borka, *et al.*, *JCAP* **11** 050 (2013).

Section 1

Classical Gravity and GR Extensions

Accretion disks around a mass with quadrupole

M. Abishev, K. Boshkayev, H. Quevedo and S. Toktarbay*
Physical-Technical Faculty, Al-Farabi Kazakh National University,
Al Farabi av. 71, 050040 Almaty, Kazakhstan
Instituto de Ciencias Nucleares, Universidad Nacional Autonoma de México,
AP 70543, México, DF 04510, Mexico
**E-mail: saken.yan@yandex.com*

We consider the stability properties of circular orbits of test particles moving around a mass with quadrupole. We show that the quadrupole modifies drastically the properties of an accretion disk made of such test particles.

Keywords: Quadrupole; compact objects; geodesics.

The simplest generalization of the Schwarzschild metric which contains a quadrupole parameter q is given by

$$ds^2 = \left(1 - \frac{2m}{r}\right)^{1+q} dt^2 - \left(1 - \frac{2m}{r}\right)^{-q}$$
$$\left[\left(1 + \frac{m^2 \sin^2\theta}{r^2 - 2mr}\right)^{-q(2+q)} \left(\frac{dr^2}{1 - \frac{2m}{r}} + r^2 d\theta^2\right) + r^2 \sin^2\theta d\varphi^2\right]. \quad (1)$$

This solution is known as the $\delta-$metric or as the $\gamma-$metric and was first obtained by Zipoy and Voorhees.[1] We propose to use the term quadrupole metric ($q-$metric) to emphasize the role of the parameter q. The $q-$metric is an axially symmetric exact vacuum solution, and reduces to the Schwarzschild metric for $q \to 0$. It is asymptotically flat with a central curvature singularity at $r = 0$ and an outer singularity at $r = 2m$ which is naked. According to the Geroch definition, the independent multipole moments are the monopole $M_0 = m(1+q)$ and the quadrupole $M_2 = -\frac{m^3}{3}q(1+q)(2+q)$. For more details, see Ref. 2.

As a first approximation, an accretion disk can be considered as a set of test particles moving along circular orbits around the central mass. In this case, the geodesic equations on the equatorial plane are equivalent to the equations for the motion in the effective potential[3]

$$V_{eff}^2(r,q) = \left(1 - \frac{2m}{r}\right)^{q+1} \left[\frac{l^2}{r^2}\left(1 - \frac{2m}{r}\right)^q + \epsilon\right], \quad (2)$$

where E and l are constants of motion. The radius and stability properties of circular orbits are completely determined by the behavior of the effective potential which, in turn, depends on the behavior of the first and second radial derivatives. We performed a detailed analysis of the behavior of the effective potential. The result of this study is summarized in Fig. 1. The region of stability determines the spatial region where an accretion disk can exist, and the radius of the last stable circular orbit is interpreted as the minimum inner radius of the disk.

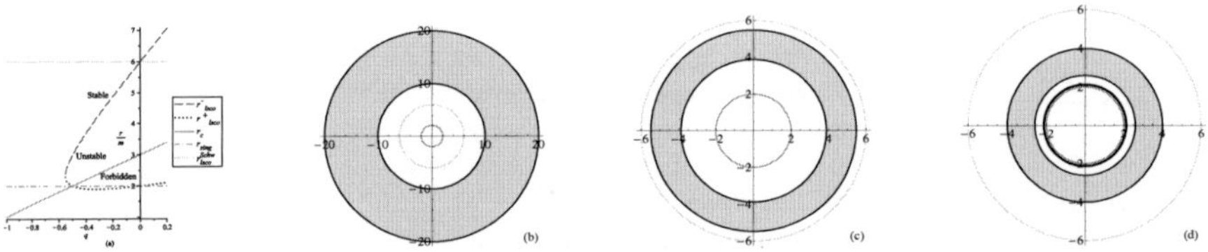

Fig. 1. (a) Radius of the last stable circular orbit as a function of the quadrupole. Below the critical radius $r_c = m(3 + 2q)$, no motion is allowed. The outer singularity $r_{sing} = 2m$ is also plotted. Accretion disks are illustrated for positive q in (b), for $-0.5 > q > 0$ in (c) and for $q = -0.52$ in (d).

We see that for positive values of q the accretion disk is always located outside the Schwarzschild radius of the last stable circular orbit $r_{lsco}^{Sch} = 6m$. For negative values of q, the disk can be completely inside the radius $r_{lsco}^{Sch} = 6m$. Finally, for values of q close to -0.5, a second inner disk appears in a region very closed to the outer singularity $r_{sing} = 2m$.

We conclude that the quadrupole parameter can modify drastically the geometric structure of accretion disks. Therefore, it should be possible to determine the value of q from the geometric properties of the disk.

We acknowledge the support through a Grant of the Target Program of the MES of the RK, Grant No. 3101/GF4 IPC-11/2015, DGAPA-UNAM, Grant No. 113514, and Conacyt, Grant No. 166391.

References

1. D. M. Zipoy, J. Math. Phys. **7** (1966) 1137; B. Voorhees, Phys. Rev. D **2** (1970) 2119.
2. H. Quevedo, Int. J. Mod. Phys. D **20**, 1779 (2011).
3. K. Boshkayev, E. Gasperin, A.C. Gutierrez-Pineres, H. Quevedo and S. Toktarbay, *Motion of test particles in the field of a naked singularity*, arXiv: gr-qc/1509.03827.

Orbital stability of the restricted three-body problem in General Relativity

M. Abishev[1], H. Quevedo[2], S. Toktarbay[1*] and B. Zhami[1]

Physical-Technical Faculty, Al-Farabi Kazakh National University, Al Farabi av. 71, 050040 Almaty, Kazakhstan

Instituto de Ciencias Nucleares, Universidad Nacional Autónoma de México, AP 70543, México, DF 04510, Mexico

**E-mail: saken.yan@yandex.com*

We consider the problem of orbital stability of the motion of a test particle in the restricted three-body problem, by using the orbital moment and its time derivative. We show that it is possible to get some insight into the stability properties of the motion of test particles, without knowing the exact solutions of the motion equations.

Keywords: Restricted three-body problem; orbital stability; vector elements.

The relativistic equation of translational motion of a test body in the field of two bodies in general relativity (GR) mechanics, corresponding to the circular restricted three-body problem was investigated in Ref. 1. The evolution equations of motion can be studied by using the asymptotic methods of adiabatic theory, according to which the vector elements $\vec{M}$ (orbital moment) and $\vec{A}$ (Laplace vector) provides information about the stability properties of a test body orbit.[2,3]

If we assume that the relativistic corrections to the motion of a test particle satisfy following limits

$$U_1 << c^2, \quad U_2 << U_1, \quad U_1/U_2 \approx v^2/c^2, \tag{1}$$

where U_1 and U_2 are the potentials of the central and second body, respectively.

The Hamilton function of the system can be written as[1]

$$H = H_0 + H_{rel}, \tag{2}$$

where H_0 and H_{rel} are the classical and relativistic Hamilton functions. We can find the evolution equations of motion of the test particle (third body), which describes the average change of its orbital momentum

$$\vec{M} = [\vec{r}_3, \vec{p}_3] = \vec{M}_0 + \vec{M}_{rel}, \tag{3}$$

and its time derivative

$$\dot{\vec{M}} = \left[\dot{\vec{r}}_3, \vec{p}_3\right] + \left[\vec{r}_3, \dot{\vec{p}}_3\right] = \dot{\vec{M}}_0 + \dot{\vec{M}}_{rel}, \tag{4}$$

where

$$\dot{\vec{r}}_3 = \frac{\partial H}{\partial \vec{p}_3}, \quad \dot{\vec{p}}_3 = -\frac{\partial H}{\partial \vec{r}_3}. \tag{5}$$

In order to obtain the evolution equations of motion one needs to integrate the Eq. (4) for a repetition period of the system configurations T (synodic period of the test body):

$$\overline{\dot{\vec{M}}} = \frac{1}{T}\int_0^T \dot{\vec{M}} dt. \tag{6}$$

It is clear that if the test particle is moving on a circular orbit, then the integral (6) is equal to zero[1]. If the particle is moving in nearly circular orbit with small values of the eccentricity, the equations of motion (5) and (6) are very complicated to be solved analytically. However, if we assume that r_{kep}

$$r_{kep} = \frac{p}{1 + e\cos\varphi} \tag{7}$$

is the approximate solution of the restricted three-body problem, then by using numerical methods, we can integrate Eq.(6) approximately. where e is the eccentricity, p is the parameter of the orbit. In this manner, we can analyze the stability properties of test particles. Thus, we avoid the difficult problem of finding exact solutions of the three-body problem and, we analyze the stability properties of an entire class of approximate solutions. In this way, we expect to get some insight into the physical properties of the motion in the field of a three-body system.

We acknowledge the support through a Grant of the Target Program of the MES of the RK, Grant No. F. 0679, 2015.

References

1. M. E. Abishev, S. Toktarbay and B. A. Zhami, *Grav.Cos.* **20**, 3 (2014).
2. L.D. Landau, E. M. Lifshitz. *The Classical Theory of Fields.* Vol. 2 (4th ed.) (1975).
3. M. M. Abdildin, *Mechanics of Einstein Theory of Gravity* (Nauka, 1988).

Rotating hot white dwarfs

K. A. Boshkayev,[1,2] J. A. Rueda[1,2] and B. A. Zhami[1,*]

[1] *Department of Physics and Technology, Al-Farabi Kazakh National University, Al-Farabi avenue 71, Almaty, 050040, Kazakhstan*

[2] *International Center for Relativistic Astrophysics Network, Piazza della Repubblica 10, Pescara, I-65122, Italy*

[*] *E-mail: zhami.bakytzhan@gmail.com*

We consider the effects of rotation and temperature on the structure of white dwarfs in order to compare them with the estimated data from observations.

Keywords: Rotating white dwarfs; hot white dwarfs; general relativity.

In this work we construct the mass-radius relation for white dwarfs (WDs) using the Chandrasekhar equation of state (EoS) within general relativity. First we perform computations for zero temperature uniformly rotating WDs at the mass shedding limit within the Hartle formalism[1] and calculate the masses and average radii. Afterwards we superpose our results with the estimated mass-radius relations obtained from the Sloan Digital Sky Survey Data Release 4 (SDSS 4) by Tremblay et al.[2] As one can see on the left panel of Fig. 1 the difference between theory and observations is quite noticeable.

In order to overcome this problem S. M. de Carvalho et al.[3] proposed to include the finite-temperature effects in the EoS. Following this idea we performed similar analysis for static WDs at finite-temperatures by solving the Tolman-Oppenheimer-Volkoff equation (see Ref. 4 for details). The results of Ref. 4 are shown on the right panel of Fig. 1. As one can see by the inclusion of the temperature in the EoS one can cover all the estimated data from observations.

It should be stressed that from the observations usually one infers the effective surface temperature and the surface gravity of WDs. All the rest parameters are estimated by using certain models. However there also exist techniques to measure the masses of WDs in close eclipsing binaries. The data inferred from close binaries are more reliable. Therefore in order to perform more realistic computations one needs to take into account the effects of rotation and temperature together for selected WDs with known

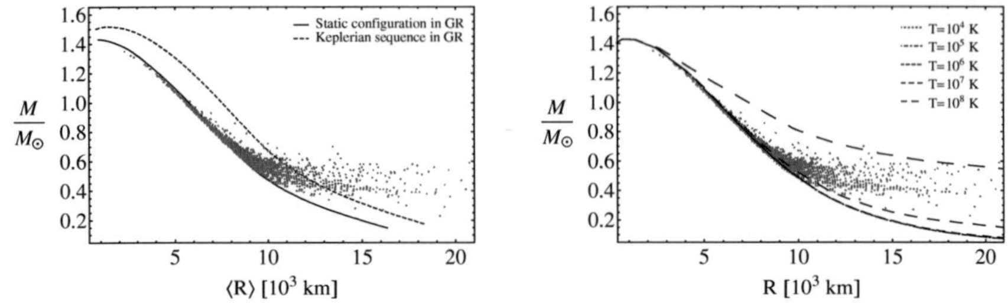

Fig. 1. Mass-radius relation of uniformly rotating WDs obtained with the Chandrasekhar EoS for T = 0 K case (left panel) and mass-radius relation of static WDs for selected finite temperatures from T $=10^4$ K to T = 10^8 K (right panel) and their superposition with the estimated masses and radii of white dwarfs taken from the SDSS 4 (blue dots).

parameters. Only after that one can make further predictions. This issue is out of the scope of the present work and will be published elsewhere.

In conclusion, we calculated the masses and radii of cold rotating and hot static white dwarfs in general relativity i.e. the effects of the rotation and finite-temperature were considered separately. We compared and contrasted our results with the estimated data from the observations of WDs. Our results cover all the data. For more detailed analysis one needs to consider both effects together and work with more or less model-indepent data.

The work was partially supported by Grant No. 3101/GF4 IPC-11/2015 of the MES of the RK.

References

1. K. A. Boshkayev, J. A. Rueda, R. Ruffini and I. Siutsou, On general relativistic uniformly rotating white dwarfs, *Astrophys. J.* **762**, 1 (2013).
2. P.-E. Tremblay, P. Bergeron and A. Gianninas, An improved spectroscopic analysis of DA White Dwarfs from the Sloan Digital Sky Survey Data Release 4, *Astrophys. J.* **730**, 128 (2011).
3. S. M. de Carvalho, M. Rotondo, J. A. Rueda and R. Ruffini, Relativistic Feynman-Metropolis-Teller treatment at finite temperatures, *Astrophys. J.* **730**, 015801 (2014).
4. K. A. Boshkayev, J. A. Rueda, B. A. Zhami, Zh. A. Kalymova and G. Sh. Balgymbekov, Equilibrium structure of white dwarfs at finite temperatures, *Int. J. Mod. Phys.: Conf. Ser.* (in press).

Spherically symmetric solution in Cartan-Weyl space with Dirac scalar field

O. V. Babourova*, B. N. Frolov*, P. E. Kudlaev and E. V. Romanova

Moscow Pedagogical State University, Institute of Physics, Technology and Information Systems,
M. Pirogovskaya ul. 29, Moscow 119992, Russian Federation
**E-mail: baburova@orc.ru, frolovbn@orc.ru*

The static spherically symmetric solutions of the field equations of the conformal theory of gravitation in Cartan–Weyl spacetime with Dirac scalar field are obtained. The metrics of this solution coincides with the Papapetou–Yilmaz–Rosen metrics. The Dirac scalar field is considered as a model of dark matter.

Keywords: Cartan–Weyl spacetime; Dirac scalar field; exterior form variational formalism; Papapetou–Yilmaz–Rosen metrics.

Let us consider a connected 4D oriented differentiable manifold $\mathcal{M}$ equipped with a metric $\hat{g}$ of the index 3, a linear connection $\Gamma^a{}_b$ and a volume 4-form η. Then a Cartan–Weyl space $\mathcal{C}W_4$ is defined as the such manifold equipped with a curvature 2-form $\mathcal{R}^a{}_b$, a torsion 2-form $\mathcal{T}^a$ and a nonmetricity 1-form $\mathcal{Q}_{ab}$ obeing the Weyl condition

$$\mathcal{Q}_{ab} = \frac{1}{4} g_{ab} \mathcal{Q} \,. \qquad (1)$$

Here $\mathcal{Q}_{ab} = -\mathcal{D} g_{ab}$, and $\mathcal{D} = \mathrm{d} + \Gamma \wedge \ldots$ is the exterior covariant differential.

In Ref. 1 the Poincaré–Weyl gauge theory of gravitation (PWTG) has been developed. The gauge field introduced by the subgroup of dilatations is named as dilatation field, its vector-potential is the Weyl 1-form, and quanta of this field can have nonzero rest masses. As it has been shown in Ref. 1, an additional scalar field $\beta(x)$ is introduced in PWTG as an essential geometrical addendum to the metric tensor, the tangent space metrics being the form,

$$g_{ab} = \beta^2 g^M_{ab} \,, \qquad (2)$$

where g^M_{ab} are the constant components of the Minkowski metric tensor.

The properties of the field $\beta(x)$ coincide with those of the scalar field introduced by Dirac and earlier by S. Deser, see Refs. 2, 3. Members of the Dirac scalar field Lagrangian have structure of the Higgs Lagrangian and can cause an appearance of nonzero rest masses of particles.

On the basis of PWTG, the conformal theory of gravitation in Cartan–Weyl spacetime with Dirac scalar has been developed[4–8] with the Lagrangian density 4-form (in exterior form formalism),[6,7]

$$\mathcal{L} = \mathcal{L}_G + \mathcal{L}_M + \beta^4 \Lambda^{ab} \wedge \left(\mathcal{Q}_{ab} - \frac{1}{4} g_{ab} \mathcal{Q} \right) , \tag{3}$$

$$\begin{aligned}\mathcal{L}_G = 2f_0 \Big[& \frac{1}{2}\beta^2 \mathcal{R}^a{}_b \wedge \eta_a{}^b - \beta^4 \Lambda \eta + \frac{1}{4}\lambda \mathcal{R}^a{}_a \wedge * \mathcal{R}^b{}_b \\ & + \rho_1 \beta^2 \mathcal{T}^a \wedge * \mathcal{T}_a + \rho_2 \beta^2 (\mathcal{T}^a \wedge \theta_b) \wedge *(\mathcal{T}^b \wedge \theta_a) \\ & + \rho_3 \beta^2 (\mathcal{T}^a \wedge \theta_a) \wedge *(\mathcal{T}^b \wedge \theta_b) + \xi \beta^2 \mathcal{Q} \wedge * \mathcal{Q} + \zeta \beta^2 \mathcal{Q} \wedge \theta^a \wedge * \mathcal{T}_a \\ & + l_1 \mathrm{d}\beta \wedge * \mathrm{d}\beta + l_2 \beta \mathrm{d}\beta \wedge \theta^a \wedge * \mathcal{T}_a + l_3 \beta \mathrm{d}\beta \wedge * \mathcal{Q} \Big] . \end{aligned} \tag{4}$$

Here $\mathcal{L}_G$ is the gravitational field Lagrange density, $\mathcal{L}_M$ is the matter Lagrange density. The first term in $\mathcal{L}_G$ is the Gilbert–Einstein Lagrangian density generalized to the Cartan–Weyl space ($\eta_a{}^b = *(\theta_a \wedge \theta^b)$, $f_0 = c^4/16\pi G$), the second term is a generalized cosmological term describing a vacuum energy (Λ is the Einstein cosmological constant). The last term in (3) contains the Weyl's condition (1) with the Lagrange multiplier Λ^{ab}.

We use the exterior form variational formalism on the base of the Lemma on the commutation rule between variation and Hodge star dualization.[9] The independent variables are the nonholonomic connection 1-form $\Gamma^a{}_b$, the basis 1-form θ^a, and the Dirac scalar field $\beta(x)$. The variational field equations are the Γ-equation, θ-equation and the β-equation. These equations can be found in Refs. 7, 8.

The variational field equations of this theory have been solved at the very early stage of evolution of universe for the scale factor $a(t)$ and the field $\beta(t)$, when a matter density has been very small.[6–8] These solutions realize exponential diminution of the field β, and thus sharp exponential decrease of physical vacuum energy (dark energy) by many orders. Thus this result can explain the exponential decrease in time at very early Universe of the dark energy describing by the effective cosmological constant. This can give a way to solving the one of the fundamental problems of the modern theoretical physics – the problem of cosmological constant (see Refs. 10–12) – as a consequence of fields dynamics at the early Universe.

Now a static spherically symmetric solution of the field equations in vacuum (in case of $\Lambda = 0,\ \lambda = 0$) is obtained for a central compact mass m.[8,13]

In the spherically symmetric case the torsion 2-form is,

$$\mathcal{T}^a = \frac{1}{3}\mathcal{T} \wedge \theta^a \qquad \mathcal{T} = *(\theta_a \wedge *\mathcal{T}^a)\,, \tag{5}$$

were $\mathcal{T}$ is a torsion trace 1-form.

After anti-symmetrization the Γ-equation on indices a, b, we obtain an equation that does not contain the Lagrange multiplier Λ^{ab}. After an exterior multiplication of this equation by θ_b, the consequence results,[7]

$$2(1 - \rho_1 + 2\rho_2)\mathcal{T} - 3(\frac{1}{4} + \zeta)\mathcal{Q} + 3(2 - l_2)\mathrm{d}\ln\beta = 0. \tag{6}$$

It can be proved that as a consequence of the condition (6) the anti-symmetrization part of the Γ-equation is fulfilled in the spherically symmetric case.

After taking the trace of the Γ-equation, another consequence of the Γ-equation is obtained,

$$(2\rho_1 - 4\rho_2 + 8\zeta)\mathcal{T} + (16\xi + 3\zeta)\mathcal{Q} + (3l_2 + 8l_3)\mathrm{d}\ln\beta = 0. \tag{7}$$

One can conclude that, as a consequence of the conditions (6) and (7), the torsion 1-form $\mathcal{T}$ and the nonmetricity 1-form $\mathcal{Q}$ can be realized as

$$\mathcal{T} = s\mathrm{d}U\,, \qquad \mathcal{Q} = q\mathrm{d}U\,, \qquad U = \log\beta\,, \tag{8}$$

where s and q are arbitrary numbers.

We shall find a static spherically symmetric solution with a metrics in the form

$$ds^2 = e^{-2U(r)}\left(e^{-\mu(r)}dt^2 - e^{\mu(r)}(dr^2 + r^2(d\theta^2 + \sin^2\theta\, d\phi^2))\right). \tag{9}$$

After calculation the Γ-, θ- and β-equations with the help of this metrics, one can conclude that these equations are reduced to the following equations,

$$\mu'' + \frac{2}{r}\mu' = 0\,, \qquad U' = \pm\frac{k}{2}\mu'\,, \qquad k = \sqrt{k^2} > 0\,, \tag{10}$$

with $q = -8$, s is arbitrary, and $k^{-2} = l_1$ under some conditions on the coupling constants of the Lagrangian density 4-form (4).

The equations (10) have solutions,

$$\mu = \frac{r_0}{r}\,, \qquad \beta(r) = \beta_0\, e^{\pm\frac{kr_0}{2r}}\,, \tag{11}$$

which leads to the metrics

$$ds^2 = e^{\mp\frac{kr_0}{r}} ds^2_{PIR}\,, \tag{12}$$
$$ds^2_{PIR} = e^{-\frac{r_0}{r}} dt^2 - e^{\frac{r_0}{r}}(dr^2 + r^2(d\theta^2 + \sin^2\theta\, d\phi^2))\,. \tag{13}$$

With the help of the conformal transformation

$$\breve{g}_{\alpha\beta} = e^{\mp\frac{kr_0}{r}} g_{\alpha\beta}\,, \tag{14}$$

the metrics (12) can be transformed to the metrics (13), the Cartan–Weyl space being transformed to the Riemann–Cartan space.

In Ref. 14 the modified variational formalism is advanced, according to which the scalar field β is not entered in the theory by "hands" irrespective of the metric (how it was carried out in the Lagrangian density 4-form (4), but this scalar field is entered in a Lagrangian density as a representation of a tangent space metric as (2) with the help of a method of Lagrange uncertain multipliers. In this case properties of this scalar field β are substantially determined by ideas of the old Weyl's gauge theory, and this field is expedient for naming as Weyl-Dirac scalar field.

Now the Lagrangian density 4-form of the theory we shall present in vacuum as

$$\mathcal{L} = \mathcal{L}_G + \Lambda^{ab} \wedge (\mathcal{Q}_{ab} - (1/4)g_{ab}\mathcal{Q}) + \lambda^{ab}(g_{ab} - \beta^2 g^M_{ab}\eta\,, \tag{15}$$
$$\begin{aligned}\mathcal{L}_G = 2f_0\,\big[&(1/2)\mathcal{R}^a{}_b \wedge \eta_a{}^b + \rho_1 \mathcal{T}^a \wedge *\mathcal{T}_a \\ &+\rho_2(\mathcal{T}^a \wedge \theta_b) \wedge *(\mathcal{T}^b \wedge \theta_a) + \rho_3(\mathcal{T}^a \wedge \theta_a) \wedge *(\mathcal{T}^b \wedge \theta_b) \\ &+16\xi \mathcal{Q}_{ab} \wedge *\mathcal{Q}^{ab} + 4\zeta \mathcal{Q}_{ab} \wedge \theta^a \wedge *\mathcal{T}^b + l \mathrm{d}\ln\beta \wedge *\mathrm{d}\ln\beta\,\big]\,.\end{aligned} \tag{16}$$

In comparison with (4) here the term with cosmological constant is omitted as the local task will consider, and also a Lagrange density of external fields is omitted as the problem in emptiness will be solved here. Last term in (16) is added for maintenance of dynamics of the scalar field β.

In order to derive the equations of a gravitational field in vacuum it is necessary to vary (15) and (16) independently with respect to the non-holonomic connection 1-form $\Gamma^a{}_b$, the basis 1-form θ^a, the tangent space metrics g_{ab}, and theWeyl– Dirac scalar field $\beta(x)$. In a static spherically symmetric case, the torsion 1-form $\mathcal{T}$ and the nonmetricity 1-form $\mathcal{Q}$ are realized as (8), and the solutions of these field equations in vacuum are the same as (11) with $k^{-2} = l$, $q = -8$, s – arbitrary, and the metrics is (13).

The metrics (13) is known as the Papapetou–Yilmaz–Rosen metrics.[15–17] This metric at large distances gives the same results as the

Schwarzshild metric, if $r_0 = 2Gm/c^2$. We can conclude that the presence of the Weyl-Dirac scalar field, which always exists, because it has an equally fundamental status as the metric, eliminates the existence of the black hole solution in the stadart form. As a consequence of the solution obtained, the Weyl–Dirac scalar field intensive concentrates near massive objects. In Ref. 8 the hypothesis has been formulated that *the Weyl–Dirac scalar field is realized itself not only as the 'dark energy', but also as one of the components of the 'dark matter'.* The Higgs particle, which has been recently found at Large Hadron Collider (CERN), maybe in reality is a quantum of the Weyl–Dirac scalar field.[8]

References

1. O. V. Babourova, B. N. Frolov, and V. Ch. Zhukovsky, *Phys. Rev. D* **74**, 064012-1-12 (2006).
2. P. A. M. Dirac, *Proc. Poy. Sos.* **A333**, 403 (1973).
3. S. Deser, *Ann. Phys.* **59**, 248 (1970).
4. O. V. Baburova, R. S. Kostkin, and B. N. Frolov, *Russian Phys. J.* **54**, 121 (2011).
5. O. V. Babourova, B. N. Frolov, ArXive: 1112.4449 [gr-qc].
6. O. V. Babourova , K. N. Lipkin, and B. N. Frolov, *Russian Phys. J.* **55**, 855 (2012).
7. O. V. Babourova, B. N. Frolov, and K. N. Lipkin, *Gravit. Cosm.* **18**, 225 (2012).
8. O. V. Babourova, B. N. Frolov, *Mathematical foundations of the modern theory of gravitation* (MPGU, Moscow, 2012) (in Russian).
9. O. V. Babourova, B. N. Frolov, and E. A. Klimova, *Class. Quantum Grav.* **16**, 1149 (1999) (gr-qc/9805005).
10. S. Weinberg, *Rev. Mod. Phys.* **61**(1), 1–23 (1989).
11. P. J. E. Peebles and B. Ratra, *Rev. Mod. Phys.* **75**, 559–606 (2003) (astro-ph/0207347v2).
12. Padmanabhan T. *Phys. Rep.* **380**, 235–320 (2003).
13. O. V. Babourova, B. N. Frolov, and E. V. Febres, *Russian Phys. J.* **57**, 1297 (2015).
14. O. V. Babourova, B. N. Frolov, and E. V. Febres, *Russian Phys. J.* **58**, 283 (2015).
15. A. Papapetrou, *Z. Phys.* **139**, 518 (1954).
16. H. Yilmaz, *Phys. Rev.* **111**, 1417 (1958).
17. N. Rosen, *Ann. Phys. (N.Y.)* **84**, 455 (1974).

Parameterized post-Newtonian limit of Horndeski's gravity theory

M. Hohmann*

Laboratory of Theoretical Physics, Institute of Physics, University of Tartu, Tartu, Tartumaa, Estonia
**E-mail: manuel.hohmann@ut.ee*
http://kodu.ut.ee/~manuel/

We present a recent result on the parameterized post-Newtonian (PPN) limit of Horndeski's gravity theory and its consistency with solar system observations.

Keywords: Horndeski gravity; scalar-tensor gravity; parameterized post-Newtonian limit.

1. Method

The starting point of our calculation are the most general second order field equations of scalar-tensor gravity,[1] whose dynamical variables are the metric $g_{\mu\nu}$ and the scalar field ϕ and whose action contains four free functions K, G_3, G_4, G_5 of ϕ and its kinetic term $X = -\nabla_\mu\phi\nabla^\mu\phi/2$. We expand the fields $g_{\mu\nu} = \eta_{\mu\nu} + h_{\mu\nu}$ and $\phi = \Phi + \psi$ around their cosmological background values $\eta_{\mu\nu}, \Phi$. Further, we expand the free functions in a Taylor series

$$K(\phi, X) = \sum_{m,n=0}^{\infty} K_{(m,n)}\psi^m X^n\,, \tag{1}$$

and analogously for G_3, G_4, G_5. We then introduce post-Newtonian orders of magnitude for the field perturbations $h_{\mu\nu}, \psi$ and expand the field equations in these orders. We finally solve the field equations order by order.

2. Results

We now display the results obtained from applying the method shown in the previous section to Horndeski's gravity theory.[2] It turns out that the only PPN parameters which potentially deviate from their observed values are γ and β. In general they depend on the distance r between the gravitating mass and the test mass and take the form

$$\gamma(r) = \frac{2\omega + 3 - e^{-m_\psi r}}{2\omega + 3 + e^{-m_\psi r}}, \tag{2}$$

$$\begin{aligned}\beta(r) = 1 + \frac{1}{(2\omega + 3 + e^{-m_\psi r})^2}\Bigg\{&\frac{\omega + \tau - 4\omega\sigma}{2\omega + 3}e^{-2m_\psi r} \\ &+a(r)\left[e^{-m_\psi r}\ln(m_\psi r) - (m_\psi r + e^{m_\psi r})\,\mathrm{Ei}(-2m_\psi r) - \frac{1}{2}e^{-2m_\psi r}\right] \\ &+b(r)\left[e^{m_\psi r}\mathrm{Ei}(-3m_\psi r) - e^{-m_\psi r}\mathrm{Ei}(-m_\psi r)\right]\Bigg\},\end{aligned} \tag{3}$$

where we used the abbreviations

$$a(r) = (2\omega + 3)m_\psi r\,, \quad b(r) = \frac{6\mu r + 3(3\omega + \tau + 6\sigma + 3)m_\psi^2 r}{2(2\omega + 3)m_\psi}, \tag{4}$$

and introduced the constants

$$m_\psi = \sqrt{\frac{-2K_{(2,0)}}{K_{(0,1)} - 2G_{3(1,0)} + 3\frac{G_{4(1,0)}^2}{G_{4(0,0)}}}}, \quad \omega = \frac{G_{4(0,0)}}{2G_{4(1,0)}^2}\left(K_{(0,1)} - 2G_{3(1,0)}\right),$$

$$\sigma = \frac{G_{4(0,0)}G_{4(2,0)}}{G_{4(1,0)}^2}, \quad \tau = \frac{G_{4(0,0)}^2}{2G_{4(1,0)}^3}(K_{(1,1)} - 4G_{3(2,0)}), \quad \mu = \frac{G_{4(0,0)}^2 K_{(3,0)}}{G_{4(1,0)}^3}. \tag{5}$$

In the limit $m_\psi \to 0$ of a massless scalar field we obtain the values

$$\gamma = \frac{\omega + 1}{\omega + 2}, \quad \beta = 1 + \frac{\omega + \tau - 4\omega\sigma}{4(\omega + 2)^2(2\omega + 3)}, \tag{6}$$

which are independent of the interaction distance r.

Acknowledgments

The author gratefully acknowledges the full financial support of the Estonian Research Council through the Postdoctoral Research Grant ERMOS115 and the Startup Research Grant PUT790.

References

1. T. Kobayashi, M. Yamaguchi and J. 'i. Yokoyama, Generalized G-inflation: Inflation with the most general second-order field equations, *Prog. Theor. Phys.* **126**, 511 (2011) [arXiv:1105.5723 [hep-th]].
2. M. Hohmann, Parameterized post-Newtonian limit of Horndeski's gravity theory, *Phys. Rev. D* **92**, 064019 (2015) [arXiv:1506.04253 [gr-qc]].

Wormhole solutions in GR with two phantom scalar fields

A. Urazalina*, V. Dzhunushaliev and A. Makhmudov

IETP, al-Farabi KazNU, Almaty, 050040, Kazakhstan
Theor. Nucl. Phys. Dept, KazNU, Almaty, 050040, Kazakhstan
**E-mail: y.a.a.707@mail.ru*

Wormhole solutions for two scalar fields in general relativity are considered. The physical characteristics of such objects are discussed.

Keywords: Phantom scalar fields; wormholes.

1. Wormhole Solutions

At the present time it is widely believed that observed acceleration of our Universe is the consequence of some special form of matter – dark energy (DE). One of the key features of the latter is its ability to violate various energy conditions. In the most extreme case. DE is modeled by matter violating the null energy condition. Such matter is called exotic. One of DE models is phantom fields. Here we use two phantom scalar fields to construct wormholes.

We consider Einstein + scalar field equations in the form

$$G_\mu^\nu = 8\pi G T_\mu^\nu, \tag{1}$$

$$\frac{1}{\sqrt{-g}}\frac{\partial}{\partial x^\mu}\left[\sqrt{-g}\, g^{\mu\nu}\frac{\partial(\varphi,\chi)}{\partial x^\nu}\right] = -\frac{\partial V}{\partial(\varphi,\chi)}. \tag{2}$$

The potential for these two phantom scalar fields is

$$V(\phi,\chi) = \frac{\lambda_1}{4}(\phi^2 - m_1^2)^2 + \frac{\lambda_2}{4}(\chi^2 - m_2^2)^2 + \phi^2\chi^2 - V_0. \tag{3}$$

The metric for our wormhole solution filled with two phantom scalar fields is

$$ds^2 = B(r)dt^2 - dr^2 - A(r)(d\theta^2 + \sin^2\theta d\phi^2), \tag{4}$$

Using this metric, one can obtain Einstein equations:

$$A''A - \frac{1}{2}\left(\frac{A'}{A}\right)^2 - \frac{1}{2}\frac{A'}{A}\frac{B'}{B} = \varphi'^2 + \chi'^2 \, , \tag{5}$$

$$\frac{A''}{A} + \frac{1}{2}\frac{A'}{A}\frac{B'}{B} - \frac{1}{2}\left(\frac{A'}{A}\right)^2 - \frac{1}{2}\left(\frac{B'}{B}\right)^2 + \frac{B''}{B} = 2\left[\frac{1}{2}(\varphi'^2 + \chi'^2) + V\right] \tag{6}$$

The corresponding field equation for φ is (the equation for χ is analogous):

$$\varphi'' + \left(\frac{A'}{A} + \frac{1}{2}\frac{B'}{B}\right)\varphi' = \varphi\left[2\chi^2 + \lambda_1(\varphi^2 - m_1^2)\right] \, . \tag{7}$$

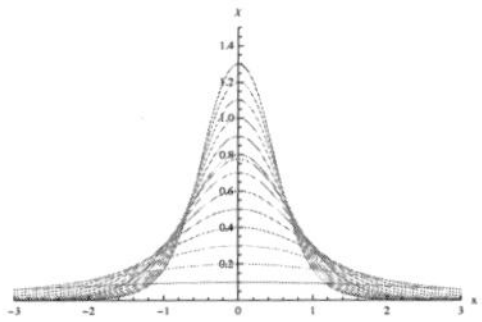

Fig. 1. $\chi(r)$

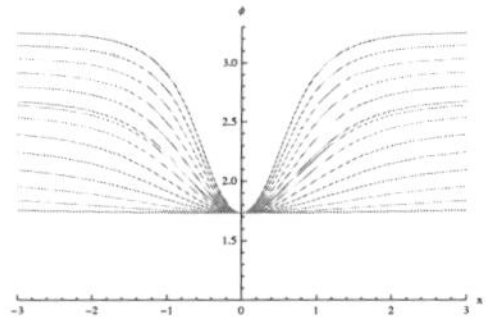

Fig. 2. $\phi(r)$

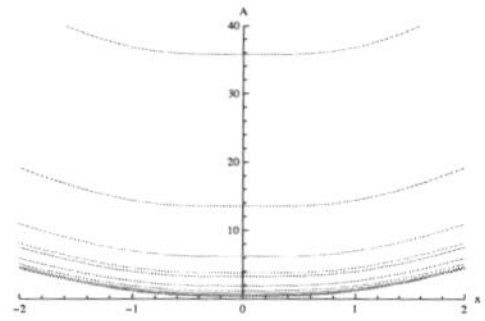

Fig. 3. $A(r)$

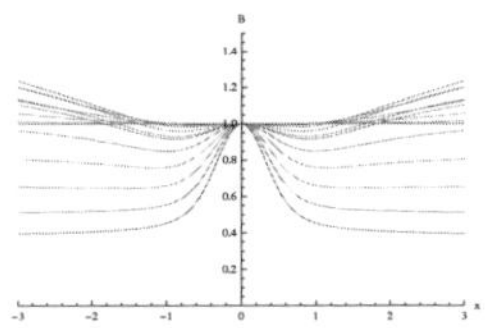

Fig. 4. $B(r)$

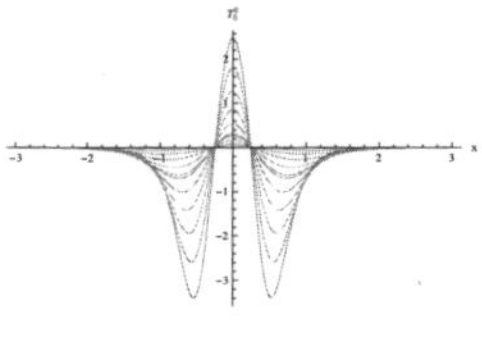

Fig. 5. The profile of energy density

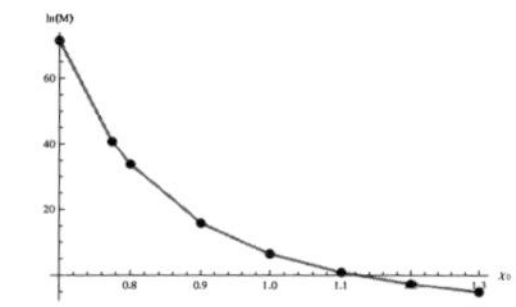

Fig. 6. The profile of the mass of wormhole vs χ_0

2. Conclusions

(a) Wormhole solutions in GR created with two phantom scalar fields are investigated; (b) solutions with different values of scalar field $\chi(0)$ at the origin are obtained; (c) the wormhole mass by different values of χ_0 is calculated.

References

1. V. Dzhunushaliev and V. Folomeev, "4D static solutions with interacting phantom fields," *Int. J. Mod. Phys. D* **17**, 2125 (2008); arXiv:0711.2840 [gr-qc].

Effective fields in the Fokker–Feynman–Wheeler scheme

S. V. Bolokhov[1] and V. V. Kolybasova[2]

Peoples' Friendship University of Russia, Miklukho–Maklaya st., 6
Moscow, 117198, Russia
[1]*E-mail: bol-rgs@yandex.ru*

Faculty of Physics, M.V. Lomonosov Moscow State University
Leninskie Gory, Moscow 119361, Russia
[2]*E-mail: kolybasova@phys.msu.ru*

We discuss the possible ways to generalize the so-called action-at-a-distance principle (initially developed for the electromagnetic interactions in the works of Fokker, Feynman, and Wheeler) on the case of gravity and non-Abelian fields. Our approach is based on the relational-statistical theory of spacetime and interactions developed by Yu. S. Vladimirov.

Keywords: Multidimensional gravity; anisotropic fluid; black hole; brane; exact solution; global structure.

1. Introduction

In recent years, various approaches was proposed to develop a background-independent theory of gravity and other interactions based on the idea of secondary nature of macroscopic spacetime (Loop Quantum Gravity, Penrose's twistor theory, causal sets theory, Vladimirov's relational approach, etc.). In such approaches, spacetime is usually supposed to be a result of more fundamental structures and processes in microscales (up to the Planckian scales).

One of the possible ways to obtain the correspondence with classical field theory in the macroscopic limit is the concept of an effective field in the spirit of the Fokker–Feynman–Wheeler "action-at-a-distance" theory initially developed for the electromagnetic interaction in Minkowski space[1,2]. In this scheme, the concept of classical field is modified in terms of direct relativistic-invariant interaction of particles on the null surface of light cone. This theory turns out to be equivalent to the Maxwell electrodynamics in the averaged classical limit.

The Fokker–Feynmann–Wheeler principle was extended on the case of linearized gravity in works of Granovsky and Pantiushin.[3] The gravitational interaction in this approach is described in terms of a geodesic motion of a system of particles in an effective Riemanian metric $\bar{g}_{\mu\nu}$ constructed from the Fokker-like action of this system. In works of Vladimirov and Turygin[4] this approach was extended to the nonlinear (Einstein) gravity via a perturbative scheme based on series in G.

We discuss the possible approaches to generalize The Fokker–Feynmann–Wheeler principle in the following cases:

(a) The linearized 5D gravity.
(b) Nonlinear 4D gravitation in the relational model of spacetime.[5]
(c) Non-Abelian fields in the framework of the Vladimirov—Turygin perturbative scheme.

2. Fokker's Action in 5D Minkowski Spacetime

Applying the principles of Granovsky and Pantiushin's approach to the five-dimensional case, it is possible to write the Fokker action of interaction for a system of N particles with masses μ_a $(a, b = 1, .., N)$ and 5-velocities u_a^A $(A = 0, 1, 2, 3, 5)$ embedded in flat spacetime $\mathbb{M}^5 = \mathbb{M}^4 \times \mathbb{R}$ with metric ${}^5g_{AB} = \begin{pmatrix} \eta_{\alpha\beta} & 0 \\ 0 & -1 \end{pmatrix}$:

$$
\begin{aligned}
{}^5S_{\text{int}} = \frac{G}{2} \sum_{a \neq b}^{N} \mu_a \mu_b \iint ds_a\, ds_b\, \delta[s^2_{(ab)}] \\
\times u_a^A u_a^B \Big({}^5g_{AC} {}^5g_{BD} + {}^5g_{AD} {}^5g_{BC} - \lambda {}^5g_{AB} {}^5g_{CD} \Big) u_b^C u_b^D, \qquad (1)
\end{aligned}
$$

where $s^2_{(ab)}$ is a square of relativistic interval between points on the world lines of particles a, b, and λ is a free parameter (for correspondence with Granovsky–Pantiushin's case one can take $\lambda = 2$). The integration is performed over world lines of the interacting particles.

It can be easily shown that the dimensional reduction $\mathbb{M}^5 \to \mathbb{M}^4$ through the metric splitting procedure leads to the following sum of contributions

$$
{}^5S_{\text{int}} \to {}^4S^{(\text{grav})}_{\text{int}} + {}^4S^{(\text{EM})}_{\text{int}} + {}^4S^{(\text{scalar})}_{\text{int}} \qquad (2)
$$

containing Granovsky–Pantiushin's action for linear gravity, well-known Fokker's 4D electromagnetism, and Fokker's 4D scalar interaction respectively.

3. Nonlinear 4D Gravitation in the Relational Scheme

As a generalization of constructions proposed in Ref. 5, one can write the following Fokker-like gravitational functional for a system of particles $\mathfrak{S}_N = \{i_1, ..., i_N\}$ with masses m_l in spacetime $\mathbb{M}^4$:

$$S(\mathfrak{S}_N) \equiv \sum_{n=2}^{4} \theta(N-n) \sum_{(i_1,...,i_n)\in\mathfrak{S}_N} \alpha_n S^{(n)}(i_1, ..., i_n). \tag{3}$$

Here the n-point contribution term is

$$S^{(n)}(i_1, ..., i_n) \equiv C_n \frac{G^{n-1}}{c^{2n-3}} \prod_{l=1}^{n} m_l \int \prod_{l=1}^{n} ds_l \Big(\det ||u_{ij}|| \sum_{\{1,...,n\}} \prod_{k=2}^{n} \delta[s^2_{(1k)}]\Big), \tag{4}$$

where α_n and C_n are constant phenomenological factors, $\det ||u_{ij}|| \equiv \det ||u^{\mu}_{(i)} u^{\nu}_{(j)} \eta_{\mu\nu}||_{i,j=1,...,n} \sim \mathrm{Vol}^2(\mathcal{H}_n)$, $\mathcal{H}_n \in T\mathbb{M}^4$ is the n-parallelotope on the 4-velocity vectors in tangent space $T\mathbb{M}^4$; notation $\{1, \ldots, n\}$ means the cyclic shift over n numbers.

The 2-point term corresponds to the Granovsky–Pantiushin case for linearized 4D gravity. The higher terms ($n > 2$) yield nonlinear corrections to the gravitational interaction picture.

Following this approach, one can consider (as a toy model) the circular motion of a test particle around a spherically-symmetric massive source in $\mathbb{M}^4$ and construct an effective Riemannian metric $\bar{g}_{\mu\nu}$ for this case in the 3-point approximation:

$$S(\mathfrak{S}_N)\big|_{n\leq 3} \longrightarrow \bar{S}^{\mathrm{eff}}[\mathrm{Riem}] = -mc \int \sqrt{\bar{g}_{\mu\nu} dx^{\mu} dx^{\nu}}. \tag{5}$$

It can be directly shown (see Ref. 6) that this procedure leads to the effective metric $\bar{g}_{\mu\nu}$ of the form

$$d\bar{s}^2 = \left(1 - \frac{2GM}{c^2 r} + \frac{2G^2M^2}{c^4 r^2}\right) dt^2 - \left(1 + \frac{2GM}{c^2 r}\right) (dr^2 + r^2 d\Omega^2) \tag{6}$$

which corresponds to the Post-Newtonian limit for the Schwarzschild-like metric in isotropic coordinates.[7]

4. Multidimensional Case and Non-Abelian Fields

Here we very briefly discuss the possible way to include non-Abelian fields into the Fokker scheme applied to the gravity.

Given the system of particles $\mathfrak{S}_N = \{i_1, ..., i_N\}$ in the $(4+d)$ dimensional manifold $\mathbb{M}^{4+d}$ with a background (maybe Minkowski) metric $\overset{0}{g}_{AB}$, consider the corresponding Fokker action using Vladimirov–Turygin's perturbative approach. In geometric terms it means the transition from the background geometry to the effective Riemannian manifold $\mathcal{M}^{4+d}[\text{Riem}]$:

$$(\mathbb{M}^{4+d},\ \overset{0}{g}_{AB}) \xrightarrow{\text{Fokker}} (\mathcal{M}^{4+d}[\text{Riem}],\ g^{\text{eff}}_{AB}). \tag{7}$$

In this approach, the existence of the non-Abelian contributions in the interaction picture corresponds to such a structure of the effective metric g^{eff}_{AB} that allows to represent $\mathcal{M}^{4+d}[\text{Riem}]$ as $\mathcal{M}^4 \times B^d$, where $B^d \simeq \mathcal{G}/H$ is a d-dimensional space whose isometry group is isomorphic to the gauge group $\mathcal{G}$ with maximal isotropy subgroup H.

The further procedure is performing the dimensional reduction $\mathcal{M}^{4+d}[\text{Riem}] \to \mathcal{M}^4$ according to the standard Kaluza–Klein receipt. As a result, the effective Yang–Mills fields $\mathbf{F}^{\mu\nu}$ appear in the structure of the effective 4-dimensional action:

$$\begin{aligned} {}^{4+d}S &\sim \int\limits_{\mathcal{M}^{4+d}} d^{4+d}x\sqrt{|g_{\text{eff}}|}R[g_{\text{eff}}] \\ &\longrightarrow {}^4S_{\text{eff}} = \int\limits_{\mathcal{M}^4} d^4x\sqrt{|{}^4g|}\left(-\frac{1}{2\kappa}R[{}^4g] - \frac{1}{4}\operatorname{Tr}\mathbf{F}^{\mu\nu}_{\text{eff}}\mathbf{F}^{\text{eff}}_{\mu\nu} + \ldots\right). \end{aligned} \tag{8}$$

References

1. A. D. Fokker, Ein invarianter Variationssatz fur die Bewegung mehrerer electrischer Massenteilchen, *Z. Phys.* **58**, 386 (1929).
2. J. A. Wheeler, R. P. Feynman, Classical Electrodynamics in Terms of Direct Interparticle Action, *Rev. Mod. Phys.* **21**, 425 (1949).
3. Ya. I. Granovsky, A. A. Pantiushin, On the relativistic gravitation theory, *Izvestija AN Kaz. SSR (Phys.-Math. Series)* **2**, 65 (1965) [in Russian].
4. A. Yu. Turygin. Fokker's Type Action at a Distance Theory of Gravitation, *Gen. Rel. Grav.* **18**, 333 (1986).
5. Yu. S. Vladimirov, Gravitation interaction in the relational approach, *Grav. Cosmol.* **14**, 41 (2008).
6. S. V. Bolokhov, A. N. Klenitsky. On the Construction of Effective Metrics in a Relational Model of Space-Time, *Grav. Cosmol.* **19**, 35 (2013).
7. C. W. Misner, K. S. Thorne, J. A. Wheeler, *Gravitation* Vol. 3 (W. H. Freeman and Company, San Francisco, USA, 1973).

Regular charged formation on an arbitrary worldline in a space with torsion

V. V. Kassandrov

Institute of Gravitation and Cosmology, Peoples' Friendship University of Russia
Moscow, Russia
E-mail: vkassan@sci.pfu.edu.ru

J. A. Rizcallah

School of Education, Lebanese University
Beirut, Lebanon
E-mail: joeriz68@gmail.com

We consider the equations of covariantly-constant vector fields (CCVF) in Weyl nonmetric geometry as well as in a conformal geometry with torsion. On a Minkowski metric background, the CCVF system alone allows us to pin down the Weyl nonmetricity vector or the torsion trace vector, interpreted as the electromagnetic potential. The electromagnetic field corresponding to the fundamental solution of the CCVF system is given by the ordinary Lienard-Wiechert field or a (regular/singular) modification thereof. The magnitude of charge is fixed to unity, "elementary", and charges of opposite signs correspond to retarded and advanced potentials respectively, suggesting a direct connection between the particle/antiparticle asymmetry and the "arrow of time".

Keywords: Covariantly-constant vector fields; Weyl geometry; torsion; Lienard-Wiehert field; quantization of electric charge.

1. Covariantly-Constant Vector Fields in Weyl Space

Consider a real-valued scalar field $\Psi(x)$ on a Weyl manifold background. Two fundamental equations for $\Psi(x)$ are the *eikonal equation* $g^{\mu\nu}\partial_\mu\Psi\partial_\nu\Psi = 0$ and the linear *wave equation* $g^{\mu\nu}\nabla_\mu\partial_\nu\Psi = 0$, where ∇_μ denotes the covariant derivative with respect to the *Weyl connection*

$$\Gamma^\rho_{\nu\mu} = \gamma^\rho_{\nu\mu} + \frac{1}{2}\left(A_\nu\delta^\rho_\mu + A_\mu\delta^\rho_\nu - A^\rho g_{\nu\mu}\right), \tag{1}$$

with Christoffel symbols $\gamma^\rho_{\nu\mu}$ and nonmetricity vector A^μ interpreted as the electromagnetic potential.

Apart from the ordinary Weyl conformal-gauge invariance, the eikonal equation possesses an additional *functional invariance.* Namely, for any solution $\Psi(x)$ of the eikonal equation, an arbitrary differentiable function $\Phi(\Psi(x))$ is also a solution. Upon this transformation, the gradient field $\partial_\mu \Psi$ gets an additional factor $\partial_\mu \Psi \mapsto \Phi' \partial_\mu \Psi, \quad \Phi' := d\Phi/d\Psi$, which, together with the gauge transformations of potentials $A_\mu \mapsto A_\mu - \frac{1}{2}\partial_\mu \ln \Phi'$, ensures the invariance of the wave equation. Thus, one has the *double gauge invariance* w.r.t. both the Weyl conformal group and that resembling the $SU(1)$ group in QFT.[1]

A *covariantly constant vector field* (CCVF) K_μ on a Weyl background, is one whose covariant derivative with respect to (1) vanishes:

$$\nabla_\nu K_\mu := \partial_\nu K_\mu - \Gamma^\rho_{\nu\mu} K_\rho = 0. \tag{2}$$

This system is overdetermined and can be used to determine both K_μ and A_μ, provided the metric is fixed. Symmetry of the connection (1) in the lower indexes results in a gradient CCVF (2), $K_\mu = \partial_\mu \Psi$, for some function $\Psi(x)$. For nontrivial EM fields $F_{\mu\nu} := \partial_\mu A_\nu - \partial_\nu A_\mu = \nabla_\mu A_\nu - \nabla_\nu A_\mu$, the norm of the CCVF K_μ must be null, and so the generating function $\Psi(x)$ necessarily satisfies the eikonal equation.

As a generating function, we may adopt the field $\tau(x)$ of retarded/advanced time-like parameter implicitly defined by $X_\mu X^\mu = 0, \quad X^\mu := x^\mu - \xi^\mu(\tau)$, along an arbitrary worldline $\xi^\mu(\tau)$ in Weyl space with constant Minkowski metric $\eta_{\mu\nu}$. Indeed, the field $\tau(x)$ obeys the eikonal equation, and the *reparametrization symmetry* corresponds to the additional gauge invariance considered above. The associated electromagnetic potential $A_\mu(x)$ can now be determined from (2) for connection (1) with $g_{\mu\nu} = \eta_{\mu\nu}$ and $\gamma^\rho_{\mu\nu} = 0$. It turns out that, up to a 4-gradient, the electromagnetic potential coincides with the *Lienard-Wiechert* ansatz

$$A_\mu = -\epsilon \xi'_\mu / X^\lambda \xi'_\lambda. \tag{3}$$

Note that for *retarded* time solution for which $X^0 = x^0 - \xi^0 > 0$ the dimensionless electric charge $\epsilon = -1$ whereas the opposite value $\epsilon = +1$ corresponds to *advanced* potentials. In either case the value of charge is fixed, "self-quantized" (the constant ϵ is the same for an arbitrary motion of a point-like source). In particular, for a charge at rest the solution reproduces the Coulomb potential[1].

2. Covariantly-Constant Vector Fields in a Space with Torsion

Now, to avoid the known difficulties in physical interpretation of Weyl geometry, we consider instead the space with the following metric compatible

$(\nabla_\rho g_{\mu\nu} = 0)$ connection:

$$\Gamma^\rho_{\mu\nu} = \gamma^\rho_{\mu\nu} + (A_\nu \delta^\rho_\mu - A^\rho g_{\mu\nu}), \tag{4}$$

whose curvature is invariant under the Weyl-type conformal transformations

$$g_{\mu\nu} \mapsto \exp{(2\lambda)} g_{\mu\nu}, \quad A_\mu \mapsto A_\mu - \partial_\mu \lambda. \tag{5}$$

Thus, as in the Weyl case, the new connection defines a conformally invariant curvature but possesses torsion $T^\rho_{[\nu\mu]}$ of the special form determined solely by its trace $T^\rho_{[\nu\rho]} = -\frac{3}{2} A_\nu$. The latter, in view of (5), may be identified with the potential 4-vector of an effective EM field.

Remarkably, the equations of PF (2) with connection (4) are form-invariant under (5) provided the PF field transforms as $K_\mu \mapsto \exp{(\lambda)} K_\mu$ so that its norm $K^2 := K_\mu K^\mu$ stays invariant. Moreover, as it follows from (2),(4), the latter is a real constant, $K^2 = \pm b^2$, $b = const$.

Maxwell's vacuum equations do not hold for every solution of the system (2), (4). Nonetheless we may consider the first and second pairs of Maxwell's equations as definitions of the EM fields through potentials and the current density 4-vector, respectively. From this standpoint, Maxwell's equations with the corresponding source will hold identically for every solution of the CCVF system. This approach to the dynamics of EM fields seems quite legitimate, as long as the CCVF system admits fundamental solutions of the Coulomb type, i.e. solutions for which the current density 4-vector asymptotically vanishes and the fields exhibit the correct multipole structure.

In particular, for a Minkowskian metric, the fundamental solution to (2),(4) is defined[2] by an arbitrary worldline $\xi_\mu(\tau)$ and given by the ansatz $K_\mu = x_\mu - \xi_\mu(\tau)$, where the field $\tau(x)$ is found from the equation of *hyperboloid* $K^2 = \pm b^2$. The associated 4-potentials $A_\mu = -\xi'_\mu/(K^\rho \xi'_\rho)$ formally correspond to the Lienard-Wiechert solution, but in fact lead to the familiar EM field strength only in the "light cone" case $b = 0$, or asymptotically far from the worldline. For $b \neq 0$ they define two topologically different solutions; one of which is everywhere regular and another – singular on the 2D "shell" $r = b$. In the simplest static case the solutions are of the form:

$$K_0 = \epsilon\sqrt{r^2 \mp b^2}, K_a = x^a; A_0 = \frac{\epsilon}{\sqrt{r^2 \mp b^2}}, A_a = 0, \epsilon = \pm 1; \tag{6}$$

at $r \to \infty$ the EM field is Coulomb-like and the electric charge is again "*self-quantized*".

Thus, we have obtained a peculiar *regular* charged field distribution localized near and moving along an arbitrary worldline (together with its singular counterpartner) . In this context, recall that A. Einstein regarded the discovery of a model with two *asymmetric* charged solutions as a crucial point in the construction of the ultimate unified field theory.

References

1. V. V. Kassandrov, J. A. Rizcallah, Gen. Rel. Grav. **46**, 1771 (2014). arXiv:1311.5423 [gr-qc]
2. V. V. Kassandrov, J. A. Rizcallah, Grav. Cosmol. **21**, No.4 (2015) (in print)

Nonpertrubative Measurements in Causal Set Dynamics

Alexey L. Krugly
Scientific Research Institute for System Analysis of the Russian Academy of Science, Nahimovskiy pr., 36, k. 1, Moscow, Russia, 117218
E-mail: akrugly@mail.ru

A causal set is locally finite partially ordered set. This is a model of a discrete spacetime on a microscopic scale. The aim of a dynamics is to predict the future of a considered system or to reconstruct its past. This means to add new parts to known causal set. A sequential growth dynamics of a causal set is an addition of new elements one by one. We consider probabilistic algorithms. These algorithms must be a consequence of some fundamental principles. Principles depend on a physical interpretation of sequential growth. We consider an addition of new element as an elementary measurement. A sequential growth is a process of reception of new information about a causal set by some ideal observer. The first principle is a causality principle. The second principle is a principle of nonpertrubative measurements. If we start from one element, this element can be any element of the universe. A probability to get some structure during sequential growth from one element must relate how often this structure is found in the universe.

Keywords: Causal set; sequential growth dynamics.

A causal set is a model of a discrete spacetime on a microscopic scale that has the same causal structure as Minkowski spacetime. (A list of references see in Ref. 1). An Alexandrov set of two points is an intersection of a future light cone of the first point and past light cone of the second point. In Minkowski spacetime any Alexandrov set is empty or consists of continuum of points. Suppose that any Alexandrov set consists of finite number of points. This is a causal set. A causal set is a locally finite partially ordered set, if a partial order is considered as a causal order. We postulate that all properties of spacetime and matter come to order relations.

Let consider a dynamics. The aim of a dynamics is to predict the future of a considered system or to reconstruct its past. This means to add new parts to known causal set. A sequential growth dynamics of a causal set is an addition of new elements one by one. We consider probabilistic algorithms. In this case, a sequential growth is a Markov chain. The aim is to get unique algorithm to calculate transition probabilities. This algorithm must be a consequence of some fundamental principles. Principles depend on a physical interpretation of sequential growth. The addition of a new element is not a real birth. We consider an addition of new element as an elementary measurement. A sequential growth is a process of reception of new information about a causal set by an observer.

The first principle is a causality principle. A ratio of the probabilities to add one or another maximal element is a function of past sets of these elements. The second principle is a principle of nonpertrubative measurements. Consider some causal sets A and B that we get at the step number N. The probability $P(A, N)$ of causal set A is a sum of probabilities of all ways S to get A from one element. The probability $P(A, N+M)$ of A at the step number $N+M$ is a sum of probabilities of all causal sets that include A as a subset.

$$P(A, N) = \sum_{S} P(A, N, S) \tag{1}$$

$$P(A, N+M) = \sum_{\{C|A \subset C\}} P(C, N+M) \geq P(A, N) \tag{2}$$

If we start from one element we can get any possible causal set. The first element can be any element of the universe. A probability to get some structure during sequential growth from one element must relate how often this structure is found in the universe. This means that we get the same distribution of causal sets' probabilities at any step of sequential growth.

$$\frac{P(A, N)}{P(B, N)} = \frac{P(A, N+M)}{P(B, N+M)} \tag{3}$$

The sequential growth is a nonpertrubative measurements.

Unfortunately we have more transition probabilities than equations. We need some additional postulates to calculate transitional probabilities. This is a subject of future investigations.

Reference

1. A. L. Krugly, A sequential growth dynamics for a directed acyclic dyadic graph, *Bulletin of Peoples' Friendship University of Russia, Series "Mathematics. Information Sciences. Physics"*, 124 (2014), arXiv: 1112.1064.

Renormalization of masses in pseudo-Finsler momentum spaces

A. V. Solov'yov

Faculty of Physics, M. V. Lomonosov Moscow State University, Moscow, 119991, Russia
E-mail: a.v.solovyov@gmail.com

This paper proposes geometrical mechanisms of renormalization and generation of particle masses. The above mechanisms are considered in the case of a 6-dimensional flat pseudo-Finsler space generated by Finsler 3-spinors over the field of real numbers and are based on dimensional reduction of this space.

Keywords: Finsler N-spinor; pseudo-Finsler space; renormalization of masses.

The theory of Finsler N-spinors (hyperspinors) over the field $\mathbb{C}$ of complex numbers was constructed in Refs. 1 and 2. Finsler N-spinors over $\mathbb{C}$ generate an N^2-dimensional flat pseudo-Finsler space with the metric defined by a homogeneous Nth degree polynomial in coordinate differentials. In Ref. 3, we developed the theory of Finsler N-spinors over the field $\mathbb{R}$ of real numbers. It was shown that Finsler N-spinors over $\mathbb{R}$ generated two flat pseudo-Finsler spaces. These spaces have $D = N(N+1)/2, N(N-1)/2$ dimensions, where $N > 1$ is a positive integer. Moreover, all of them are equipped with the "nth root metrics" $(dS)^n = G_{A_1 \ldots A_n} dX^{A_1} \cdots dX^{A_n}$, where $n = N, N/2$ and $A_1, \ldots, A_n = 0, \ldots, D-1$. Lagrangian mechanics of a classical particle in the above spaces was considered in Ref. 4. However, problems of dimensional reduction of such mechanics require an additional discussion, which is presented here.

In this paper, we consider the simplest nontrivial case of the theory when the corresponding pseudo-Finsler space has six dimensions and the cubic metric $(dS)^3 = G_{A_1 A_2 A_3} dX^{A_1} dX^{A_2} dX^{A_3}$ (see Ref. 3 for details). Let $X^A(S)$ be a world line of a particle in this space and $A = 0, 1, \ldots, 5$. The 6-velocity of the particle is defined by the natural formula $U^A = dX^A/dS$. By using the explicit expression for the metric from Ref. 3, we can easily show that the 6-momentum $P^A = MU^A$ of the particle satisfies the following "mass shell" equation

$$[(P^0)^2-(P^1)^2-(P^2)^2]P^5-P^0[(P^3)^2+(P^4)^2] \\ +2P^1P^3P^4+P^2[(P^3)^2-(P^4)^2]=M^3, \quad (1)$$

where M is the 6-dimensional particle "mass".

In Ref. 3, we proved that $SL(3,\mathbb{R})$ transformations of Finsler 3-spinors over $\mathbb{R}$ induced linear transformations of coordinates X^A, which preserved the metric $(dS)^3$. Therefore, Eq. (1) is invariant under such transformations. In $SL(3,\mathbb{R})$, we select a subgroup isomorphic to $SL(2,\mathbb{R})$. Direct calculations show that $SL(2,\mathbb{R})$ transformations of Finsler 3-spinors over $\mathbb{R}$ induce the split transformations of X^A:

$$X'^{\alpha}=L^{\alpha}_{\beta}X^{\beta} \quad (\alpha,\beta=0,1,2), \quad (2)$$

$$X'^{a+2}=M^{a}_{b}X^{b+2} \quad (a,b=1,2), \quad (3)$$

$$X'^{5}=X^{5}. \quad (4)$$

Here Eq. (2) is a 3-dimensional Lorentz transformation, Eq. (3) is a transformation of a Majorana 2-spinor, and Eq. (4) expresses a scalar character of X^5. It is evident that the same is valid for P^A.

Now, we can perform the dimensional reduction of Eq. (1). First of all, it is necessary to find the relation between dS and the 3-dimensional line element $ds=\sqrt{(dX^0)^2-(dX^1)^2-(dX^2)^2}$. By using the explicit expression for dS, we obtain

$$dS=\sqrt[3]{\dot{X}^5-g_{\alpha\beta}u^{\alpha}\dot{\xi}^{\top}\sigma^{\beta}\dot{\xi}}\,ds, \quad (5)$$

where $(g_{\alpha\beta})=\mathrm{diag}(1,-1,-1)$ is the matrix of components of the metric tensor of the 3-dimensional Minkowski space in a pseudo-orthonormal basis, $u^{\alpha}=dX^{\alpha}/ds$ is the 3-velocity of the particle, $\dot{\xi}^{\top}=(\dot{X}^3,\dot{X}^4)$ is a row and $\dot{\xi}=(\dot{X}^3,\dot{X}^4)^{\top}$ is a column of components of the Majorana 2-spinor $\dot{X}^{a+2}=dX^{a+2}/ds$, σ^{β} are the identity matrix and the two real Pauli matrices, $\dot{X}^5=dX^5/ds$ is a 3-scalar. Transformation properties of the above quantities follow immediately from Eqs. (2), (3), and (4).

Let us insert Eq. (5) into Eq. (1). This leads to the 3-dimensional Lorentz invariant relation

$$g_{\alpha\beta}P^{\alpha}P^{\beta}=m^2, \quad (6)$$

where

$$m=\frac{M}{\sqrt[3]{\dot{X}^5-g_{\alpha\beta}u^{\alpha}\dot{\xi}^{\top}\sigma^{\beta}\dot{\xi}}} \quad (7)$$

is the 3-dimensional mass of the particle and $P^\alpha = mu^\alpha$ is its 3-momentum. Thus, from the viewpoint of a 3-dimensional observer, the mass m of the particle is *variable*.

Equation (6) is the energy-momentum-mass relation for a particle in the 3-dimensional Minkowski space. It should be noted that P^α and m are the only observable quantities here. Therefore, we can interpret Eq. (7) as the peculiar renormalization of the 6-dimensional "mass" M.

When $M = 0$, we should start directly with Eq. (1). Then the analog of Eq. (6) is

$$g_{\alpha\beta}P^\alpha P^\beta = \frac{P^0[(P^3)^2 + (P^4)^2] - 2P^1P^3P^4 - P^2[(P^3)^2 - (P^4)^2]}{P^5}. \quad (8)$$

It is possible to interpret the right side of Eq. (8) as the square of an effective mass. This is a mechanism of generating the particle mass from extra dimensions.

In the present paper, we have shown that the simplest multidimensional pseudo-Finsler geometry can propose the mechanisms of renormalization and generation of particle masses. There are similar mechanisms in the Kaluza–Klein models. Probably, this is a common feature of all multidimensional physical theories.

References

1. D. Finkelstein, S. R. Finkelstein, and C. Holm, Hyperspin Manifolds, *Int. J. Theor. Phys.* **25**, 441 (1986).
2. A. V. Solov'yov and Yu. S. Vladimirov, Finslerian N-spinors: Algebra, *Int. J. Theor. Phys.* **40**, 1511 (2001).
3. A. V. Solov'yov, Finsler N-spinors with real components, *Theor. Math. Phys.* **183**, 756 (2015).
4. A. V. Solov'ev, General solution of equations of motion for a classical particle in 9-dimensional Finslerian space, *Russ. J. Math. Phys.* **13**, 466 (2006).

Relational justification of the cosmological redshift

Yu. S. Vladimirov* and A. B. Molchanov**

Lomonosov Moscow State University
Moscow, Russia
* *E-mail: yusvlad@rambler.ru*
** *E-mail: alexeybm2009@gmail.com*

There is suggested a new interpretation of the cosmological redshift in terms of relational approach to the nature of the world. And there are provided quantitative comparisons of the energy densities, which are corresponding to radiation and universal expansion to justify this effect in the framework named above.

Keywords: Cosmological redshift; relational approach.

1. Introduction

Research in theoretical physics is carried out within three physical (metaphysical) paradigms: field theory paradigm (nowadays dominated), geometrical, on which general relativity and it's native extensions are based, and relational, which consists of two parts: 1) relational interpretation of classical space-time concepts and 2) description of physical interactions in terms of action at a distance.

Thus it is interpreted the cosmology in terms of general relativity, i.e. in the framework of geometrical paradigm. There are described the evolution of the Universe by ΛCDM-model. As it is known, this model includes two hypothetical substances (dark matter and dark energy), which are considered as fundamental problems in theoretical physics.

So, there is suggested a relational approach to cosmology, which allows to bypass these problems. The first step on this way is the relational justification of the cosmological redshift, as one of the main observable cosmological effects. Since in this approach there is no space-time background, on which photons can propagate, then the emission of the photon means assignment of the world matrix of relations between the emitter and all possible receivers.[1] In this sense, photon's energy is located in all possible absorbers. But there are a lot of photons and not less absorbing objects in the Universe. It means, that all objects around any observer should have some additional energy.

To demonstrate the validity of this statement, compare the energy of the "sea" of photons, existing in the Universe with the energy of the matter "expansion" in observable part of it.

2. Energy Density of Electromagnetic Radiation in the Universe

The overall density of electromagnetic radiation consists of two main parts: relict radiation and other radiation of stars.

In accordance with modern data,[2] the temperature of relict radiation is $T_{rel} = 2{,}72548 \pm 0{,}00057K$. It corresponds to the radiation density in g/cm^3 $\rho_{rel} = \frac{4\sigma T^4}{c^3} = 4{,}644 \times 10^{-34}$ g/cm^3.

Energy density of the overall stellar radiation is known not so precisely (A. Eddington): $\rho_{star} \simeq 7{,}67 \times 10^{-13}$ erg/cm^3 $\rightarrow$ $\rho_{star} \simeq 8{,}58 \times 10^{-34}$ g/cm^3. So, the overall matter density, corresponding to radiation energy in the Universe, can be estimated as $\rho_{sum} = \rho_{rel} + \rho_{star} \simeq 13{,}22 \times 10^{-34}$ g/cm^3.

The unabsorbed radiation energy, which is located in all possible absorbers (according to relational approach), should be interpreted, as an additional energy of matter, which reveals in the expansion of the Universe.

3. Energy Density of the Universe Expansion

In evaluating of the Universe expansion energy we restrict our analysis to a separate part of it, in which we can confidently talk about the density of matter and it's motion (the volume, in which the Hubble's law is linear, $z < 0{,}1 \pm 0{,}01$).[3] Moreover, we will take into account only ordinary (baryonic) matter with observable density, which is characterized by the ratio of the critical density $\Omega_b = 0{,}0456 \pm 0{,}0016$.[2] As a result we obtain: $\rho_E = \frac{3}{10} z^2 \Omega_b \frac{3H_0^2}{8\pi G} \simeq (12 \pm 3{,}1) \times 10^{-34}$ g/cm^3.

So, the matter density, which corresponds to the energy of the Universe expansion, coincides with the matter density, which corresponds to the energy of the electromagnetic radiation.

4. Conclusions and Comments

When we use relational approach to describe cosmology, it remains a number of questions. For example, we need to justify the non-linear part of Hubble's law. This can be done in relational framework by using ideas of P. Rashevskiy about the statistical nature of geometrical concepts and about the possible variation of arithmetics on large distances.[4] One can come to this paying attention on a symmetry between coordinate and momentum in classical mechanics. This symmetry has acquired greater significance in quantum mechanics. But there is a limit value in the velocity space – speed of light c. This allows to put forward the hypothesis, that there may exist a limit value of distance (or another features) in coordinate space. And, as it was showed by V. Rvachev,[5] it leads to the redshift effect, which corresponds to non-linear part of Hubble's law.

References

1. Yu.S. Vladimirov, *Action at a distance physics* (LIBROKOM Publ., Moscow, 2012).
2. P.A.R. Ade et. al, Planck 2013 results. I. Overview of products and scientific results, *Astronomy and Astrophysics* **571**, A1 (2014).
3. S. Perlmutter, Supernovae, Dark Energy and Accelerating Universe, *Physics Today* **56**, 53 (2003).
4. P.K. Rashevskiy, On the dogma of the natural numbers, *Uspekhi Matematicheskikh Nauk* **28**, 243 (1973).
5. V.L. Rvachev, *Static objects in the far space have redshifts of their spectra*, (Pre-print SA of Ukraine. Institute of mechanical engineering problems, 377, 1994).

Relational-statistical nature of metric

Yu. S. Vladimirov* and D. A. Tereshchenko

Faculty of Physics of Moscow State University,
Institute of gravitation and cosmology, PFUR
Moscow, Russia
** E-mail: yusvlad@rambler.ru*
dima91ter@yandex.ru

Based on relationally-statistical approach to nature of space-time and interactions we consider the theory of hydrogen atom. The condition of stationarity for interaction between proton and electron was obtained on the basis of theory of binary systems of complex relations.

Keywords: nature of metric; theory of atom; relational statistical approach; complex relations binary systems.

This paper proposes substantiation of $O(4)$-symmetry of solutions of Schrodinger equation for the problem of hydrogen-like atoms discovered by V. A. Fock[1] and E. A. Hyleraas[2] on the basis of relational statistical approach to nature of space-time and interactions. This is done on the basis of the mathematical apparatus of minimal ranks $(2,2)$, $(3,3)$ and $(4,4)$ complex relations binary systems[3] (CRBS).

In this theory, particles are described by pairs of set elements, and each of these elements is characterized by a two-component spinor. Since a hydrogen atom is a bound state of a proton and electron, there exists a relation between proton–electron elements. This relation is equivalent to the given three-dimensional hypersphere in four-dimensional Euclidean space with a signature $(++++)$. The hypersphere equation is derived from the corresponding $SL(2,C)$ transformation invariant.

When one make a conformal transformation on the set elements, there appears a conformal factor for each relation between elements. The relations between set elements of atom particles and particles of the surrounding world can be represented as dots on the surface of a hypersphere. Distribution of conformal factors for these relations on a hypersphere surface satisfies a stationary condition which corresponds to eigenvalue problem for Laplace operator on 3-dimensional hypersphere, which reduced to Laguerre equation. Laguerre equation is a radial part of both Schrodinger equation for hydrogen-like atoms and Klein–Fock–Gordon equation.

Thus, based on purely algebraic equations initially not related to the classical theory, both nonrelativistic and relativistic quantum equations were deduced. Also, a classical space-time idea was not used at all.

This problem was solved within a scope of more fundamental problem: derivation of classical space-time relations from the concepts and laws of microscopic physics instead of using already existing space-time in conventional theoretical constructions. In this regard, the theory of bound states is a source of metric relations.

References

1. V. Fock, The hydrogen atom and non-Euclidean geometry, *Proceedings of the AS USSR* **2**, (1935), p. 169 [in Russian].
2. E. Hylleraas, Wellengleichung des Kepler-Problems im Impulsraum, *ZS. f. Phys.* **74**, (1932), p. 216.
3. Yu. Vladimirov, *Physics of action at a distance. Nature of space-time* (Editorial URSS, Moscow, 2012) [in Russian].

Relational statistical spacetime: A new approach to description of gravitation

V. V. Aristov

Dorodnicyn Computing Centre of Russian Academy of Sciences,Federal Research Centre of Informatics and Rule of Russian Academy of Sciences
Moscow, 119333, Russia
E-mail: aristovvl@yandex.ru
www.ccas.ru

The concept of the relational statistical spacetime allows us to elaborate an apparatus in which both quantum and gravitational effects can be described.

Keywords: Relational statistical spacetime; gravitation; quantum mechanics.

1. Introduction

Contemporary problems of gravitation: Dark Matter, Dark Energy, necessity of reconciling General Relativity (GR) with quantum physics, etc. require to seek for a more general theory. We develop the relational statistical (RS) spacetime concept, which can be compared with the concept by Yu. Vladimirov.[1] In our approach properties of time and space are treated as properties of physical clocks (see Ref 2) and scale rules (see Ref 3) respectively. It is assumed that a basis for an origin of physical equations is a structure of spacetime. A postulated time increment is as follows

$$d\tau^2 = \frac{a^2}{N}\sum_{i=1}^{N}(dr_i - \frac{1}{N}\sum_{j=1}^{N} dr_j)^2, \tag{1}$$

where N is the number of particles in a system under consideration (e.g. Metagalaxy). Here changing instantaneous pictures (snapshots) obtained by an ideal camera with particle coordinates provides the coordinate intervals dr_i. It implies that synchronisation is defined for two events in the same light ray. In fact a sort of absolute frame is used according to discussion by Smolin[4] if one takes into account that a value $dr_{absi} = dr_i - \frac{1}{N}\sum_{j=1}^{N} dr_j$ is invariant relatively shifting transformation. The appropriate transition to the Einstein synchronisation can be made and all relations of special relativity are deduced, including the motion equations, see Ref. 2. A constant $a = 1/c$, where c is speed of light. From (1) for the light ray $d\tau = 0$. This leads to the equation for any r_i characterising a trajectory of the ray $dr = dr_i = \frac{1}{N}\sum_{j=1}^{N} dr_j$. We consider a simple model of identical particles ("nucleons"). The RS space is expressed through configuration of masses. This realizes a projection of a discrete set of particles onto the physical instrument for measuring space which consists of discrete elements (particles). It is natural to use the formalism of graphs in which particles are associated with vertices of graphs, see Ref. 3. Hence there is the minimum length r_e associated with a single mass m_e, of a particle, namely $r_e = bm_e$, where b is expressed through fundamental constants.

2. Microscopic Limit: Quantum Relationships

For a discrete medium for small distances when $r \sim r_e$, there are two or more straight lines (geodesics) between two points (particles), and one can interpret this as manifestation of indeterminism in quantum mechanics. For large distances for the uniform discrete distribution the straight lines tend to the unique straight line. Let us obtain restrictions of measuring (in fact the analogue of the Heisenberg inequality). Taking into account Eq. (1) and the restriction for the spatial intervals we have for a finite time interval $\Delta\tau \geq ad_\tau r_e$, where $d_\tau \sim 1$ if we use a natural assumption for Metagalaxy concerning a random distributions of vectors Δr_i which are not equal to an average value. When all $|\Delta r_i|$ tend to a limiting value r_e, an interval $\Delta\tau$ does not vanish, because vectors have different directions and tend to the value of the order τ_e. Thus the velocity should be treated as the limit of the ratio of the spatial and the temporal increments, the first tends to r_e and the second tends to the finite value which can possess different magnitudes. The relative error of the velocity is of the order of unity, see Ref. 3. Then $\Delta u_x \sim r_e/(r_e a) = 1/a = c$. We have for a free particle $\Delta p_x \Delta x = m_e \Delta u_x \Delta x \sim m_e r_e/a = m_e r_e c$. If we assume the equality of this value and the Planck constant, we find $m_e r_e c = \hbar$ and a constant $b = \hbar/(m_e^2 c)$. For deriving the Schrödinger equation we use the Nelson formalism[5] in which an assumption ad hoc has been proposed that a particle of mass m_e is subject to Brownian motion with diffusion coefficient $v = \hbar/2m_e$. In RS spacetime an effective "diffusion coefficient" appears when a small finite coordinate increment can be written in terms of a certain average velocity u_{mx}, multiplied by the time increment with the addition term which supposedly is a random variable:

$$\Delta x(\tau) = u_{mx}(\tau)\Delta\tau + (1/2)u'_{mx}(\Delta\tau)^2 \sim u_{mx}(\tau)\Delta\tau + (1/2)r_e$$

and we have $\Delta x(\tau) \sim u_{mx}(\tau)\Delta\tau + \hbar/(2m_e c)$, where u'_{mx} is an certain average derivative of u_{mx}. The "diffusion coefficient" is $\hbar/(2me)$ and we are in the framework of Nelsons formalism. The ordinary velocity loses its sense but the average "forward" and "back" derivatives can be defined by a positive increment and negative increment of time respectively. So we obtain not an ordinary motion equation but a pair of equations for the sum and the difference of forward and back velocities (u and v respectively)

$$\frac{\partial u}{\partial t} = -(\hbar/(2m_e)\nabla(divv) - \nabla(vu),$$

$$\frac{\partial v}{\partial t} = (1/m_e)F - (v\nabla)v + (u\nabla)u + (\hbar/(2m_e)\Delta u.$$

Introducing the notation $u = \hbar\nabla R/m_e, v = \hbar\nabla S/m_e$ and combining the equations for u and v we obtain the equation with complex variables (the potential U is related to the force F) and introducing a function $\Psi = exp(R + iS)$, we obtain the Schrödinger equation

$$\frac{\partial \Psi}{\partial t} = (i\hbar/(2m_e)\Delta\Psi - i/\hbar U\Psi.$$

3. Macroscopic Limit of RS Spacetime: Gravitation

If there is a nonuniformity in the distribution of particles (a massive body exists), the nonuniqueness of straight line will also appear even for the limiting images of the discrete medium. For large scales one can obtain a deviation from a flat metric and treat that as manifestation of gravitation. Deflection of light with the presence of a massive body can be associated with changing the statistical sum which describes the light-like geodesics. If a mass of the body is M then in the sum describing a light ray (see Introduction) a part (M) of vectors r_i related to this nonuniformity are deflected in comparison with the symmetry uniform situation, this denotes that r is deflected in the direction to a massive body; so gravitational lensing is predicted as well. We compare a real distribution of moving masses with the appropriate RS sums for representing "real" time and space intervals ($d\tau$, dr) and ideal uniform distributions for representing "ideal" RS time and space intervals ($d\tau_c$, dr_c). The origin of the equations is in fact mathematical relationships. For the system of the uniform distributed moving particles the mentioned values are equal and one can substitute dr_{ci} and $d\tau_c$ into Eq.

(1) and obtain the equation in which all values are measured by ordinary instruments. In a case with a massive body the uniformity is disturbed and the connection between these values is determined by analogues of the metric coefficients g_{ij}. For example for connections of the increments of time we have $d\tau^2 = g_{00} d\tau_c^2$. One can prove that $d\tau < d\tau_c$ (and also $dr > dr_c$). It is a natural way for introducing the gravitational potential by taking into account slowing down of time when a massive body exists. So we consider a metric coefficient g_{00} in which a gravitational potential should appear. A form of the Newton gravitational potential is obtained when we compare two dimensionless sums $\sum_i \frac{u_i^2}{c^2}$ and $\sum_i \frac{bm_e}{r_i}$ in view of the fact that the distance is measured by masses and using the limit theorems of the probability theory. The fundamental constants will appear in the potential if we consider parameters of Metagalaxy (in the framework of a variant of the Mach principle). Explanation of some cosmological coincidences is obtained, and the analogue of the Schwarzschild metric is derived and all effects of General Relativity are reproduced with the accuracy of the modern experiments for weak gravitation. However the RS concept predicts the deviations from these classical effects for possible experiments with a more high accuracy. The problem of Dark Matter could be explained because the different distribution of masses in Metagalaxy results in the different value of the gravitational potential.

References

1. Yu. S. Vladimirov, *Metaphysics* (Binom, Moscow, 2008).
2. V. V. Aristov, Statistical model of clocks in a physical theory, *Physics-Doclady (Doclady of Russian Acad.Sci.)* **39**, 45 (1994).
3. V. V. Aristov, *On the relational statistical space-time concept*, in *The Nature of Time: Geometry, Physics and Perception*, eds. R. Bucchery et al., (Kluwer Academic Publishers, Dordrecht, 2003), pp. 221–229.
4. Lee Smolin, *Time reborn* (Spin Networks Ltd, 2013).
5. E. Nelson, Derivation of the Schrödinger equation from Newtonian mechanics, *Phys. Rev.* **150**, 1079 (1966).
6. V. V. Aristov, The gravitational interaction and Riemannian geometry based on the relational statistical space-time concept, *Gravitation and Cosmology* **17**, 166 (2011).

Relational and geometric approaches to justifying the magnetic fields of astrophysical objects

I. A. Babenko

Institute of gravitation and cosmology of Peoples' Friendship University of Russia, Miklukho-Maklaya str.6, Moscow, 117198, Russia
E-mail: inna.babenko@list.ru

We propose justification of the Sutherland hypotheses about origin of the magnetic fields of the Earth, Sun and other astrophysical objects as a part of the relational theory of space-time and interactions ("binary geometrophysics") and multidimensional geometrical models of physical interactions (like the Kaluza-Klein theories).

Keywords: Magnetic fields of astrophysical objects; Sutherland hypotheses; relational theory of space-time; multidimensional geometrical models.

Two options for justification of Sutherland's hypothesis about magnetic fields origin of astrophysical objects are offered within geometrical and relational approaches to the physical reality description. This hypothesis is other than nowadays dominating dynamo hypothesis and is based on the assumption about very small (outside a resolution by the modern equipment) distinction of heavy positively and light negatively charged particles. It leads to positive volume charge of astrophysical enough massive objects which Coulomb field is compensated by the stuck negative particles. But as astrophysical objects rotate, because of different laws of opposite signs charges distribution, the magnetic fields generated by them differ, as is shown in the form of observed magnetic fields. For this hypothesis acceptance it was necessary to prove distinction of the electrons and protons electric charges forming objects substance.

Two ways of Sutherland's hypothesis justification are offered in this work. The first of them is geometrical leans on 6-dimensional version of Kaluza-Klein theory in which two types of $G_{5\alpha}$ and $G_{4\alpha}$ mixed a component of a metric tensor are interpreted through vector potentials of the electromagnetic field A_α. But as in version of the 5-dimensional Klein's theory a source of a field is a mass, and in version of the 5-dimensional theory of Kaluza is a charge, this distinction of sources leads to Sutherland's hypothesis.

In relational approach, elementary particles mass are treated proportional to squares of their electric charges that allows to assume that the observed charge is presented in the form of a number of decomposition on value of an inoculating charge. The proportionality coefficients determined by the accounting of Mach's principle for two types particles are different that also leads to Sutherland's hypothesis.

Thus, two approaches (geometrical and relational) lead to identical results.

References

1. W. Sutherland, *Solar magnetic fields and the cause of terrestrial magnetism.* (Terr. Mag. Planet Sci. **5**, 73 (1900); **8**, 49 (1903); **9**, 167 (1904); **13**, 155 (1908)).
2. B. M Yanovsky, *Earth's magnetism* (Leningrad, Leningrad State University, 1964).
3. Yu. S Vladimirov, *Relational theory of space-time and interactions. Part 2* (Moscow, Moscow State University, 1998).
4. Yu. S Vladimirov, *The origin the magnetic field of astrophysical objects* (Moscow, Moscow State University, 2000).

Conformal gravity, gravitational bubbles and all that

V. A. Berezin[1,*], V. I. Dokuchaev[1,2] and Yu. N. Eroshenko[1]

[1] *Institute for Nuclear Research of the Russian Academy of Sciences, 60th October Anniversary Prospect 7a, Moscow, 117312, Russia*
**E-mail: berezin@inr.ac.ru*

[2] *National Research Nuclear University MEPhI, Kashirskoe sh. 31, Moscow, 115409, Russia*

We are reporting on some general ideas and first results in studying the spherical symmetric conformal gravity.

Keywords: Gravitation; conformal gravity; cosmology; black holes.

The conformal gravity was invented by H.Weyl in 1918. His motivation was to combine the gravitational and electromagnetic fields into one unified theory. Then such a theory was rejected by A.Einstein and H.Weyl because it was recognized that the conformal invariance allows only for massless particles to exist. But, nowadays, this obstacle can be overcomed by means of the Higgs mechanism for generating particle masses. The conformal gravity came again into play in 1989 due to the paper by P.D. Mannheim and D. Kazanas, who obtained the static spherically symmetric vacuum solution, depending on three arbitrary constants and generalizing the Schwarzschild solution in General relativity. What was really new is the appearance of the term linear in radius, what forced to remember the Mach's principle.

All the homogeneous and isotropic space-times (i.e., described by the Robertson-Walker line element) have zero Weyl tensor. This means that they are solutions to the vacuum conformal gravity field equations. Such an empty high symmetric space-time is a good candidate for the creation of the universe "from nothing" (the very possibility of which was first proposed by A. Vilenkin). The idea that the initial state of the universe should be conformal invariant, is advocated also by R. Penrose and G.'t Hooft. But, this empty universe should be filled with the particles (necessarily massless prior to the spontaneous breaking of the conformal symmetry). How? - Due to the quantum fluctuations that will cause the local deviation of the Weyl tensor from zero. This, in turn, will cause the creation of massless particles (as wa shown first by Ya.B. Zel'dovich, I.D. Novikov and A.A. Starobinsky) which, moving with the speed of light, will deviate the Weyl tensor form zero in much more large region, and so on, and so forth. For how long this could last, and what could be the result? As was shown by Ya.B. Zel'dovich, the process of particle creation can make the universe isotropic. The isotropization and homogenization will lead to decreasing of both the creation rate and the square of the Weyl tensor. But, in conformal gravity such solution are the vacuum ones! This controversy can be removed by taking in account not only the energy-momentum tensor of the already created particles, but also the quantum corrections coming from the vacuum polarization and trace anomaly. In one-loop approximation the corresponding counter-terms can be incorporated into the action integral as the curvature scalar (manifesting the renormalization/emergence of Newton's constant), the cosmological constant and the term, quadratic in Ricci tensor. This was foreseen by A.D. Sakharov in 1967. Thus, while the Weyl tensor is gradually vanishing, the counter-term, linear in scalar curvature is becoming dominant. This is how the General relativity may come onto play. Note, that for this to happen, the very process of particle creation plays the central role with spontaneously broken conformal symmetry.

First, we found the general vacuum spherically symmetric solution with varying two-dimensional conformally transformed scalar curvature. This appears to be the one-parameter family (up to the overall conformal factor equal to the square of the radius). For the specific choice of the dependence of the radius on this scalar curvature it reduces to the famous Mannheim-Kazanas metrics. Moreover, we showed that the Schwarzschild-(anti)de Sitter of General Relativity is also the member of out family, and the vacuum solution exhibits the effective cosmological constant.

Second, we obtained the solution for the radiation (a la Vaidya) and the electro-vacuum (with the spherical Coulomb field) space-time, and showed that they differ from the corresponding counterparts in General Relativity.

Finally, we found two different types of the pure vacuum solutions, completely source-free. One of them has zero Weyl tensor, and it is conformally related to the homogeneous and isotropic cosmological models (but empty!). That one which is compact (with positive spatial curvature) we called "the gravitational bubble", it is quite suitable for the "creation from nothing". Another one, with non-zero Weyl tensor, is the best candidate for the vacuum fluctuations, the seeds for all the subsequent evolution.

The reported study was supported by RFBR grant 15-02-05038 a.

Proof of the entropy principle in Einstein–Maxwell theory

Sijie Gao
Physics Department, Beijing Normal University,
Beijing, 100875, China
E-mail: sijie@bnu.edu.cn

We consider a self-gravitating charged perfect fluid in a static spacetime. We assume that the Einstein constraint equation is satisfied. Then we prove that the extrema of the total entropy of fluid implies other components of Einstein's equation. Conversely, if Einstein's equation is satisfied, we show that the total entropy achieves an extremum. This work suggests that the maximum entropy principle is consistent with Einstein's equation when an electrostatic field is taken into account.

Keywords: Self-gravitating fluid; entropy principle; Einstein's equation.

The distribution of a self-gravitating fluid in a curved spacetime can be determined in two ways. One way is solving Einstein's equation while the other way is to configure the fluid such that its total entropy achieves a maximum. The equivalence of the two methods has been discussed for the past decades (Ref. 1– 7). Previously, we have shown that the extremum of the total entropy of a perfect fluid and the Einstein constraint equation imply the other components of Einstein's equation.[8] It is interesting to know whether this result can be extended to charged fluid. The following theorem shows that the entropy principle is consistent with not only Einstein's equation, but also Maxwell's equation.

Theorem:

Consider a uniformly charged perfect fluid in a static spacetime (M, g_{ab}). Let C be a region on a three-dimensional hypersurface Σ and $\bar{C}$ be the boundary of C. Let h_{ab}, Φ and χ be the induced metric on Σ, electrostatic potential and redshift factor, respectively. Assume that the temperature of the fluid obeys Tolman's law. Assume that the Einstein constraint equation and Maxwell's equation are satisfied in C. Then the other components of Einstein's equation are implied by the extrema of the total fluid entropy for all variations of h_{ab}, Φ and χ in C with fixed values on $\bar{C}$.

Detailed proof can be found in Ref. 8. The sketch of the proof is given as follows. The total entropy is

$$S = \int_C \sqrt{h} s(\rho, n) \tag{1}$$

where s, ρ and n are the entropy density, energy density and number density. Applying the first law of thermodynamics, we find the variation of S

$$\delta S = \int_C s\delta\sqrt{h} + \sqrt{h}\left(\frac{1}{T}\delta\rho - \frac{\mu}{T}\delta n\right) \tag{2}$$

Using the Einstein constraint equation

$$\rho = \frac{1}{16\pi}R^{(3)} - \frac{1}{4\pi}(F_{ac}F_b{}^c u^a u^b + \frac{1}{4}F_{ab}F^{ab}) \,. \tag{3}$$

and Maxwell's equation

$$\nabla_b F^{ab} = 4\pi\rho_e u^a, \tag{4}$$

where ρ_e is proportional to n for uniformly charged fluid, we may express δS as

$$\delta S = \int_C \delta L_\Phi + \delta L_\chi + \delta L_h \tag{5}$$

where δL_Φ, δL_χ and δL_h denote the terms proportional to $\delta\phi$, $\delta\chi$ and δh_{ab}, respectively. It can be shown that δL_Φ and δL_χ vanish identically. Since $\delta S = 0$ by the assumption of the theorem, we have $\delta L_h = 0$, which gives exactly the projection of Einstein's equation on Σ:

$$8\pi p h^{ab} = R_{cd}h^{ac}h^{bd} - \frac{1}{2}Rh^{ab} - 2(F_{ce}F_d{}^e - \frac{1}{4}g_{cd}F_{ef}F^{ef})h^{ac}h^{bd} \,. \tag{6}$$

The inverse of the theorem is also true: If Einstein's equation and Maxwell's equation are satisfied in C, then the total entropy of the fluid achieves an extrema for all variations.

References

1. W. J. Cocke, Ann. Inst. Henri Poincaré **2**, 283 (1965).
2. R. D. Sorkin, R. M. Wald and Z. J. Zhang, *General Relativ. Gravit.* **13**, 1127 (1981).
3. S. Gao, *Phys. Rev. D* **84,** 104023 (2011); **85,** 027503 (2012).
4. Z. Roupas, *Classical Quantum Gravity* **30**, 115018 (2013).

5. L. M. Cao, J. Xu, and Z. Zeng, *Phys. Rev. D.* **87**, 064005 (2013).
6. Z. Roupas, arXiv:1305.4851.
7. S. R. Green, J. S. Schiffrin and R. M. Wald, *Classical Quantum Gravity* **31** 035023 (2014).
8. X. Fang and S. Gao, *Phys. Rev. D.* **92,** 024044 (2015)

On one Hawking's hypothesis

M. Yu. Konstantinov*

Department of Physics, N.E. Bauman Moscow State Technical University
5, 2nd Bauman Street, 105005, Moscow, Russia
**E mail: konst@bmstu.ru*

The Hawking hypothesis about Euclidean nature of space–time is discussed.

Keywords: Space-time; general relativity; polymetric space-time theories; dark matter.

For the improvement of the continual integrals convergence in quantum field theory and quantum gravity the so called Wick rotation $x^0 \to ix^0$ which changes signature of space–time metric from the Lorentzian one to the Euclidean one is often used. Because of high efficiency of this procedure Hawking supposed that both quantum and classical space–time is really Euclidean (i.e. has Euclidean signature) and the interpretation of classical space–time as pseudo–Riemannian (Lorentzian) is a result of peculiarities of our perception.[1]

To apply Hawking's supposition to the classical physics the correspondence between Riemannian and pseudo–Riemannian spaces must be realized in the form[2,3]

$$g_{\mu\nu} = 2u_\mu u_\nu - G_{\mu\nu}, \tag{1}$$

instead of Wick rotation. Here metrics $g_{\mu\nu}$ has Lorentzian signature $(+,-,-,-)$, $G_{\mu\nu}$ is Riemannian metric with signature $(+,+,+,+)$ and u_μ is some unit vector field i.e. $G^{\mu\nu}u_\mu u_\nu = 1$. The correspondence between contravariant tensors $G^{\mu\nu}$ and $g^{\mu\nu}$ is similar to equation (1), $G^{\mu\nu}u_\mu u_\nu = g^{\mu\nu}u_\mu u_\nu = 1$ and $G = \det\|G_{\mu\nu}\| = -\det\|g_{\mu\nu}\| = -g$[3].

In accordance with Hawking's hypothesis classical space–time must be considered as a Riemannian space with metrics $G_{\mu\nu}$ and the action integral must be written in the following general form

$$S = \int (\kappa R_G + L_m)\sqrt{G} d^4x, \tag{2}$$

where κ is a constant, R_G is the Ricci scalar corresponding to $G_{\mu\nu}$, L_m is a matter (source fields) Lagrangian.

Because of Hawking's supposition the kinetic term of the matter Lagrangian L_m must be defined by Lorentzian metrics $g_{\mu\nu}$. By means of (1) action (2) may be transformed to the following equivalent form

$$S = \int (\kappa R_g + L_m + F(u_\mu, u_{\mu,\nu}))\sqrt{-g}d^4x. \tag{3}$$

Among of the usual matter is described by the Lagrangian L_m action (3) contains additional term $F(u_\mu, u_{\mu,\nu})$ which describes additional source field u_μ which may be considered as a dark matter field.

So, Hawking's supposition about Euclidean nature of space–time leads to the existence of dark matter. It is clear that this result does not exclude existence of dark matter of another nature.

Another consequence of Hawkings hypothesis is a possible coexistence of several Lorentzian structures on the same Riemannian manifold which corresponds to different kinds of matter (polimetric space-time). For the first time such supposition was made by author in 1985[4] and was repeatedly made recently by Geroch.[5] Several consequences from polymetric space–time models were considered in Refs. 6, 7.

References

1. S. W. Hawking, Euclidian Quantum Gravity, *in Recent Developments in Gravitation*, ed. S. Deser, Plenum Press, 1978.
2. S. W. Hawking, G. F. R. Ellis, *The Large Scale Structure of Space Time* (Cambridge University Press, 1973).
3. N. V. Mitzkevich, The map of pseudo-Riemannian spaces on Riemannian in general relativity, *in: Gravitation and the theory of relativity*, Kazan, 1982 (in Russian).
4. M. Yu. Konstantinov, Topology change in classical gravitation theory: scalar-tensor formalism, *in: The problems of the gravitation theory and elementary particles N 16*, (Moscow, Energoatom, 1985), p. 148. (in Russian)
5. R. Geroch, Faster Than Light?, arXiv: 1005.1614v1 [gr-qc].
6. M.Yu. Konstantinov, Superlight particles in polymetric space–time theories *Izvestiya vuzov: Fizika* **5**, 65 2012 (in Russian).
7. M.Yu. Konstantinov, On the metric structure of space–time *in: Herald of the Bauman Moscow State Technical University, Ser. Natural Sciences* **1**(58), 63 (2015).

Dynamics of test bodies in scalar-tensor theory and equivalence principle

Yuri N. Obukhov

Theoretical Physics Laboratory, Nuclear Safety Institute,
Russian Academy of Sciences, B.Tulskaya 52, 115191 Moscow, Russia
E-mail: obukhov@ibrae.ac.ru

Dirk Puetzfeld

ZARM, University of Bremen,
Am Fallturm, 28359 Bremen, Germany
E-mail: dirk.puetzfeld@zarm.uni-bremen.de

How do test bodies move in scalar-tensor theories of gravitation? We provide an answer to this question on the basis of a unified multipolar scheme. In particular, we give the explicit equations of motion for pointlike, as well as spinning test bodies, thus extending the well-known general relativistic results of Mathisson, Papapetrou, and Dixon to scalar-tensor theories of gravity. We demonstrate the validity of the equivalence principle for test bodies.

1. Introduction

Scalar-tensor theories are considered to be close and viable generalizations of Einstein's general relativity theory. Since their introduction in Refs. 17, 18 and 26 they attracted a lot of attention after the works of Brans and Dicke[2,3,5,7,8] with the scalar field interpreted as a variable gravitational coupling – for an overview of the history and results of scalar-tensor theories see Refs. 4, 11, 12 and 24.

However, little attention was paid to the study of motion of extended test bodies in such theories. Following Refs. 1, 3 and 27, the issue was thoroughly analysed[6] in the framework of the post-Newtonian formalism. Here we present the test body dynamics for a very large class of scalar-tensor theories.

2. Scalar-Tensor Theory

We study a class of scalar-tensor theories (along the lines of Ref. 6) with the action $I = \int d^4x \stackrel{J}{\mathfrak{L}}$ on the spacetime manifold with the metric $\stackrel{J}{g}_{ij}$. The

Lagrangian density $\overset{J}{\mathfrak{L}} = \sqrt{-\overset{J}{g}}\overset{J}{L}$ has the following form

$$\overset{J}{L} = \frac{1}{2\kappa}\left(-F^2 R(\overset{J}{g}) + \overset{J}{g}{}^{ij}\overset{J}{\gamma}_{AB}\partial_i\varphi^A\partial_j\varphi^B - 2\overset{J}{U}\right) + L_{\rm m}(\psi, \partial\psi, \overset{J}{g}_{ij}). \quad (1)$$

This extends the Brans-Dicke theory[28,29] to the case with a multiplet of scalar fields φ^A (capital indices $A, B, C = 1, \ldots, N$ label the components of the multiplet). Here $\kappa = 8\pi G/c^4$ is Einstein's gravitational constant and

$$F = F(\varphi^A), \qquad \overset{J}{U} = \overset{J}{U}(\varphi^A), \qquad \overset{J}{\gamma}_{AB} = \overset{J}{\gamma}_{AB}(\varphi^A). \quad (2)$$

The Lagrangian $L_{\rm m}(\psi, \partial\psi, \overset{J}{g}_{ij})$ depends on the matter fields ψ.

The metric $\overset{J}{g}_{ij}$ determines angles and intervals in the *Jordan frame.* The Riemannian curvature scalar $R(\overset{J}{g})$ is constructed from the Jordan metric. With the help of the conformal transformation

$$\overset{J}{g}_{ij} \longrightarrow g_{ij} = F^2\overset{J}{g}_{ij} \quad (3)$$

we obtain the metric in the *Einstein frame.*

In the Einstein frame the Lagrangian density in the scalar-tensor theory reads $\mathfrak{L} = \sqrt{-g}L$ with

$$L = \frac{1}{2\kappa}\left(-R + g^{ij}\gamma_{AB}\partial_i\varphi^A\partial_j\varphi^B - 2U\right) + \frac{1}{F^4}L_{\rm mat}(\psi, \partial\psi, F^{-2}g_{ij}). \quad (4)$$

The scalar curvature $R(g)$ is constructed from the Einstein metric g_{ij}, and

$$\gamma_{AB} = \frac{1}{F^2}(\overset{J}{\gamma}_{AB} + 6F_{,A}F_{,B}), \qquad U = \frac{1}{F^4}\overset{J}{U}. \quad (5)$$

The metrical energy-momentum tensor of matter is defined by $\sqrt{-g}t_{ij} := 2\partial(\sqrt{-g}L_{\rm mat})/\partial g^{ij}$. The Noether theorem yields the conservation law

$$\nabla_j t^{kj} = \frac{1}{F}\left(4t^{kj} - g^{kj}g_{mn}t^{mn}\right)\partial_j F = -V_{ij}{}^k t^{ij}. \quad (6)$$

For details see Ref. 20. Here $V_{ij}{}^k = A_j\delta_i^k - \frac{1}{4}g_{ij}A^k$, and $A_i := \partial_i \log F^{-4}$.

3. Equations of Motion

We derive the equations of motion in the Mathisson-Papapetrou-Dixon[9,10,19,21] approach by integrating the conservation law (6) using the geodesic expansion technique of Synge.[25] With the world function σ and the parallel propagator by $g^y{}_x$, we introduce integrated moments to an arbitrary order $n = 0, 1, 2, \ldots$ by:

$$p^{y_1 \cdots y_n y_0} := (-1)^n \int\limits_{\Sigma(s)} \sigma^{y_1} \cdots \sigma^{y_n} g^{y_0}{}_{x_0} \sqrt{-g} t^{x_0 x_1} d\Sigma_{x_1}, \tag{7}$$

$$k^{y_2 \cdots y_{n+1} y_0 y_1} := (-1)^n \int\limits_{\Sigma(s)} \sigma^{y_1} \cdots \sigma^{y_n} g^{y_0}{}_{x_0} g^{y_1}{}_{x_1} \sqrt{-g} t^{x_0 x_1} w^{x_2} d\Sigma_{x_2}. \tag{8}$$

We use a condensed notation so that y_n denotes indices at the point y which we associate with the world-line $y(s)$ of a test body, parametrized by the proper time s. The integrals are performed over spatial hypersurfaces $\Sigma(s)$.

4. Pole-Dipole Equations of Motion

In the pole-dipole approximation, an extended body is characterized by the multipole moments $p^a, p^{ab}, k^{ab}, k^{abc}$. Using the general multipolar scheme[23] we derive the equations of motion for these moments:

$$0 = k^{(a|c|b)} - v^{(a} p^{b)c}, \tag{9}$$

$$\frac{D}{ds} p^{ab} = k^{ba} - v^a p^b - V_{dc}{}^b k^{acd}, \tag{10}$$

$$\frac{D}{ds} p^a = -V_{cb}{}^a k^{bc} - V_{dc}{}^a{}_{;b} k^{bcd} - \frac{1}{2} R^a{}_{cdb} \left(k^{bcd} + v^d p^{bc} \right). \tag{11}$$

Here $v^a := dy^a/ds$ denotes the normalized four-velocity of a body. Since $k^{a[bc]} = 0$, we can solve (9) to find explicitly

$$k^{abc} = v^a p^{cb} + v^c p^{[ab]} + v^b p^{[ac]} + v^a p^{[bc]}. \tag{12}$$

Plugging this into (10) and (11), we obtain the generalized Mathisson-Papapetrou-Dixon system

$$\frac{D\mathcal{P}^a}{ds} = \frac{1}{2} R^a{}_{bcd} v^b \mathcal{J}^{cd} - \xi \nabla^a F^{-4} - \xi^b \nabla_b \nabla^a F^{-4}, \tag{13}$$

$$\frac{D\mathcal{J}^{ab}}{ds} = -\, 2 v^{[a} \mathcal{P}^{b]} - 2\xi^{[a} \nabla^{b]} F^{-4}. \tag{14}$$

Here, following Refs. 20, 22 and 23, we introduce the generalized total energy–momentum 4-vector and the generalized total angular momentum by

$$\mathcal{P}^a := F^{-4} p^a + p^{ba} \nabla_b F^{-4}, \tag{15}$$

$$\mathcal{J}^{ab} := F^{-4} L^{ab}. \tag{16}$$

The orbital angular moment is defined by $L^{ab} := 2p^{[ab]}$, and we denoted

$$\xi^a = -\,\frac{1}{4} g_{bc} k^{abc}, \qquad \xi = -\,\frac{1}{4} g_{ab} k^{ab}. \tag{17}$$

5. Monopolar Equations of Motion

At the monopolar order, the only nontrivial moments are p^a, and k^{ab}. The system (9)-(11) then reduces to

$$0 = k^{ba} - v^a p^b, \tag{18}$$

$$\frac{Dp^a}{ds} = -\, V_{cb}{}^a k^{bc}. \tag{19}$$

Making use of $k^{[ab]} = 0$, the first equation yields $v^{[a}p^{b]} = 0$, hence we have

$$p^a = M v^a \tag{20}$$

with the mass $M := v^a p_a$. Substituting (18) and (20) into (19) we find $\xi = -\frac{v^a p_a}{4} = -\frac{M}{4}$ and

$$\frac{Dv^a}{ds} = -\,(g^{ab} - v^a v^b)\frac{\nabla_b F}{F}. \tag{21}$$

Quite remarkably, the dynamics of an extended test body in the monopole approximation is independent of body's mass. For a trivial coupling function F, equation (21) reproduces the general relativistic result. Interestingly, the mass of a body is not constant: $M = F^3 M_0$ with $M_0 = \text{const}$.

6. Conclusions

Our main result is the system (13)-(14) that describes the dynamics of extended test bodies in scalar-tensor gravity. In the monopolar case, our analysis revealed a surprisingly simple equation of motion (21). In contrast to geodesic motion in General Relativity, freely falling massive test bodies in scalar-tensor gravity experience an additional force, determined by the new scalar degrees of freedom encoded in the function F.

A remarkable feature of (21) is the prediction that all massive test bodies move in the same way, independently of their mass. We thus demonstrate the validity of the equivalence principle in scalar-tensor gravity. This is consistent with the previous observation[15,16] that the total scalar charge of a body is equal to its mass when the scalar field self-interactions are neglected. The latter is in agreement with the test body assumption that underlies the Mathisson-Papapetrou-Dixon approach. When one goes beyond the test body approximation, however, the scalar charge is no longer equal to the mass and a further study is needed to fix their relation.[13,14]

Acknowledgements

D.P. was supported by the Deutsche Forschungsgemeinschaft (DFG) through the grant SFB 1128 (geo-Q).

References

1. P. Bergmann, *Int. J. Theor. Phys.*, **1**:25, 1968.
2. C. Brans, *Phys. Rev.*, **125**:2194, 1962.
3. C. Brans, *Phys. Rev.*, **125**:388, 1962.
4. C. Brans, *arXiv:gr-qc/0506063*, 2005.
5. C. Brans and R. H. Dicke, *Phys. Rev.*, **124**:925, 1961.
6. T. Damour and G. Esposito-Farèse, *Class. Quantum Grav.*, **9**:2093, 1992.
7. R. H. Dicke, *Phys. Rev.*, **125**:2163, 1962.
8. R. H. Dicke, *The theoretical significance of experimental relativity* (Gordon and Breach: New York, 1964).
9. W. G. Dixon, *Nuovo Cimento*, **34**:317, 1964.
10. W. G. Dixon, *Phil. Trans. R. Soc. Lon. A*, **277**:59, 1974.
11. Y. Fujii and K. Maeda, *The scalar-tensor theory of gravity* (Cambridge University Press: Cambridge, 2003).
12. H. Goenner, *Gen. Rel. Grav.*, **44**:2077, 2012.
13. S. E. Gralla, *Phys. Rev. D*, **81**:084060, 2010.
14. S. E. Gralla, *Phys. Rev. D*, **87**:104020, 2013.
15. L. Hui and A. Nicolis, *Phys. Rev. Lett.*, **105**:231101, 2010.
16. L. Hui, A. Nicolis, and C. W. Stubbs, *Phys. Rev. D*, **80**:104002, 2009.
17. P. Jordan, *Schwerkraft und Weltall* (Vieweg: Braunschweig, 1955).
18. P. Jordan, *Z. Phys.*, **157**:112, 1959.
19. M. Mathisson, *Acta Phys. Pol.*, **6**:163, 1937.
20. Y. N. Obukhov and D. Puetzfeld, *Phys. Rev. D*, **90**:104041, 2014.
21. A. Papapetrou, *Proc. Roy. Soc. Lond. A*, **209**:248, 1951.
22. D. Puetzfeld and Y. N. Obukhov, *Phys. Rev. D*, **88**:064025, 2013.
23. D. Puetzfeld and Y. N. Obukhov, *Phys. Rev. D*, **90**:084034, 2014.
24. T. P. Sotiriou, *Lect. Notes Phys.*, Ed. E. Papantonopoulos, **892**:3, 2015.
25. J. L. Synge, *Relativity: The general theory* (North-Holland: Amsterdam, 1960).
26. Y. Thiry, *J. Math. Pures et Appl., sér. 9*, **30**:275, 1951.
27. R. V. Wagoner, *Phys. Rev.*, **12**:3209, 1970.
28. C. M. Will, *Theory and experiment in gravitational physics* (Cambridge University Press: Cambridge, 1993).
29. C. M. Will, *Living Reviews in Relativity*, 17(4), 2014.

Averaging the Schwarzschild spacetime

I. Drobov and S. Tegai*

Siberian Federal University,
Krasnoyarsk, Russia
**E-mail: tegai_s_f@inbox.ru*

We explicitly perform the averaging of the Schwarzschild spacetime using Weitzenbock connection for the parallel transport. We show that choosing either the connection or the curvature as the objects for the averaging has it's own problems within the considered approach.

Keywords: Averaging Einstein equations; macroscopic gravity.

1. Introduction

Likewise the Maxwell equations the gravitational field equations are inherently microscopic. In spite of that they are often used for macroscopic problems and therefore require the proper averaging scheme as was first pointed out by Shirokov and Fisher.[1] Since then various approaches to the averaging were proposed.[2,3] However due to the difficulties with non-linear and tensorial character of the gravitational theories the problem of averaging still remains an open subject. Recently Brannlund, van den Hoogen and Coley[4] suggested a new averaging procedure and we would like to take a close look at it. As a test object we choose the Schwarzschild spacetime because it is a very important solution on the one hand and relatively simple on the other.

2. Averaging the Schwarzschild Spacetime

Let the parallel transport of a vector v^α to be performed by

$$v^\alpha(x) = P^\alpha_{\alpha'}(x, x')v^{\alpha'}(x'), \tag{1}$$

where $P^\alpha_{\alpha'} = e^\alpha_{(\beta)}e^{(\beta)}_{\alpha'}$ and $e^{(\beta)}_\alpha$ is a local frame co-basis. The average of a tensor field $T^{\alpha_1\ldots\alpha_n}_{\beta_1\ldots\beta_m}$ over a region Σ is then defined as[4]

$$\overline{T}^{\alpha_1\ldots\alpha_n}_{\beta_1\ldots\beta_m} = \frac{1}{V_\Sigma}\int\limits_\Sigma P^{\alpha_1}_{\alpha'_1}\ldots P^{\alpha_n}_{\alpha'_n}P^{\beta'_1}_{\beta_1}\ldots P^{\beta'_m}_{\beta_m}T^{\alpha'_1\ldots\alpha'_n}_{\beta'_1\ldots\beta'_m}e'd^4x', \tag{2}$$

where V_Σ is the volume of the region measured by the unaveraged frame field and e' is the determinant of the frame field.

Following the above rule we perform an explicit averaging of the Schwarzschild spacetime. In the Cartesian–Lemaitre coordinates it's non-trivial connection forms are given by

$$\omega^{(i)}_{\ (0)} = \frac{\gamma_1(r)}{r} x^i dt + \left(\frac{\gamma_2(r)}{r^2} x^i x^j + \gamma_3(r) \delta_{ij} \right) dx^j \tag{3}$$

where $\gamma_1 = m/r^2$, $\gamma_2 = 3\sqrt{2m}/2r^{3/2}$, $\gamma_3 = -\sqrt{2m}/r^{3/2}$.

According to the first variant of the proposed scheme the average of the connection form $\overline{\omega}$ is declared to be the connection form of the averaged spacetime. With the 4-cylinders $\{-T < t < T; |r - r'| < R\}$ as the domains of averaging $\overline{\omega}$ takes the same structure as (3) but with

$$\begin{aligned} \gamma_3 = \frac{2\sqrt{2m}}{15 r^3 R^3} \Big[& \left(2R^4 - R^3 r - 6R^2 r^2 - R r^3 + 2r^4\right) \sqrt{R + r} \\ & - \left(2R^4 + R^3 r - 6R^2 r^2 + R r^3 + 2r^4\right) \sqrt{|R - r|} \Big] \end{aligned} \tag{4}$$

and $\gamma_1 = mr/R^3$ for with $r < R$, $\gamma_1 = m/r^2$ for $r > R$ and $\gamma_2 = r d\gamma_3/dr$.

The averaged spacetime admits the same symmetries as the original one. However for $m/R \lesssim 0.85$ the Killing vector $\partial/\partial t$ is timelike everywhere and there is no horizon. The Ricci tensor has the Segre type [(11)(1, 1)] and remains singular.

The second variant suggests that the average of the curvature form $\overline{\mathcal{R}}$ is the curvature form of the averaged spacetime. This time the exterior averaged region $r > R$ coinsides with the original Schwarzschild spacetime. The curvature form for the interior $r < R$ does not correspond to any spacetime because the second Cartan structure equations are inconsistent.

References

1. M. F. Shirokov and I. Z. Fisher, *Astron. Zh.* **39**, 899 (1962).
2. R.M. Zalaletdinov, *Gen. Rel. Grav.* **24**, 1015 (1992); **25**, 673 (1993).
3. T. Buchert, *Gen. Rel. Grav.* **32**, 105 (2000); **33**, 1381 (2001).
4. J. Brannlund, R. van den Hoogen, A. Coley, *Int. J. Mod. Phys.* **D 19**, 1915 (2010).

The tetradic approach and BF-gravity

A. O. Shishanin

Department of Physics, Bauman Moscow State University,
Moscow, Russia
E-mail: shishandr@rambler.ru

Here we discuss a limit of the gravitational constant G goes to zero for the Einstein gravity. It is convenient to use for this a tetradic approach (using tetrads and spin-connection) and the Palatini formalism. Also some external field B is entered, which is 2-form. In this limit an equation for the spin-connection is obtained. Such an equation was obtained earlier in the MacDowell-Mansouri-Stelle-West gravity.

Keywords: Tetrads; spin connections; differential forms; the Palatini formalism; BF-gravity.

1. Introduction

For several reasons it is convenient to consider Einstein gravity in the tetradic approach.[1] Usualy on Pseudo-Riemannian manifold M with signature (p, q) can be defined the metric

$$ds^2 = g_{\mu\nu} dx^\mu dx^\nu.$$

Here, instead of metric, the tetrads are introduced (naively "the square root" of metric), and instead of Christoffel symbols the spin-connection is entered. The metric tensor can be expressed in terms of tetrads (vielbein)

$$g_{\mu\nu} = e^a_\mu e^a_\nu.$$

Let us define the 1-forms of tetrads

$$e^a = e^a_\mu dx^\mu$$

Covariant deravitive via the spin connection is

$$\nabla_\mu = \partial_\mu + \omega^{ab}_\mu M^{ab}$$

and 1-form spin connection is

$$\omega^{ab} = \omega^{ab}_\mu dx^\mu,$$

here M^{ab} are generators of $so(p,q)$. It is well known that in the metric approach it is impossible to determine the covariant derivative for fermions. Using the spin connection, one can define covariant derivative for fermions. As usual the curvature 2-form is

$$R = d\omega + \omega \wedge \omega.$$

2. BF-Theory

It is well-known that Yang-Mills theory at the limit $g \to 0$ describes BF-model. Recall that the action of Yang-Mills theory is

$$S_{YM} = \frac{1}{4g^2}\int_M trF \wedge *F.$$

Here F is 2-form of curvature of the principial bundle at and $F = dA + A \wedge A$ where A is connection of this bundle. We may rewrite this action as

$$S_{YM} = \int_M trF \wedge B - 2g^2 \int_M trB \wedge *B.$$

Here B is some external field ($(d-2)$-form). When $g \to 0$ here BF-theory is obtained

$$S_{BF} = \int_M trF \wedge B.$$

3. The Limit G Goes to 0

It is natural to ask what happens to gravity when the gravitational constant G goes to 0. Let us using the first order formalism or the Palatini formalism. Then

$$S_P = \int_M tr(\frac{1}{2\kappa} e \wedge e \wedge R - \frac{\Lambda}{\kappa} e \wedge e \wedge e \wedge e)$$

Here $\kappa = 8\pi G_N$, Λ is the cosmological constant.

For n-dimensional case

$$S_P = \int_M tr(\frac{1}{2\kappa} \underbrace{e \wedge ... \wedge e}_{n-2} \wedge R - \frac{\Lambda}{\kappa} \underbrace{e \wedge ... \wedge e}_{n}).$$

Let us consider some 4d-action

$$\int_M tr(\alpha B \wedge B + (\beta R + \gamma e \wedge e) \wedge B)$$

An equation of motion for B is

$$2\alpha B + \beta R + \gamma e \wedge e = 0$$

Then one can obtaine the following action

$$S = \frac{1}{4\alpha}\int_M tr(\beta^2 R \wedge R + 2\beta\gamma R \wedge e \wedge e + \gamma^2 e \wedge e \wedge e \wedge e)$$

Let $\beta = 1$. Then

$$\frac{\gamma}{2\alpha} = \frac{1}{2\kappa} \quad \frac{\gamma^2}{4\alpha} = -\frac{\Lambda}{\kappa}$$

and the solution is $\alpha = -4\Lambda\kappa$, $\gamma = -4\Lambda$.

When G_N goes to 0 we can obtain

$$S_{G_N \to 0} = \int_M tr(R \wedge B - 4\Lambda e \wedge e \wedge B). \tag{1}$$

Here equations of motion have the following form

$$d_\omega B = dB + \omega \wedge B = o, \quad e \wedge B = 0,$$

and varying by B it is obtained equation for spin connection

$$d\omega + \omega \wedge \omega = 4\Lambda e \wedge e. \tag{2}$$

When $\Lambda = 0$ then a simplest solution of the equation (2) is the Minkowski space. The simplest solutions of this equation when Λ is positive is dS_4-space, and if Λ is negative then solution is AdS_4-space. Similar to (2) obtained by the equation MacDowell-Mansouri-Stelle-West gravity.[2–4] There is unified connection $\mathcal{A}$ at MMSW-gravity

$$\mathcal{A} = \omega + \frac{1}{l}e.$$

Here $l = \sqrt{\frac{\pm 1}{4\Lambda}}$. The case '+' corresponds to dS-space ($\mathcal{A}$ is $so(1,4)$-connection) and '-' to AdS-space. Then the curvature 2-form is

$$\mathcal{F} = R - 4\Lambda e \wedge e + d_\omega e.$$

The last term is nothing but a form of torsion $T = d\omega + \omega \wedge \omega$. Then the equation (2) is condition of null curvature for MacDowell-Mansouri-Stelle-West connection with zero torsion.

I would like to thank D. A. Polyakov for correspondance.

References

1. T. Eguchi, P. B. Gilkey, A. J.Hanson, Gravitation, gauge theories and differential geometry *Phys. rept.* **66**, 213 (1980).
2. S. W. MacDowell, F. Mansouri, Unified geometric theory of gravity and supergravity *Phys. Rev.Lett.* **38**, 739 (1979)
3. K. S. Stelle, P. C. West. Spontaneously broken de Sitter symmetry and the gravitational holonomy group *Phys.Rev.* **D21**, 1466 (1980)
4. X. Bekaert, S. Cnockaert, C. Iazeolla, M. A. Vasiliev. Nonlinear higher spin theories in various dimensions. hep-th/0503128

Electrodynamics of oriented point as a consequence of the real relativity principle

E. A. Gubarev

Intersectoral Science-Technical Centre of Venture Non-Conventional Technologies, Moscow, Russia
E-mail: e.gubarev.21@gmail.com

The equations of electrodynamics of oriented point are formulated based on the principle of real relativity. The resulted equations are analyzed in particular cases. The most important is the case of generating the electromagnetic field by the charge in a circular nonrelativistic motion.

Keywords: Oriented point; real relativity principle; electrodynamics.

1. The Principle of Real Relativity and Electrodynamics of the Oriented Point

In Ref. 1, the author put forward *the principle of real relativity*, with its first part stating the equivalency of real reference frames for the description of physical events (such as registering the flashes of light from a distant source): *All the real reference frames are equal to each other as a means of describing events.*

The second part of the principle sets the requirement for the basic equations describing the laws of nature: *It is required of the equation expressing locally the law of nature to be invariant under transformations of coordinates of events between the real reference frames (L-covariance).*

The equations for electrodynamics of oriented point in the L-covariant form are formulated uniquely.[2] The first pair of equations appears as

$$\forall_c F_{ab} + \forall_b F_{ca} + \forall_a F_{bc} = 0, \tag{1}$$

where $\forall_c$ is the absolute local derivative with respect to the local coordinate[1]

$$\forall_c Z^a = Z^a{}_{,c} - T^a_{bc} Z^b, \quad \forall_c Z_a = Z_{a,c} + T^b{}_{ac} Z_b. \tag{2}$$

In these formulas, T^a_{bc} are Ricci rotation coefficients.[1] The second pair of equations of electrodynamics of the oriented point and the continuity equation form the following relations

$$\nabla_b F^{ab} = -\frac{4\pi}{c} j^a, \quad \nabla_a j^a = 0. \tag{3}$$

2. Final Provisions

The equations of new electrodynamics are formulated in $4D$ form for an arbitrary proper reference frame of the charge through the components of L-tensor of the electromagnetic field. The more common case of exercising the proper reference frame of the charge — in its circular nonrelativistic motion — appeared to be more interesting in view of emerging qualitatively new effects.

It is shown that there is a solution for a free electromagnetic field in the form of a set of plane electromagnetic waves. In a particular case of the plane wave propagating perpendicular to the plane of rotation of the charge, the dispersion relation between the wave vector and the frequency of the wave is obtained, which has four solutions (two for each direction of propagation). In the forward direction, two monochromatic waves with different wavelengths and, consequently, with different phase velocities are generated at each fixed frequency: $v_{phase} < c$, holds for the slow wave, $v_{phase} > c$ holds for the fast wave. As a result, the remote observer at rest will register *the effect of the dual signal*, which involves different times of arrival of the fast and slow waves, emitted at the same time.

A permanent free electromagnetic field in the form of *a plane monochromatic wave of zero frequency* is shown to be possible.

It is shown that a quasi-static free electromagnetic field has special properties: it is noninductive and does not exert any electromotive force in a plane perpendicular to the wave vector. This property is different from a classical free electromagnetic field, which should vary in time and have an inductive electromotive force in the transverse plane. The examined quasi-static field should have a high penetrating ability in conducting media (that is be *superpenetrative*) because it does not perform any work with free charges and, therefore, does not dissipate in conductors.

References

1. E.A. Gubarev, *Theory of real relativity* (NEW CENTRE publishing, Moscow, 2010).
2. E.A. Gubarev, *Electrodynamics of oriented point* (NEW CENTRE publishing, Moscow, 2013).

Intrinsic time in Wheeler–DeWitt conformal superspace

A. E. Pavlov* and V. N. Pervushin

Bogoliubov Laboratory of Theoretical Physics,
Joint Institute for Nuclear Research, Dubna, 141980, Russia
**E-mail: alexpavlov60@mail.ru*

In Geometrodynamics, the York's extrinsic time, constructed of the tensor of extrinsic curvature and the Misner's intrinsic time, built of the spatial metric tensor, coexist. In our paper, we prove the preference of selecting the internal time. To extract the intrinsic time, we generalize the Dirac's mapping of transition to conformal variables. In Friedmann cosmology, the many-fingered intrinsic time obtains a sense of a global time of the Universe. An accounting of metric scalar linear perturbations leads to adding some corrections not dominated to the effective energy density in the Hubble law. The metric vector and tensor perturbations do not influence the internal time in linear approximation.

Keywords: Misner's time; York's time; scaled Dirac's mapping; conformal gravitation; conformal cosmological model parameters.

The extraction of internal time in General Relativity is achieved by generalization of Dirac's mapping. The constructed local time is a scalar relative to diffeomorphisms of changing of spatial coordinates. In contrast with the York's extrinsic time, constructed of the tensor of extrinsic curvature, it is built of the metric tensors of the space in spirit of the intrinsic Misner's time. Friedmann equation in cosmology has an interpretation of the formula, connecting two time intervals: intrinsic time (a function of the scaling factor) and of conformal time (time of a photon's life). The Hubble diagram demonstrates near-linear character of connection between the times. As far as the conformal time is a physical observable as coordinate distance, so the intrinsic time is also a physical magnitude.

The choice of the intrinsic time leads inevitably to the Conformal gravitation, where conformal metric is postulated to be a metric of physical space. York's conformal transverse-traceless method of solving the Cauchy problem in General Relativity looks quite natural in conformal variables. If we wish accept the York's extrinsic time, we get the Friedmann equation as algebraic one. Hence, the connection between temporal intervals (geometrical coordinate time t in the pseudoRiemannian space and intrinsic one D in the Wheeler – DeWitt superspace) is lost.

In Geometrodynamics, the many-fingered intrinsic time is a scalar field. The spatial metric γ_{ij} carries information about the inner time. According to the scaled Dirac's mapping, the metric γ_{ij} is factorized in the exponent function of inner time D and the conformal metric $\tilde{\gamma}_{ij}$. So, the extracted internal time has the spatial geometric origin. Cauchy problem was successfully solved after mathematically formal transition to conformal variables not by chance. The point is that we have come to the physical variables.

In Friedmann cosmology, the many-fingered intrinsic time in General Relativity has a sense of a global time of the Universe. The problem of time and the problem of energy associated with it is a key one in the mathematical General Relativity. Researchers encountered to the problems since the creation of the theory. In frame of Conformal cosmology it is meaningful to speak of the energy of the Universe that was lost in Standard cosmology. The question of the origin of initial small values of the particle masses is much simpler than the unresolved problem of initial singularity of the Universe (notorious the Big Bang problem).

Recent observations of the so-called "standard candles" supernovae demonstrated that they are turned out to be too far away from us. Cosmologists had to introduce one more inflation in the Universe's evolution scenario, not explaining the reason. The Conformal cosmological model allows us to describe the supernova data without Lambda term, because in this model the observed distances are identified with conformal longer intervals. The evolution of the lengths in the Standard cosmology is replaced by the evolution of the masses in the Conformal cosmology. Weierstrass functions $\wp$ and ζ, used for a long time in classical mechanics and astronomy, are useful in cosmology for solving the Friedmann equation. Calculations demonstrate that linear metric perturbations do not influence the intrinsic time D chosen.

Reference

1. V. Pervushin, A. Pavlov, *Principles of Quantum Universe* (Lambert Academic Publishing, Saarbrücken, 2014).

A perfect-fluid spacetime for a slightly deformed mass

M. Abishev, K. Boshkayev, H. Quevedo and S. Toktarbay*

Physical-Technical Faculty, Al-Farabi Kazakh National University,
Al Farabi av. 71, 050040 Almaty, Kazakhstan
Instituto de Ciencias Nucleares, Universidad Nacional Autónoma de México,
AP 70543, México, DF 04510, Mexico
**E-mail: saken.yan@yandex.com*

We present approximate exterior and interior solutions of Einstein's equations which describe the gravitational field of a static deformed mass distribution. The deformation of the source is taken into account up to the first order in the quadrupole.

Keywords: Quadrupole; compact objects; perfect fluid.

To describe the gravitational field of a static axially symmetric mass distribution in general relativity, it is necessary to consider the multipole moments of the source. From a physical point of view, one expects that the quadrupole is the largest contributor and higher multipoles can be neglected in a first approximation. In this case, to describe the exterior field one can use, for instance, the exact quadrupole metric ($q-$metric).[1,2]

$$ds^2 = A^{1+q}dt^2 - A^{-q} \\ \times \left[\left(1 + \frac{m^2 \sin^2\theta}{r^2 A}\right)^{-q(2+q)} \left(\frac{dr^2}{A} + r^2 d\theta^2\right) + r^2 \sin^2\theta d\varphi^2 \right], \quad (1)$$

with $A = 1 - 2m/r$, which has been shown to be the simplest generalization of the Schwarzschild metric containing a quadrupole parameter q. Considering the quadrupole up to the first order only, we obtain

$$ds^2 = A\,(1 + q\ln A)\,dt^2 - r^2 \sin^2\theta\,(1 - q\ln A)\,d\varphi^2 \\ - \left[1 + q\ln A - 2q\ln\left(A + \frac{m^2}{r^2}\sin^2\theta\right)\right]\left(\frac{dr^2}{A} + r^2 d\theta^2\right) . \quad (2)$$

This is an approximate solution of Einstein's vacuum equations up to the first order in q. The total mass of the spacetime turns out to be $M_0 = m(1+q)$ and the quadrupole moment is $M_2 = -(2/3)qm^3$.

The interior solution can be generated by using the method proposed recently in Ref. 3. We obtain

$$ds^2 = e^{2\psi_0}(1+2\tilde{q}\psi_0)dt^2 - e^{-2\psi_0}(1-2\tilde{q}\psi_0)$$
$$\times\left[e^{2\gamma_0}(1+4\tilde{q}\gamma_0+\tilde{q}\gamma_1)\left(\frac{dr^2}{r^2f^2(r)}+r^2(\sin^2\theta-\tilde{q}\sin\theta\cos\theta)d\varphi^2\right)\right], \quad (3)$$

$$e^{2\psi_0} = \frac{3}{2}f(R)-\frac{1}{2}f(r),\ f(r)=\sqrt{1-\frac{2\tilde{m}r^2}{R^3}},\ e^{\gamma_0}=re^{2\psi_0}\ , \quad (4)$$

$$\gamma_1 = -2\int\frac{1+4\pi\sin^2\theta r^2p_0}{rf^2(r)(1+r\psi_{0,r})\sin^2\theta+\frac{r}{1+r\psi_{0,r}}\cos^2\theta}dr+\kappa\ , \quad (5)$$

$$\psi_{0,r} = \frac{2\tilde{m}r}{R^3f(r)[3f(R)+f(r)]}\ , \quad (6)$$

where $\tilde{m}$, $\tilde{q}$, R and κ are real constants. This is an interior solution up to the first order in $\tilde{q}$ for a perfect fluid with density and pressure

$$\rho = \rho_0[1+\tilde{q}(1+\psi_0-4\gamma_0-\gamma_1)],\ \rho_0 = const. \quad (7)$$

$$p = p_0[1+\tilde{q}(1+\psi_0-4\gamma_0-\gamma_1)],\ p_0 = \rho_0\frac{f(r)-f(R)}{3f(R)-f(r)}, \quad (8)$$

respectively. In the limiting case $\tilde{q}\to 0$, the metric (3) represents a perfect fluid with constant density ρ_0 and pressure p_0 as given in Eq.(8). If $\tilde{m}=m$, this particular solution can be matched with the exterior Schwarzschild metric along a sphere of radius R.

In the general case $\tilde{q}\neq 0$, a more detailed analysis must be carried out in order to match the above approximate interior solution with the approximate exterior $q-$metric given in Eq.(2). First, the matching surface must be established. Then, the matching conditions must be imposed for all metric components. This would imply a relationship between the exterior parameters m and q and the interior parameters $\tilde{m}$, $\tilde{q}$, ρ_0 and κ. This result will be presented elsewhere.

We acknowledge the support through a Grant of the Target Program of the MES of the RK, Grant No. 1597/GF3 IPC-30, DGAPA-UNAM, Grant No. 113514, and Conacyt, Grant No. 166391.

References

1. H. Quevedo, *Int. J. Mod. Phys. D* **20**, 1779 (2011).
2. D. M. Zipoy, *J. Math. Phys.* **7** (1966) 1137; B. Voorhees, *Phys. Rev. D* **2** (1970) 2119.
3. H. Quevedo and S. Toktarbay, *J. Math. Phys.* **56**, 052502 (2015).

Section 2

Classical and Quantum Cosmology, Dark Matter and Dark Energy

Breaking of conformal symmetry in cosmology

A. B. Arbuzov[a,*], V. N. Pervushin[a], R. G. Nazmitdinov[a,b] and A. F. Zakharov[a,c]

[a] *Bogoliubov Laboratory of Theoretical Physics, Joint Institute for Nuclear Research, Dubna, 141980, Russia*

[b] *Department de Física, Universitat de les Illes Balears, Palma de Mallorca, E-07122, Spain*

[c] *Institute of Theoretical and Experimental Physics, 117218, Moscow, Russia*

[*] *E-mail: arbuzov@theor.jinr.ru*

Spontaneous breaking of the conformal symmetry via quantum anomalies both in the General Relativity and the Standard Model is introduced. Dynamics of a conformal cosmological model is considered at the initial moment of the Universe evolution. It is shown that the relation between the Planck mass and the electroweak energy scale is provided by the conformal weights of these quantities.

Keywords: Spontaneous symmetry breaking; quantum anomalies; conformal symmetry.

We suggest to start from the standard Hilbert action and apply the formalism developed by Dirac[1] in order to reveal the hidden conformal symmetry of General Relativity (GR). The Goldstone mode which appears due to spontaneous breaking of the conformal symmetry is identified with the scalar dilaton field. Various cosmological models based on the conformal symmetry are popular now, see *e.g.* Refs. 2–4 and references therein. Here we follow the version of conformal GR in which the zero mode of the dilaton plays the role of the intrinsic time in the Wheeler–DeWitt superspace of events. Our main task here is to define the origins of the conformal symmetry breaking and of the corresponding energy scale.

Our version of the conformal GR was presented in Refs. 5, 6. It was shown that starting from the Hilbert action without introduction of any extra field or interaction, one can construct a model of GR and Cosmology based on the conformal and affine symmetries. Obviously, the conformal symmetry in Nature is broken. But this breaking is very likely of a spontaneous (soft) kind. In fact, there are many indirect indications in favor of the hypothesis that the conformal symmetry is one of the true underlying fundamental symmetries. In particular, it would be very natural to get the observed Lorentz symmetry from spontaneous breaking the conformal one, since the former is a subgroup of the latter.

Various mechanisms of spontaneous symmetry breaking are possible. In our previous works[5,6] the idea to break the scale invariance in GR and Cosmology by a finite size of the universe horizon was proposed. We have also shown that a natural hierarchy of energy scales appears at the present day if at the beginning there was just one scale (the Planck mass).

From the other side in Refs. 7, 8, a simple reduction of the Standard Model (SM) to a conformal symmetric construction was suggested. It was shown that the standard Brout-Englert-Higgs mechanism can be realized with the help of a dimensional transmutation due to a conformal anomaly without the explicit breaking of the conformal symmetry by the tachyon-like mass term in the Higgs sector. The aim of this study is to establish a link between the mentioned above mechanisms of the conformal symmetry breaking in Cosmology and in the SM.

Let us construct a conformal-invariant action by summing up the actions of the Conformal Standard Model (CSM) and the Conformal General Relativity (CGR) $W = W_{\rm CSM} + W_{\rm CGR}$. Here the first term is obtained[7] from the SM by neglecting the tachyon mass term in the Higgs sector. The second one is the conformal version of the General Relativity[1,9]

$$W_{\rm CGR} = W_{\rm GR}[g_{\mu\nu} = e^{-2D}\widetilde{g}_{\mu\nu}], \tag{1}$$

where D is the dilaton field associated with the metric determinant. The action of the CGR is obtained directly from the Hilbert one where the conformal interval $\widetilde{ds}^2_{\rm Dirac} = \widetilde{g}_{\mu\nu}dx^\mu dx^\nu$ is used instead of the Einstein's one $ds^2_{\rm Einstein} = g_{\mu\nu}dx^\mu dx^\nu$. In CGR, the dimensionful Newton constant is absorbed by the dilaton field. Thus, the classical equations of the theory do not contain any dimensionful constant parameter.

By definition, the Planck epoch corresponds to the initial value of the redshift z_I for which the primordial Planck mass $M^*_{I\rm Pl} \equiv M^*_{\rm Pl}(z_I) = M^*_{0\rm Pl}(1 + z_I)^2$ coincides with the the primordial Hubble constant $H_I \equiv H(z_I) = M^*_{I\rm Pl} = M^*_{0\rm Pl}(1 + z_I)^2$, where the we used the laws for the dependence on redshift, which emerge in the conformal cosmological model under consideration. Thus, the primordial redshift is of the order $z_I \simeq 10^{15}$.

It is quite natural to assume that at the initial moment of a quantum universe creation there is just a single energy scale introduced as the initial data. The evolution of an energy scale of a physical quantity from the initial moment to the present day is obviously defined by the corresponding dimension d and conformal weight w. In particular we can consider the evolution of the Hubble constant, of the Planck mass, and of a particle mass which takes today the value m_0:

$$H_0 \simeq H_I \cdot (1 + z_I)^{-2}, \quad \{d = 1, w = 2\}, \tag{2}$$

$$m_0 \simeq H_I \cdot (1 + z_I)^{1}, \quad \{d = 1, w = -1\}, \tag{3}$$

$$M^*_{\text{Pl }0} \simeq H_I \cdot (1 + z_I)^{2}, \quad \{d = 1, w = -2\} \tag{4}$$

Using the known values of H_0 and M_{Planck} together with the defined above magnitude of z_I, we see that $m_0 \simeq 300$ GeV, *i.e.* it is just of the order of the electroweak energy scale. But we still have to demonstrate how does a massive particle emerge in the conformally symmetric model and how its mass is related to the *primary* energy scale H_I.

Obviously, the electroweak scale is related both to the Higgs boson mass (and vacuum expectation value) and to the top quark mass. Here we assume that the mechanism of dimensional transmutation[10] is working in the sector of the SM, which contains the most intensive interactions of this model, see details in Refs. 7, 8. Then at the Planck epoch the renormalization energy scale is naturally given by H_I. The Coleman-Weinberg mechanism then provides non-zero masses and condensates for both scalar and fermion fields. Since the two coupling constants are both of the order of unity, the emerged masses and condensates are of the same order H_I. In particular using Eq. (2), we can relate the top quark mass with the Hubble constant and the Planck mass: $m_t \simeq H_0[1 + z_I]^3$ and $m_t \simeq M^*_{\text{Pl }0}[1 + z_I]^{-1}$. On the other hand, the top quark condensate should have the same energy scale: $\langle \bar{t}t \rangle = -\gamma_t m_t^3$ where γ_t is a constant of the order of unity. To get a concrete number for γ_t, we have to regularize the corresponding divergent integral. It can be done by introduction of a simple distribution function:

$$\gamma_t = -\frac{\langle \bar{t}\, t \rangle}{m_t^3} = 4N_c \int \frac{d^3\mathbf{p}}{(2\pi)^3} \frac{1}{2\sqrt{\mathbf{p}^2 + 1}} f_{1,+}(\mathbf{p}) \approx 0.39,$$

$$f_{1,\pm}(\mathbf{p}) = \left[1 \pm \exp\left(\sqrt{\mathbf{p}^2 + 1} - 1\right)\right]^{-1}. \tag{5}$$

This distribution function reflects the degeneration of the vacuum state over momenta. That allows to predict the Higgs particle mass in the tree-level approximation

$$m_{0,h}^2 = -\frac{3 y_t \langle \bar{t}\, t \rangle}{\sqrt{2}\, v} \approx 131 \text{ GeV}. \tag{6}$$

All the rest of standard Brout-Englert-Higgs remains unchanged except that the value of the self-coupling constant λ is $3/2$ times less than in the

SM. Note that this quantity is not yet measured experimentally.

The presented model for the definition of the Planck epoch is based on the assumption that the conformal symmetry is a fundamental property of Nature. A spontaneous breaking of this symmetry due to a quantum conformal anomaly provides a single energy scale. We have shown the energy scales of the Hubble constant, the Planck mass, and the SM can be obtained with the cosmological evolution from the single scale taking into account different conformal weights of the corresponding quantities.

Some other aspects of the discussed above model were considered in our earlier papers.[5,6,11]

References

1. P. A. M. Dirac, Long range forces and broken symmetries, *Proc. Roy. Soc. Lond. A* **333**, 403 (1973).
2. D. Behnke, D. B. Blaschke, V. N. Pervushin, D. Proskurin, Description of supernova data in conformal cosmology without cosmological constant, *Phys. Lett. B* **530**, 20 (2002).
3. E. E. Flanagan, The Conformal frame freedom in theories of gravitation, *Class. Quant. Grav.* **21**, 3817 (2004).
4. D. Blas, M. Shaposhnikov, D. Zenhausern, Scale-invariant alternatives to general relativity, *Phys. Rev. D* **84**, 044001 (2011).
5. A. B. Arbuzov *et al.* Conformal Hamiltonian dynamics of general relativity, *Phys. Lett. B* **691**, 230 (2010).
6. V. N. Pervushin *et al.* Conformal and affine Hamiltonian dynamics of general relativity, *Gen. Relativ. Gravit.* **44**, 2745 (2012).
7. V. Pervushin *et al.*, Condensate mechanism of conformal symmetry breaking, *PoS (Baldin ISHEPP XXI)* 023 [arXiv:1209.4460 [hep-ph]].
8. A. B. Arbuzov, V. N. Pervushin, R. G. Nazmitdinov *et. al.* Spontaneous radiatively induced breaking of conformal invariance in the Standard Model, *arXiv:1411.5124* [hep-ph].
9. A. B. Borisov and V. I. Ogievetsky, Theory of Dynamical Affine and Conformal Symmetries as Gravity Theory, *Theor. Math. Phys.* **21**, 1179 (1975).
10. S. R. Coleman and E. J. Weinberg, Radiative Corrections as the Origin of Spontaneous Symmetry Breaking, *Phys. Rev. D* **7**, 1888 (1973).
11. A. B. Arbuzov, B. M. Barbashov, A. Borowiec, V. N. Pervushin, S. A. Shuvalov and A. F. Zakharov, General relativity and the standard model in scale-invariant variables, *Grav. Cosmol.* **15**, 199 (2009).

High-frequency gravitational waves in exact inflationary models

I. V. Fomin

Department of Physics,
Bauman Moscow State Technical University,
Moscow, 105005, Russian Federation
E-mail: ingvor-@inbox.ru
http://www.bmstu.ru

This work is devoted to the new method of obtaining the exact solutions of the scalar field equations in spatially flat FRW cosmology. The power spectrum and energy density of the high-frequency relic gravitational waves is defined.

Keywords: Inflation; gravitational waves; exact solutions.

1. The Method of Exact Solutions of the Scalar Field Equations

The cosmological perturbations are a source of the evolution of large-scale structure of the Universe. Relic gravitational waves or tensor perturbations are a valuable source of information about the early stage. State of accelerated expansion of the universe in the early stages describes the theory of inflation.

The equations of a scalar field dynamics in the flat Friedman-Robertson-Walker Universe are written as follows[1]:

$$3H^2 = \frac{1}{2}\dot{\phi}^2 + V(\phi) \tag{1}$$

$$\ddot{\phi} + 3H\dot{\phi} = -\frac{dV(\phi)}{d\phi} \tag{2}$$

The exact solutions of these equations are required to compare the cosmological parameters, obtained from different models of inflation, with observational data .

From (1) – (2) we can write the equation:

$$8\left(\frac{dH(\phi)}{d\phi}\right)^2 - r(\phi)H^2(\phi) = 0, \tag{3}$$

where $r(\phi) = -4\frac{dH}{d\phi}\frac{\dot{\phi}}{H^2}$ is the tensor-to-scalar ratio.

The solution of the equation (3) allows getting the Hubble parameter $H(\phi)$, the potential $V(\phi)$ and the scalar field ϕ from $r(\phi)$:

$$H(\phi) = A\exp\left(\int \sqrt{r(\phi)}d\phi\right) \tag{4}$$

$$V(\phi) = A^2\left[3 - 8r(\phi)\right]\exp\left(2\int \sqrt{r(\phi)}d\phi\right) \tag{5}$$

$$\dot{\phi} = -4A\sqrt{r\left(\phi\right)}\exp\left(\int \sqrt{r(\phi)}d\phi\right) \tag{6}$$

The exact solutions of the scalar field dynamical equations can be obtained for different cosmological models by choosing of the $r(\phi)$. The power spectrum $\mathcal{P}_{\mathcal{G}}$ and the spectral index $n_{\mathcal{G}}$ of tensor perturbations can be calculated on the basis of these exact solutions.

2. The Power Spectrum of High-Frequency Gravitational Waves

We consider the calculation of the relic gravitational waves parameters from exact solutions of the scalar field's dynamical equations, following.[2] The power spectrum and the spectral index are calculated as:

$$\mathcal{P}_{\mathcal{G}}(k) = \frac{H^2}{2\pi^2}\left(\frac{k}{aH}\right)^{n_{\mathcal{G}}} \tag{7}$$

$$n_{\mathcal{G}} = \frac{H^2}{\dot{H} + H^2} \tag{8}$$

The energy density of the high-frequency gravitational waves is determined as[3]:

$$\Omega_{GW}(k) = \frac{k^2}{H_0^2}T_g^2(t,k)\mathcal{P}_{\mathcal{G}}(k), \tag{9}$$

where $T_g(t,k)$ is the transfer function on the time of observation.

References

1. A. R. Liddle and D. H. Lyth, *Phys. Rep.* **231**, 1 (1993).
2. S. V. Chervon, I. V. Fomin, About the calculation of cosmological parameters in exact models of inflation, *Grav & Cosm*, Vol.14, **2** (2008).
3. A.Riazuelo, J. Uzan, Quintessence and Gravitational Waves, *Phys.Rev.*, **62**, 083506 (2000).

Symmetry in Cartan language for geometric theories of gravity

M. Hohmann

Laboratory of Theoretical Physics, Institute of Physics, University of Tartu, Tartu, Tartumaa, Estonia
E-mail: manuel.hohmann@ut.ee
http://kodu.ut.ee/~manuel/

We present a recent definition of symmetry generating vector fields on manifolds equipped with a first-order reductive Cartan geometry. We apply this definition to a number of spacetime geometries used in gravity theories and show that this definition agrees with the usual notions of symmetry of affine, Riemann-Cartan, Riemannian, Weizenböck and Finsler spacetimes.

Keywords: Cartan geometry; symmetry; spacetime model.

1. Definition

Let M be a manifold and $\varphi : \mathbb{R} \times M \to M$ a one-parameter group of diffeomorphisms generated by a vector field ξ on M. On the general linear frame bundle

$$\mathrm{GL}(M) = \bigcup_{x \in M} \{\text{linear bijections } f : \mathbb{R}^n \to T_xM\} \tag{1}$$

we define a one-parameter group of diffeomorphisms $\bar{\varphi} : \mathbb{R} \times \mathrm{GL}(M) \to \mathrm{GL}(M)$ by $\bar{\varphi}_t(f) = \varphi_{t*} \circ f$. This one-parameter group is generated by a vector field $\bar{\xi}$ on $\mathrm{GL}(M)$, which we call the frame bundle lift of ξ.

Let G be a Lie group with closed subgroup $H \subset G$, $\pi : P \to M$ a principal H-bundle with $P \subset \mathrm{GL}(M)$ and $A \in \Omega^1(P, \mathfrak{g})$ a Cartan connection which is first order reductive, i.e., the adjoint representation of H on the Lie algebra $\mathfrak{g}$ splits into subrepresentations $\mathfrak{g} = \mathfrak{h} \oplus \mathfrak{z}$ and the resulting representation on $\mathfrak{z}$ is faithful, and such that the $\mathfrak{z}$-valued part $e \in \Omega^1(P, \mathfrak{z})$ of A is the solder form on P. We call the Cartan geometry $(\pi : P \to M, A)$ *invariant*[1] under a vector field ξ on M if and only if the frame bundle lift $\bar{\xi}$ is tangent to P and its restriction to P preserves A, i.e., $\mathcal{L}_{\bar{\xi}} A = 0$.

2. Applications

We apply the notion of invariance defined above to a number of common spacetime geometries used in various models of gravity. These geometries can be written as first-order reductive Cartan geometries. In particular, we obtain the following notions of invariance under a vector field ξ:

- Affine geometry with connection Γ:

$$\mathcal{L}_\xi \Gamma = 0\,. \tag{2}$$

- Riemann-Cartan geometry with metric g and torsion T:

$$\mathcal{L}_\xi g = 0 \quad \wedge \quad \mathcal{L}_\xi T = 0\,. \tag{3}$$

- Riemannian geometry with metric g:

$$\mathcal{L}_\xi g = 0\,. \tag{4}$$

- Weizenböck geometry with tetrad e:

$$\mathcal{L}_\xi e = \lambda e\,, \tag{5}$$

  where λ is a constant infinitesimal Lorentz transformation.
- Finsler geometry[2] with Finsler length function F:

$$\mathcal{L}_{\hat{\xi}} F = 0\,, \tag{6}$$

  where $\hat{\xi}$ is the tangent bundle lift of ξ, i.e., the vector field on TM which is defined in analogy to the frame bundle lift $\bar{\xi}$, but replacing the diffeomorphisms $\bar{\varphi}_t$ with $\hat{\varphi}_t = \varphi_{t*}$.

These notions of invariance agree with the standard notions of invariance on the respective spacetime models.

Acknowledgments

The author gratefully acknowledges the full financial support of the Estonian Research Council through the Postdoctoral Research Grant ERMOS115 and the Startup Research Grant PUT790.

References

1. M. Hohmann, Spacetime and observer space symmetries in the language of Cartan geometry, *arXiv:1505.07809 [math-ph]*, to appear.
2. M. Hohmann, Extensions of Lorentzian spacetime geometry: From Finsler to Cartan and vice versa, *Phys. Rev. D* **87**, 124034 (2013) [arXiv:1304.5430 [gr-qc]].

On the Universe rotation effect for different cosmological epochs

L. M. Chechin

Fesenkov Astrophysical Institute
National centre for cosmic researches and technologies
050020, Almaty, Kazakhstan
E-mail:chechin-lm@mail.ru

Keywords: Universe rotation; cosmology; redshift; spin.

We argue that our Universe rotates and rotates differentially, i.e. at different distances its angular velocity describes by different regularities. Our argumentation is done by three steps. First the concept of spacetime foam with spin was attracted. That is why it is reasonable to base on the statement that any quantum bubble from the spacetime foam initially possesses a spin and corresponds to a point spinning particle. If it so, the suggestion that our very early Universe is described by the Dirac equation may be forwarded.

Second, basing on the generalized Ehrenfest theorem that allows reduce the Dirac equation to the Papapetrou equation in classical limit, it may result that from classical viewpoint the very early Universe describes by Papapetrous equations.

Third, for its application to searching the Universe rotation it is necessary to have in mind that the Universe may peform the rotational movement only. As a result, later on we'll use the spinning part of Papapetrou equations and show that rotation radically depends on the properties of concrete epoch of the Universe expansion. Resemble, that Papapetrous equations describe a spinning (rotating) particle in the external gravitational field. But gravitational field for the Universe may be the internal only. If so, it is pertinently to take advantage of the Fock way by dividing general relativistic equations of motion on the external and internal types whose form are equal each other. Following this way we'll apply the spinning part of Papapetrou equations for describing the Universe rotation in its own gravitational field.

By setting Pirani condition on spin $u_\gamma S^{\alpha\gamma} = 0$ and having in mind the absence of Universe translational movement, we get the space-like equation for spin

$$\frac{dS^{kl}}{dx^0} + \left(\Gamma^k_{m0} S^{ml} - \Gamma^l_{m0} S^{mk}\right) = 0. \tag{1}$$

For the Friedmann model the mixed Christoffel symbols are - $\Gamma^k_{m0} = \dfrac{1}{a(x^0)} \dfrac{da(x^0)}{dx^0} \delta^k_m = H(x^0)\delta^k_m$. So, we find the final equation which describes the rotating and evolving Universe (here and after c=1) in the form $\dfrac{dS^{kl}}{dt} + 2H(t) \cdot S^{kl} = 0$, where $H(t)$- is Hubble parameter. The dimensionless variable Hubble parameter for the flat space (k=0) reads

$$h(z) = \frac{H(z)}{H_0} = \left(\Omega_M \cdot (1+z)^3 + \Omega_V \cdot (1+z)^3 \cdot exp\left(-3\omega_a \frac{z}{1+z}\right)\right)^{\frac{1}{2}}, \tag{2}$$

where $z = a^{-1}$ is the redshift, ω_M and ω_V are the mass and vacuum density parameters, accordingly. Next, it is advisable to leave the traditionally assumed statement that spin is directly proportional to angular velocity and constant moment of inertia. In fact, for the spherical-symmetric object as our Universe spin is $S^{kl}_{Un} = J_{Un}\omega^{kl}_{Un}$, where the moment of inertia $J_{Un} = \dfrac{2}{5}Ml^2 \approx \dfrac{1}{2}\pi\rho_{Un}a^5$ can't be the constant value in expanding space-time. So, combining $S^{kl}_{Un} = J_{Un}\omega^{kl}_{Un}$ with previous results we get

$$\omega^{kl}(z) = \omega^{kl}_0 \cdot$$

$$\cdot exp\left\{7 \cdot \left(\int_0^z \frac{\sqrt{\Omega_M \cdot (1+z)^3}}{(1+z)^2 \cdot \sqrt{1 + (\Omega_M + \Omega_V)z}} dz + \right.\right.$$

$$\left.\left. + \int_0^z \frac{\sqrt{\Omega_V \cdot (1+z)^3 \cdot exp(-3\omega_a \frac{z}{1+z})}}{(1+z)^2 \cdot \sqrt{1 + (\Omega_M + \Omega_V)z}} dz\right)\right\}, \tag{3}$$

where ω_0^{kl} is the initial spin of the Universe. We performed the approximate integration of Eq. (3) for three cosmological epochs: $\omega^{kl}(z) \approx \omega_{Ob}^{kl}(1 + 7\Omega_V^{\frac{1}{2}} z)$ for the present-days epoch ($\Omega_M \ll \Omega_V$); $\omega^{kl}(z) \approx \omega_{Ob}^{kl} \cdot z^7$ for the early Universe ($\Omega_M \gg \Omega_V$) and $\omega^{kl}(z) \approx \omega_{Ob}^{kl}\left\{\sqrt{2\Omega_{tr}(1+z)} + \sqrt{1 + 2\Omega_{tr} z}\right\}^{14}$ for the transition epoch ($\Omega_M \cong \Omega_V$).

So, it was shown that our Universe rotates and rotates differentially, i.e. its rotation radically depends on the properties of concrete epoch of the Universe evolution. At the vacuum dominated epoch and for the small red-shifts ($z \ll 1$) the Universe angular velocity will be about $\omega_{Un} \sim 10^{-19} Hz$.

For the transitional epoch (redshifts $z \approx 1$) it is possible to set that $\Omega_M \sim \Omega_V$. So, we get that $\omega_{Un} \sim 10^{-12}$ Hz.

For the matter dominated epoch, i.e. for the essentially large redshifts ($z \gg 1$), the Universe was rotated with very large angular velocity. As the redshift for recombination epoch is $z \sim 10^3$, than at such distances $\omega_{Un} \sim 10^2$ Hz. Note that at such frequencies and redshifts the Universe linear velocity $\nu_{Un} = \omega_{Un} \cdot a(z) = \omega_{Un} \cdot a_0/(1+z) \approx 1.3 \cdot 10^{27} cm/sec$, ten times smaller than the speed of light that doesnt contradict with the relativity principle.

These results mean the essential increasing of Universe rotation at early epochs of its evolution with respect to its observable magnitude today. Such consequence allows estimate the Universe reasonable physical sizes. Really, assuming that $\omega_{Un} < \omega_{Pl} \sim \frac{1}{t_{Pl}} \sim 10^{-43}$ Hz, where $t_{Pl} = \sqrt{\frac{Gh}{c^5}} \approx 10^{-43} sec$ is the Planck time, we obtain the corresponding redshift $z \sim 10^9$ for the moment of the Universe birth.

As the article total summarizing I conclude the following. From the history of the standard relativistic cosmology forming it is well-known situation when idea about the Universe evolving was the subject of numerous discussions, though such possibility is the natural consequence from the logical structure of the Universe epistemological basis development. Analogously the Universe rotating concept is the normal result created from the pure logical foundation. So, the aim of this article was arguing the correct effort to justify this statement within the framework of standard cosmological model.

Oscillating universe with two sources of torsion

A. M. Galiakhmetov

Department of Physics, Donetsk National Technical University,
Kirova street 51, 84646, Gorlovka, Ukraine
E-mail: agal17@mail.ru

Closed oscillating universes filled with stiff fluid and two sources of torsion are considered in the framework of the Einstein–Cartan theory.

Keywords: Oscillating universes; stiff fluid; two sources of torsion; the Einstein–Cartan theory.

Closed Fiedmann cosmologies with stiff fluid are considered in the framework of the Einstein–Cartan theory. The variant of this theory, which simultaneously takes into consideration two sources of torsion, namely a perfect fluid with the vacuum equation of state and a nonminimally coupled canonical scalar field is investigated.

The Lagrangian Ł of the model is chosen as the sum of Lagrangians: gravitational, scalar field, and perfect fluids[1]

$$Ł = -\ R/2\kappa + (1/2)\left[\Phi_{,k}\Phi^{,k} + \xi R\Phi^2\right] - \mathrm{V}(\Phi) - \rho(1 + \Pi(\rho, \mathrm{e}))$$

$$+\ \ k \overset{\Gamma}{\nabla}_i (\rho u^i) + k_1(u_i u^i - 1) + k_2 u^i \partial_i X + k_3 u^i \partial_i e + Ł_{fl(2)}\ . \quad (1)$$

Here, $R(\Gamma)$ is the curvature scalar obtained from the full connection $\Gamma^k_{ij} = \{^{\,k}_{ij}\} + S_{ij\cdot}^{\ \ k} + S^k_{\cdot ij} + S^k_{\cdot ji}$; $\{^{\,k}_{ij}\}$ are Christoffel symbols of the second kind; $S_{ij\cdot}^{\ \ k} = \Gamma^k_{[ij]}$ is the torsion tensor; $\kappa = 8\pi G$, G is the Newtonian constant; $V(\Phi)$ is the scalar field potential; ρ is the perfect fluid mass density; $\Pi(\rho, e)$ is the internal energy; k, k_1, k_2, k_3 are the Lagrange multipliers; X is the Lagrangian coordinates of the matter particles; e is the entropy per volume; u^i is the four-velocity; $\overset{\Gamma}{\nabla}_i$ is the covariant derivative of the Riemann – Cartan space. The Lagrangian $Ł_{fl(2)}$ for the stiff fluid is not indicated since, for it, the derivative of the term responsible for particles number conservation does not contain the torsion vector.

For closed FRW cosmologies we have for the perfect fluid with the vacuum equation of state, which induces torsion, and stiff fluid, respectively:

$$P_{fl(1)} = -\varepsilon_{\mathrm{fl}(1)} = -C_{\mathrm{fl}(1)} \ , \quad \Theta = C_{\Theta} a^{-3} \ , \quad \varepsilon_{\mathrm{fl}(2)} = P_{\mathrm{fl}(2)} = C_{\mathrm{fl}(2)} a^{-6} \ , \tag{2}$$

where $C_{\mathrm{fl}(1)}$, C_{Θ} and $C_{fl(2)}$ are positive integration constants, $\Theta = k\rho$.

The scalar field potential $V(\Phi)$ is chosen in the form:

$$V(\Phi) = \frac{\mu^2}{2}\Phi^2 + \frac{\lambda}{4!}\Phi^4 \ , \tag{3}$$

where μ and λ are constants.

The exact solution to the Einstein–Cartan equations for $\xi < 0$ can be presented in the form of quadratures:

$$a(t) = AW \ , \qquad \Phi(t) = (D/\kappa|\xi|)^{1/2} W^{-1} |\sin \lambda t| \ ,$$

$$S^2(t) = \frac{9\kappa^2\lambda^2}{4W^2}\Big(\frac{C_{\Theta}}{2\sqrt{\kappa|\xi|}} - \frac{D\sin(2\lambda t)}{2\kappa W}\Big)^2 \ , \tag{4}$$

where $D = \mathrm{const}$, $S^2 = S_k S^k$ is the squared trace of torsion and

$$A = \left(\frac{\kappa(D + 2C_{\mathrm{fl}(2)})}{2(1+\chi)}\right)^{1/4} \ , \qquad W = (D\cos^2 \lambda t + 2C_{\mathrm{fl}(2)})^{1/2} \ ,$$

$$\chi = \frac{3\kappa C_{\Theta}^2}{2(D + 2C_{\mathrm{fl}(2)})} \ , \quad \lambda = \left(\frac{8\xi^2}{\kappa(D + 2C_{\mathrm{fl}(2)})}\right)^{1/4} (1+\chi)^{-3/4}. \tag{5}$$

This solution exits provided that

$$8 = \kappa^3 C_{\mathrm{fl}(1)}^2 [D + 2C_{\mathrm{fl}(2)} + (3\kappa/2)C_{\Theta}^2] \ . \tag{6}$$

The universes show oscillatory behavior in which all cycles have the same amplitude and duration:

$$T = \tilde{T}(1+\chi)^{3/4} \ , \qquad \tilde{T} = \pi\left(\frac{\kappa(D + 2C_{\mathrm{fl}(2)})}{8\xi^2}\right)^{1/4} \ , \tag{7}$$

where $\tilde{T}$ — the duration of the one cycle for the models, which contain the scalar-torsion field and stiff fluid. Thus the presence of two sources of torsion leads to an increased duration of the one cycle for the cosmological models.

Reference

1. A. M. Galiakhmetov, Cosmology in the Einstein–Cartan theory with two sources of torsion, *Grav. Cosmol.* **7**, 327 (2001).

Construction of integrable cosmological models with non-minimally coupled scalar fields

A. Yu. Kamenshchik[1,2], E. O. Pozdeeva[3], A. Tronconi[1], G. Venturi[1] and S. Yu. Vernov[3]

[1] *Dipartimento di Fisica e Astronomia and INFN, Via Irnerio 46, 40126, Bologna, Italy,*
[2] *L.D. Landau Institute for Theoretical Physics of the Russian Academy of Sciences, Kosygin str. 2, 119334, Moscow, Russia,*
[3] *Skobeltsyn Institute of Nuclear Physics, Lomonosov Moscow State University, Leninskie Gory 1, 119991, Moscow, Russia*

We consider the relation between integrable cosmological models having minimally and non-minimally coupled scalar fields. We show how to obtain general solutions for some flat Friedmann universes filled with a scalar field in induced gravity models and models including the Hilbert–Einstein curvature term plus a scalar field conformally coupled to gravity. The explicit forms of the self-interaction potentials for seven exactly solvable models are presented here.

Keywords: Cosmology; integrable models; scalar fields; modified gravity.

1. Introduction

Cosmological models with scalar fields have acquired a great popularity during the last decades due to the development of inflationary cosmology and to the appearance of different quintessence models of dark energy, which should explain the phenomenon of cosmic acceleration.

Exact solutions to the Einstein equations play an important role in cosmology. The number of integrable cosmological models based on scalar fields is rather limited. First of all, let us mention the spatially flat Friedmann model with a minimally coupled scalar field and a self-interaction exponential potential.[1,2] Other interesting classes of models with a general exact solutions are that wherein the potential is a combination of exponential functions or hyperbolic functions.[3,4] In Ref. 5, the general classification of integrable cosmological models based on scalar fields was suggested and studied in great detail. Examples of integrable models with non-minimal coupled scalar fields are given in Refs. 6 and 7.

Models with a scalar field non-minimally coupled to gravity attract a growing interest nowadays. Induced gravity models, wherein the curvature arises as a quantum effect have found diverse applications in cosmology. The recent discovery of the Higgs particle makes the model where the role of the inflaton is played by the Higgs field non-minimally coupled to gravity especially attractive.

It is known that a gravitational model with a non-minimal coupling can be transformed into a model with a minimal coupling, by using a properly chosen conformal transformation of the metric, combined with some parametric transformation of the scalar field. Such a procedure is usually called the transition from the Jordan frame to the Einstein frame.

The goal of our investigation is to show how the knowledge of an integrable system with a minimally coupled scalar field allows one to obtain the integrable system with a non-minimally coupled one and vice versa. We have constructed the explicit form of the self-interaction potentials for seven integrable induced gravity models and seven integrable models with the Hilbert–Einstein curvature term and a scalar field conformally coupled to gravity.[8,9] We also shown that this correspondence assists to obtain the general solutions explicitly.

2. Relations between General Exact Solutions in the Models with Minimally Coupled and Non-Minimally Coupled Scalar Fields

Let us consider a cosmological model, described by the following action

$$S = \int d^4x\sqrt{-g}\left[U(\sigma)R - \frac{1}{2}g^{\mu\nu}\sigma_{,\mu}\sigma_{,\nu} - V(\sigma)\right], \tag{1}$$

where $U(\sigma)$ and $V(\sigma)$ are differentiable functions of the scalar field σ. If $U(\sigma)$ is a constant: $U(\sigma) = U_0$, then the scalar field is minimally coupled to gravity, otherwise we have a model with non-minimal coupling.

In a spatially flat FLRW universe, with the interval

$$ds^2 = N^2(\tau)d\tau^2 - a^2(\tau)d\vec{l}^{\,2}, \tag{2}$$

where $a(\tau)$ is the scale factor and $N(\tau)$ is the lapse function, the Friedmann equations, derived by variation of the action (1), have the following form[8]:

$$\frac{6U\dot{a}^2}{a^2} + \frac{6U'\dot{a}\dot{\sigma}}{a} = \frac{1}{2}\dot{\sigma}^2 + N^2V, \tag{3}$$

$$\frac{4U\ddot{a}}{a} + \frac{2U\dot{a}^2}{a^2} + \frac{4U'\dot{a}\dot{\sigma}}{a} - \frac{4U\dot{a}\dot{N}}{aN} + 2U''\dot{\sigma}^2 + 2U'\ddot{\sigma} - \frac{2U'\dot{\sigma}\dot{N}}{N} = -\frac{1}{2}\dot{\sigma}^2 + N^2V. \tag{4}$$

where differentiation with respect to time τ is denoted by a dot, a "prime" means a derivative with respect to σ. Variation of the action (1) with respect to σ yields to the equation that is a consequence of system (3)–(4). Our goal is to find such functions U and V so that system (3)–(4) is integrable.

Let us make the conformal transformation of the metric $g_{\mu\nu} = \frac{U_0}{U}\tilde{g}_{\mu\nu}$, where U_0 is a constant. We also introduce a new scalar field ϕ such that

$$\frac{d\phi}{d\sigma} = \frac{\sqrt{U_0(U+3U'^2)}}{U} \quad \Rightarrow \quad \phi = \int \frac{\sqrt{U_0(U+3U'^2)}}{U}d\sigma\,. \tag{5}$$

Finally we transform the action (1) to the following action for a minimally coupled scalar field:

$$S = \int d^4x\sqrt{-\tilde{g}}\left[U_0 R(\tilde{g}) - \frac{1}{2}\tilde{g}^{\mu\nu}\phi_{,\mu}\phi_{,\nu} + W(\phi)\right], \quad \text{where} \quad W = \frac{U_0^2 V}{U^2}.$$

The FLRW metric (2) becomes $ds^2 = \tilde{N}^2 d\tau^2 - \tilde{a}^2 \vec{dl}^2$, where the new functions are defined as $\tilde{N} = \sqrt{\frac{U}{U_0}}N$, $\tilde{a} = \sqrt{\frac{U}{U_0}}a$.

Let us suppose that for some potential W we know the general exact solution of the system of equations in the Einstein frame: $\phi(\tau)$, $\tilde{a}(\tau)$, $\tilde{N}(\tau)$. We also assume that the function $\sigma(\phi)$ is known explicitly. In this case, we can also find the general solution of the system of equations (3)–(4) with the potential

$$V(\sigma) = \frac{U^2(\sigma)W(\phi(\sigma))}{U_0^2}, \tag{6}$$

which is given by $\sigma(\phi(\tau))$, $a(\tau) = \sqrt{\frac{U_0}{U(\sigma(\phi))}}\tilde{a}(\tau)$, $N(\tau) = \sqrt{\frac{U_0}{U(\sigma(\phi))}}\tilde{N}(\tau)$.

If we choose the induced gravity case, *i.e.*

$$U_i(\sigma) = \frac{\gamma}{2}\sigma^2, \tag{7}$$

where γ is a constant, then we have from Eq. (5)

$$\phi = \sqrt{\frac{2U_0(1+6\gamma)}{\gamma}}\ln\left[\frac{\sigma}{\sigma_0}\right] \quad \text{and, inversely,} \quad \sigma = \sigma_0 e^{\sqrt{\frac{\gamma}{2U_0(1+6\gamma)}}\phi}.$$

We assume that the coupling is not conformal one: $\gamma \neq -1/6$. Another interesting choice is

$$U_c(\sigma) = U_0 - \frac{\sigma^2}{12}, \tag{8}$$

i.e. the case when the coupling is conformal and a nonzero Einstein–Hilbert term is also present. In this case

$$\phi = \sqrt{3U_0}\ln\left[\frac{\sqrt{12U_0}+\sigma}{\sqrt{12U_0}-\sigma}\right] \quad \text{and} \quad \sigma = \sqrt{12U_0}\tanh\left[\frac{\phi}{\sqrt{12U_0}}\right].$$

In our papers,[8,9] we considered those integrable cases, mentioned in Ref. 5, for which the explicit solutions for $\tilde{N}(\tau)$, $\tilde{a}(\tau)$ and $\phi(\tau)$ can be found at least in quadratures. To make the comparison with the models analyzed in paper[5] more convenient, we choose $U_0 = 1/4$ and set $\sigma_0 = 1$.

W (minimal)	V_i (induced)	V_c (conformal)
$c_0 e^{2\sqrt{3}\lambda\phi}$	$\tilde{c}_0\sigma^{4+\lambda\tilde{\Gamma}}$	$c_0\Theta\Upsilon^{3\lambda}$
$c_0 + c_1 e^{\sqrt{3}\phi} + c_2 e^{-\sqrt{3}\phi}$	$\tilde{c}_0\sigma^4 + \tilde{c}_1\sigma^{4+\frac{\tilde{\Gamma}}{2}} + \tilde{c}_2\sigma^{4-\frac{\tilde{\Gamma}}{2}}$	$\Theta\left[c_0 + c_1\Upsilon^{\frac{3}{2}} + c_2\Upsilon^{-\frac{3}{2}}\right]$
$c_1 e^{2\sqrt{3}\lambda\phi} + c_2 e^{\sqrt{3}(\lambda+1)\phi}$	$\tilde{c}_1\sigma^{4+\lambda\tilde{\Gamma}} + \tilde{c}_2\sigma^{4+\frac{\lambda+1}{2}\Gamma}$	$\Theta\left[c_1\Upsilon^{3\lambda} + c_2\Upsilon^{\frac{3}{2}(\lambda+1)}\right]$
$c_1 e^{2\sqrt{3}\phi} + c_2$	$\sigma^4\left[\tilde{c}_1\sigma^{\tilde{\Gamma}} + \tilde{c}_2\right]$	$\Theta\left[c_1\Upsilon^3 + c_2\right]$
$c_0\phi e^{2\sqrt{3}\phi}$	$\frac{\sqrt{3}}{6}\tilde{\Gamma}\tilde{c}_0\sigma^{4+\tilde{\Gamma}}\ln(\sigma)$	$\frac{\sqrt{3}}{2}c_0\Theta\Upsilon^3\ln(\Upsilon)$
$c_1 e^{2\sqrt{3}\lambda\phi} + c_2 e^{\frac{2\sqrt{3}}{\lambda}\phi}$	$\sigma^4\left[\tilde{c}_1\sigma^{\lambda\tilde{\Gamma}} + \tilde{c}_2\sigma^{\frac{\tilde{\Gamma}}{\lambda}}\right]$	$\Theta\left[c_1\Upsilon^{3\lambda} + c_2\Upsilon^{\frac{3}{\lambda}}\right]$
$c_1\cosh\left(\sqrt{3}\beta\phi\right)^B +$ $+c_2\sinh\left(\sqrt{3}\beta\phi\right)^B$	$\frac{2^{-B}\sigma^4}{\sigma^{\tilde{\Gamma}(1-\beta)}}\left[\tilde{c}_1\left(\sigma^{\beta\tilde{\Gamma}}+1\right)^B +\right.$ $\left. + \tilde{c}_2\left(\sigma^{\beta\tilde{\Gamma}}-1\right)^B\right]$	$c_1\frac{\left[(\sqrt{3}+\sigma)^{3\beta}+(\sqrt{3}-\sigma)^{3\beta}\right]^B}{2^B 9(3-\sigma^2)^{1-3\beta}} +$ $+c_2\frac{\left[(\sqrt{3}+\sigma)^{3\beta}-(\sqrt{3}-\sigma)^{3\beta}\right]^B}{2^B 9(3-\sigma^2)^{1-3\beta}}$

In the Table we present the list of the potentials, for which the corresponding cosmological models are integrable,

$$\Theta \equiv \left[1-\frac{\sigma^2}{3}\right]^2, \quad \Upsilon \equiv \frac{\sqrt{3}+\sigma}{\sqrt{3}-\sigma}, \quad \tilde{\Gamma} \equiv 6\sqrt{\frac{1+6\gamma}{6\gamma}}, \quad B \equiv \frac{2(1-\beta)}{\beta}.$$

The constants c_0, c_1, and c_2 are arbitrary, $\tilde{c}_i = 4\gamma^2 c_i$. The parameter λ is an arbitrary number, but $\lambda \neq \pm 1, \lambda \neq 0$.

3. Conclusions

We have shown in Refs. 8 and 9 that given the knowledge of explicit general cosmological solutions for flat Friedmann models with minimally coupled scalar fields one can construct the general solutions for induced gravity models and models having a Hilbert–Einstein curvature term plus a scalar field conformally coupled to gravity. The latter are connected with the minimally coupled model by the combination of a conformal transformation and a transformation of the scalar field. We present the forms of the self-interacting potentials for seven explicitly integrable models.

Research of E.P. and S.V. is supported in part by the RFBR grant 14-01-00707, A.K. was partially supported by the RFBR grant 14-02-00894.

References

1. D.S. Salopek and J.R. Bond, Nonlinear evolution of long wavelength metric fluctuations in inflationary models, *Phys. Rev.* D **42**, 3936 (1990).
2. A.A. Andrianov, F. Cannata, and A.Yu. Kamenshchik, General solution of scalar field cosmology with a (piecewise) exponential potential *JCAP* **1110,** 004 (2011) [arXiv:1105.4515].
3. R. de Ritis, G. Marmo, G. Platania, C. Rubano, P. Scudellaro and C. Stornaiolo, New approach to find exact solutions for cosmological models with a scalar field, *Phys. Rev.* D **42,** 1091 (1990).
4. I. Bars, S.H. Chen, P.J. Steinhardt and N. Turok, Complete Set of Homogeneous Isotropic Analytic Solutions in Scalar-Tensor Cosmology with Radiation and Curvature, *Phys. Rev. D* **86,** 083542 (2012) [arXiv:1207.1940].
5. P. Fré, A. Sagnotti, and A.S. Sorin, Integrable Scalar Cosmologies I. Foundations and links with String Theory, *Nucl. Phys.* B **877,** 1028 (2013) [arXiv:1307.1910].
6. A. Borowiec, S. Capozziello, M. De Laurentis, F.S.N. Lobo, A. Paliathanasis, M. Paolella, and A. Wojnar, Invariant solutions and Noether symmetries in Hybrid Gravity, *Phys. Rev.* D **91,** 023517 (2015) [arXiv:1407.4313].
7. B. Boisseau, H. Giacomini, D. Polarski, and A.A. Starobinsky, Bouncing Universes in Scalar-Tensor Gravity Models admitting Negative Potentials, *JCAP* **1507,** 002 (2015) [arXiv:1504.07927].
8. A.Yu. Kamenshchik, E.O. Pozdeeva, A. Tronconi, G. Venturi, and S.Yu. Vernov, Integrable cosmological models with non-minimally coupled scalar fields, *Class. Quant. Grav.* **31**, 105003 (2014) [arXiv:1307.1910].
9. A.Yu. Kamenshchik, E.O. Pozdeeva, A. Tronconi, G. Venturi, and S.Yu. Vernov, Interdependence between integrable cosmological models with minimal and non-minimal coupling, arXiv:1509.00590.

Cosmological models with non-minimally coupled scalar fields and renormalization-group inflationary scenarios

E. Elizalde[1], E.O. Pozdeeva[2] and S. Yu. Vernov[2]

[1] *Instituto de Ciencias del Espacio (ICE/CSIC) and Institut d'Estudis Espacials de Catalunya (IEEC) Campus UAB, Carrer de Can Magrans, s/n 08193 Cerdanyola del Vallès (Barcelona), Spain*

[2] *Skobeltsyn Institute of Nuclear Physics, Lomonosov Moscow State University, Leninskie Gory 1, 119991, Moscow, Russia*

The possibility to construct an inflationary scenario for renormalization-group improved potentials corresponding to the Higgs sector of grant unification models is investigated. Taking into account quantum corrections to the renormalization-group potential which sums all leading logs of perturbation theory is essential for a successful realization of the inflationary scenario, with very reasonable parameters values. The inflationary models thus obtained are seen to be in good agreement with the most recent and accurate observational data.

Keywords: Cosmology; inflation; scalar fields; non-minimal coupling; renormalization group.

1. Introduction

Models of cosmic inflation yield accurate quantitative predictions for a number of observable quantities. Recent observations confirm that the single-field inflationary models are realistic. At the same time, the predictions of the simplest inflationary models with a minimally coupled scalar field lead to sufficiently large values of the tensor-to-scalar ratio of the density perturbations r, and are therefore ruled out by Planck data.[1] Inflationary scenarios with a minimally coupled scalar field can be improved by adding a tiny non-minimal coupling of the inflaton field to gravity. Note that generic quantum corrections to the action of the scalar field minimally coupled to gravity include a non-minimally coupling term,[2] proportional to $R\phi^2$, and quantum corrections must definitely be taken into account.

Quantum field theory in curved space-time is an unavoidable, extremely important concept at the very early universe, where curvature is large and typical energies are very high. In the absence of a consistent theory of quantum gravity, all classical theories should better be treated as quantum field theories in curved space-time.[3,4] some recent results by the Planck collaboration[1] seem to point towards the Grand Unified Theory (GUT) scale, what is quite a remarkable connection between GUTs and inflation. So, GUTs can be treated as quantum field theories in the curved space-time corresponding to the very early universe (for a complete description, see Refs. 3–6).

As a consequence, it is natural to start by addressing the issue of the renormalization-group improved effective potential for an arbitrary renormalizable massless gauge theory in curved space-time.[4] Note that it is enough to work in the linear curvature approximation, because these linear curvature terms are expected to give dominant contribution in the discussion of the inflationary effective potential corresponding to GUTs. By generalizing the flat space-time Coleman–Weinberg potential, the explicit form of the renormalization group (RG) improved effective potentials in curved space finite gauge models were obtained in Ref. 6.

In our papers (Refs. 7,8) we showed that for scalar electrodynamics, finite gauge and for the $SU(5)$ RG-improved models inflationary scenarios are possible and are in good agreement with the most recent astronomical data,[1] provided some reasonable values were taken for the parameters. In Ref. 7 we checked the possibility to construct inflationary models using RG-improved potentials, considering inflation based on de Sitter solutions, their instability providing a graceful exit from inflation. In Ref. 8 we showed that for the finite gauge models, the inflationary scenario can be realized without having an exact de Sitter solution, but indeed from a quasi-de Sitter solution with a slowly changing Hubble parameter. In the paper Ref. 8 it has been also shown that RG-corrections can generate the Hilbert–Einstein term in the action. In other words, we do not need to include this term by hand, as is done in Higgs-driven inflation,[9] since we get it naturally, as part of the quantum corrections to the induced gravity term.

There are many different models of inflation, some of them giving very close predictions to each other, for the observable inflationary parameters. For instance, two different models, as R^2 Starobinsky inflation[10] and Higgs-driven inflation,[9] yield very similar values for the spectral index of the curvature perturbations n_s and for the tensor-to-scalar ratio of the density perturbations r, respectively. In recent papers (Ref. 11) it has been shown that there are several classes of inflationary models such that all models within a given class predict the same values of n_s and r in the leading approximation in $1/N_e$, where N_e is a number of e-foldings. The model considered in Ref. 8 yield the same inflationary parameters as the R^2 and Higgs-driven inflationary models and does belong to the same important class of inflationary models which goes under the name of "cosmological attractors".

2. Inflationary Models with RG-Improved Potentials

Let us consider a gravity model with a non-minimally coupled scalar field, described by the action

$$S = \int d^4x\sqrt{-g}\left[U(\phi)R - \frac{1}{2}g^{\mu\nu}\phi_{,\mu}\phi_{,\nu} - V(\phi)\right], \tag{1}$$

where $U(\phi)$ and $V(\phi)$ are differentiable functions of the scalar field ϕ, g is the determinant of the metric tensor $g_{\mu\nu}$, and R is the scalar curvature.

We make the conformal transformation of the metric $\tilde{g}_{\mu\nu} = 2\kappa^2U(\phi)g_{\mu\nu}$, where $\kappa^2 = 8\pi M_{\rm Pl}^{-2}$, $M_{\rm Pl}$ is the Planck mass, and get a model for a minimally coupled scalar field, described by the following action

$$S_E = \int d^4x\sqrt{-\tilde{g}}\left[\frac{1}{2\kappa^2}R(\tilde{g}) - \frac{1}{4\kappa^2U}\left[1+\frac{3U'^2}{U}\right]\tilde{g}^{\mu\nu}\phi_{,\mu}\phi_{,\nu} - V_{\rm E}\right], \tag{2}$$

where $V_{\rm E} = V/(4\kappa^4U^2)$.

In the model considered, the tree-level functions of the scalar field V and U are[4,6–8]:

$$V^{(0)} = \tilde{a}\lambda\phi^4, \qquad U^{(0)} = b\xi\phi^2, \tag{3}$$

where $\tilde{a}$ and b are positive constants and ξ is the conformal coupling. The corresponding potential in the Einstein frame $V_{\rm E}$ is a constant and is not suitable for inflation. To realize the inflationary scenario we use the RG-improved potentials.[4,6] The functions U and V, obtained in the linear curvature approximation (see Refs. 6,8 for details), are given by

$$V = \tilde{a}\kappa_1 g^2 f^4(\phi)\phi^4, \qquad U = b\xi(\phi)f^2(\phi)\phi^2, \tag{4}$$

$$f = \left(\frac{\phi}{\mu}\right)^{-C_1g^2}, \qquad \xi = \frac{1}{6} + \left(\xi_0 - \frac{1}{6}\right)\left(\frac{\phi}{\mu}\right)^{Cg^2}.$$

where $\tilde{a}$, b, and κ_1, μ, C, and C_1 are some constants which depend on the gauge parameter and on the features of the theory. It is important that both functions U and V should be modified.[4,6]

The function $U^{(0)}$ corresponds to the induced gravity, in other words, the term proportional to ϕ^2R plays the role of the Hilbert–Einstein term in the action. It is quite interesting that for some values of the parameters the renormalization-group corrections can yield the standard Hilbert–Einstein term. Indeed, if $C_1 = \frac{1}{g^2} + \frac{C}{2}$, then function U acquires the following form

$$U = \frac{1}{6} b\mu^2 \left(6\xi_0 - 1 + \left[\frac{\phi}{\mu}\right]^{-Cg^2} \right). \tag{5}$$

Thus, we come to the interesting conclusion that the Hilbert–Einstein term can arise, in the induced gravity action, as a renormalization-group correction. In this case, we also obtain

$$V = a\kappa_1 g^2 \mu^4 \left(\frac{\phi}{\mu}\right)^{-2Cg^2} = ag^2\kappa_1\mu^4 \left(\frac{6}{b\mu^2} U + 1 - 6\xi_0 \right)^2. \tag{6}$$

This relation between the functions V and U means that the corresponding inflationary model is a cosmological attractor.[8,11]

3. Conclusions

It has been shown in our paper Ref. 8 that the inflationary scenario for finite gauge theory belongs to the class of $\alpha - \beta$ cosmological attractors, following the convenient classification given in the literature.[11] The form of the potential in the Einstein frame is quite close to the forms of the corresponding potentials for Higgs-driven inflation[9] and for Starobinsky's R^2 inflation.[10] As is known, these inflationary scenarios have parameters which are compatible with the astronomical data[1] and very close to one another.[12] This yields even more value to the fact that, at the same time, there is this remarkable difference that distinguishes our model from both R^2-inflation and Higgs-driven inflation. The scalar field belongs to the matter sector; moreover, in clear distinction to Higgs-driven inflation, in the model constructed the Hilbert–Einstein term arises as a result of unavoidable quantum corrections at the one-loop approximation and needs not be imposed by hand. In our model[8] the principles of quantum field theory lead to the necessary Hilbert–Einstein term in a generic and natural way.

Acknowledgements

We would like to thank Sergei Odintsov for collaboration and fruitful discussions. E.E. are supported in part by MINECO (Spain), project FIS2013-44881-P, and by the CPAN Consolider Ingenio Project. E.O.P. and S.Yu.V. are supported in part by RFBR grant 14-01-00707.

References

1. P. A. R. Ade, *et. al.* [Planck Collaboration], Planck 2015 results. XX. Constraints on inflation, arXiv:1502.02114.
2. N. A. Chernikov, E. A. Tagirov, Quantum theory of scalar fields in de Sitter space-time, *Annales Poincare Phys. Theor.* A **9,** 109 (1968).
3. I. L. Buchbinder, S. D. Odintsov and I. L. Shapiro, *Effective Action in Quantum Gravity* (IOP Publishing, Bristol and Philadelphia, 1992).
4. E. Elizalde and S. D. Odintsov, Renormalization group improved effective potential for gauge theories in curved space-time, *Phys. Lett.* B **303,** 240 (1993) [arXiv:hep-th/9302074];
   E. Elizalde and S. D. Odintsov, Renormalization group improved effective Lagrangian for interacting theories in curved space-time, *Phys. Lett. B* **321,** 199 (1994) [arXiv:hep-th/9311087].
5. I. L. Buchbinder, S. D. Odintsov, Effective potential and phase transitions induced by curvature in gauge theories in curved spacetime, *Class. Quant. Grav.* **2**, 721 (1985).
6. E. Elizalde and S. D. Odintsov, Renormalization group improved effective potential for finite grand unified theories in curved space-time, *Phys. Lett.* B **333,** 331 (1994) [arXiv:hep-th/9403132].
7. E. Elizalde, S. D. Odintsov, E. O. Pozdeeva, and S. Yu. Vernov, Renormalization-group inflationary scalar electrodynamics and $SU(5)$ scenarios confronted with Planck2013 and BICEP2 results, *Phys. Rev.* D **90,** 084001 (2014) [arXiv:1408.1285].
8. E. Elizalde, S. D. Odintsov, E. O. Pozdeeva, and S. Yu. Vernov, Cosmological attractor inflation from the RG-improved Higgs sector of finite gauge theory, arXiv:1509.08817.
9. F. L. Bezrukov and M. Shaposhnikov, The Standard Model Higgs boson as the inflaton, *Phys. Lett.* B **659,** 703 (2008) [arXiv:0710.3755];
   A. O. Barvinsky, A. Y. Kamenshchik, and A. A. Starobinsky, Inflation scenario via the Standard Model Higgs boson and LHC, *J. Cosmol. Astropart. Phys.* **0811,** 021 (2008) [arXiv:0809.2104];
   A. De Simone, M. P. Hertzberg and F. Wilczek, Running Inflation in the Standard Model, *Phys. Lett.* B **678,** 1 (2009) [arXiv:0812.4946].
10. A. A. Starobinsky, A New Type of Isotropic Cosmological Models Without Singularity, *Phys. Lett.* B **91,** 99 (1980).
11. M. Galante, R. Kallosh, A. Linde and D. Roest, The Unity of Cosmological Attractors, *Phys. Rev. Lett.* **114,** 141302 (2015) [arXiv:1412.3797].

12. F. L. Bezrukov and D. S. Gorbunov, Distinguishing between R^2-inflation and Higgs-inflation, *Phys. Lett.* B **713,** 365 (2012) [arXiv:1111.4397]

Section 3

Black Holes and Wormholes

Spin and mass of the supermassive black hole at the Galactic center

V. I. Dokuchaev[1,2]

[1] *Institute for Nuclear Research of the Russian Academy of Sciences, 60th October Anniversary Prospect 7a, Moscow 117312, Russia*

[2] *National Research Nuclear University MEPhI, Kashirskoe sh. 31, Moscow 115409, Russia*

E-mail: dokuchaev@inr.ac.ru

The promising observational signatures for the measurement of black hole mass and spin are the azimuthal and latitudinal oscillation frequencies of the bright spots in the accretion flow and also the frequency of black hole event horizon rotation. Interpretation of the known quasi-periodic oscillations data by dint of a signal modulation from the hot spots in the accreting plasma reveals the Kerr metric rotation parameter, $a = 0.65 \pm 0.05$, and mass, $M = (4.2 \pm 0.2)10^6 M_\odot$, of the supermassive black hole in the Galactic center. The observed first 11.5 min quasi-periodic oscillation period is identified with a period of the black hole event horizon rotation, and, respectively, the second 19 min period is identified with a latitudinal oscillation period of hot spots in the accretion flow.

Keywords: Galactic center; black hole.

1. Introduction

The major problem with observations of the supermassive black hole SgrA* in the Galactic center is that it is a 'dormant' quasar, i. e., nearly completely inactive with a very rare splashing of activity. Nevertheless, there are serendipitous observations of Quasi-Periodic Oscillations (QPOs) from SgrA* with a sufficiently high statistical significance in the near-infrared[1] and X-rays.[2] We describe here the QPO interpretation, related to the oscillation frequencies of the numerous hot spots in the accretion plasma,[3] which are independent of the accretion model and defined completely by the properties of the black hole gravitational field. We use units with $G = c = 1$ and define the spin parameter of Kerr black hole as $a = J/M^2$, where M and J are, respectively, the black hole mass and angular momentum. Equations of motion for test particles (e. g., planets or clumps of hot plasma) with a mass μ in the Kerr metric in the Boyer–Lindquist coordinates (t, r, θ, φ) are[4]:

$$\rho^2 \frac{dr}{d\lambda} = \pm\sqrt{V_r}, \quad \rho^2 \frac{d\theta}{d\lambda} = \pm\sqrt{V_\theta}, \tag{1}$$

$$\rho^2 \frac{d\varphi}{d\lambda} = L \sin^{-2}\theta + a(\Delta^{-1}P - E), \tag{2}$$

$$\rho^2 \frac{dt}{d\lambda} = a(L - aE\sin^2\theta) + (r^2 + a^2)\Delta^{-1}P, \tag{3}$$

where $\lambda = \tau/\mu$, τ — is the proper time of a particle and

$$V_r = P^2 - \Delta[\mu^2 r^2 + (L - aE)^2 + Q], \tag{4}$$

$$V_\theta = Q - \cos^2\theta[a^2(\mu^2 - E^2) + L^2 \sin^{-2}\theta], \tag{5}$$

$$P = E(r^2 + a^2) - aL, \quad \rho^2 = r^2 + a^2\cos^2\theta, \quad \Delta = r^2 - 2r + a^2. \tag{6}$$

The motion of a test particle is completely defined by three integrals of motion: the total particle energy E, the azimuthal component of the angular momentum L and the Carter constant Q, related to the total angular momentum of the particle and non-equatorial motion. It is useful to choose the dimensionless variables and parameters: $t \Rightarrow t/M$, $r \Rightarrow r/M$, $a \Rightarrow a/M$, $E \Rightarrow E/\mu$, $L \Rightarrow L/(M\mu)$, $Q \Rightarrow Q/(M^2\mu^2)$. The radius of the black hole event horizon is $x_{\rm h} = 1 + \sqrt{1 - a^2}$.

Fig. 1. The numerically calculated in the Boyer–Lindquist coordinates trajectories of the "planet" (thin curve) and photon (thick curve) infalling into the black hole with a spin $a = 0.9982$ and event horizon radius $x_{\rm h} = 1.063$. The both trajectories are "winding up" with the angular velocity $\Omega_\varphi \to \Omega_{\rm h}$, by approaching the black hole horizon.)

Fig. 1 shows the numerically calculated "plunging" trajectory of the test "planet" and photon infalling into the rotating black hole. From equations of motion (2) and (3) follows that any trajectory is quasi-periodically "winding up" with an azimuthal angular velocity $\Omega_\varphi = d\varphi/dt$, coming close to the angular velocity of the black hole horizon $\Omega_{\rm h}$, where

$$\Omega_{\rm h} = \left.\frac{d\varphi}{dt}\right|_{x\to x_{\rm h}} = \frac{a}{2(1+\sqrt{1-a^2})}. \tag{7}$$

As a result, the angular velocity of black hole event horizon Ω_φ must be inevitably imprinted to the QPO signal from the accreting black holes. Any source of radiation, e. g., clump of the hot plasma, approaching the event horizon of the rotating black hole will be viewed by the distant observer in a relativistic "synchrotron mode" as short splashes of radiation, beamed and boosted forward into the narrow solid angles,[5–8] and repeated quasi-periodically with a frequency very near to $\nu_{\rm h} = \Omega_{\rm h}/2\pi$. The oscillation with $\Omega_{\rm h}$ from (7) is a first observational signature for revealing the spin of the supermassive black hole in the Galactic center.

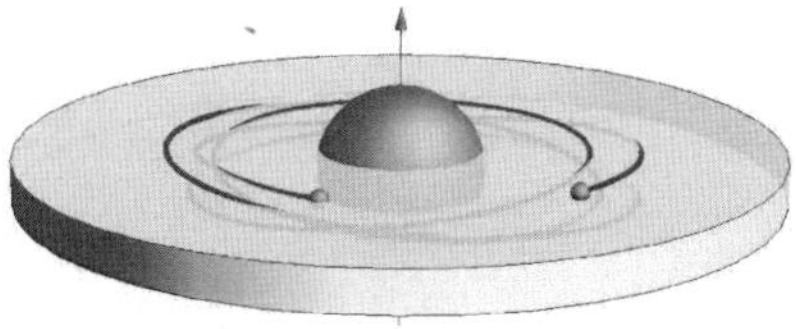

Fig. 2. The non-equatorial orbit of the bright clump of plasma with $Q = 0.1$, $E = 0.91$, $L = 2.715$, $x_{\rm p} = 3.85$, $x_{\rm a} = 5.01$ and $\theta_{\rm max} = 6.6°$, oscillating in the thin opaque accretion disk around the black hole with $a = 0.65$, $x_{\rm h} = 1.76$ and $x_{\rm ms} = 3.62$.

The corresponding second signature is related with the latitudinal QPOs of non-equatorial bound orbits in the accretion flow. The latitudinal angular velocity of the near equatorial circular orbits, Ω_θ, in the limit $Q \to 0$ is equal[11]

$$\Omega_\theta = \frac{2\pi}{T_\theta} = \frac{\sqrt{x^2 - 4ax^{1/2} + 3a^2}}{x(a+x^{3/2})}\frac{1}{M}. \tag{8}$$

This angular velocity describes the latitudinal oscillation of the hot spot or clump of plasma in the thin accretion disk. It is assumed that the brightest hot spots in the accretion flow are located near the radius of the marginally stable circular orbit,[5] $x = x_{\rm ms}$:

$$x_{\rm ms} = 3 + Z_2 - \sqrt{(3-Z_1)(3+Z_1+2Z_2)}, \tag{9}$$

where

$$Z_1 = 1 + (1-a^2)^{1/3}[(1+a)^{1/3} + (1-a)^{1/3}], \quad Z_2 = \sqrt{3a^2 + Z_1^2}. \tag{10}$$

See in Fig. 2 the example of the oscillating clump of plasma in the thin accretion disk around the moderately fast rotating black hole. The latitudinal oscillation with an angular velocity Ω_θ from (8), estimated at the radius $x = x_{\rm ms}$, is the second requisite observation signature of the spinning black hole in the Galactic center. At the same time the values of azimuthal angular oscillations of hot spots in the accretion disk in the accretion disk are spread in a wide range and would not produce any prominent features in the spectrum of QPOs.

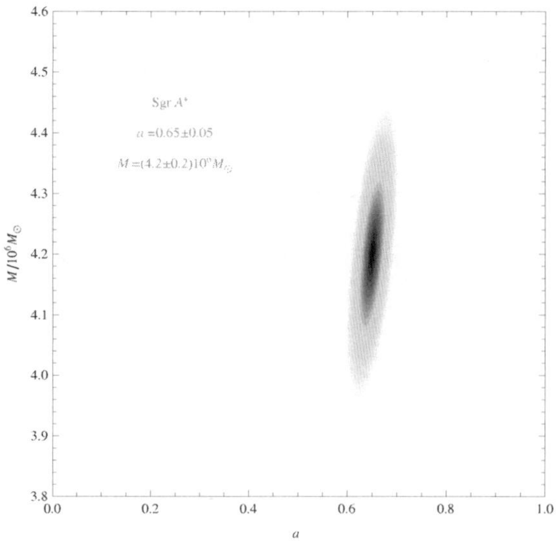

Fig. 3. The 1-sigma error (M, a)-region for the joint resolution of equations (7) and (8) with the observed 11.5 min QPO1 period, identified with $T_{\rm h}$, and, respectively, the 19 min period QPO2, identified with T_θ.

Fig. 3 shows the 1-sigma error (M, a)-region for the joint resolution of equations (7) and (8) with the observed 11.5 min QPO1 period, identified with $T_{\rm h}$, and, respectively, the 19 min period QPO2, identified with T_θ. This 1-sigma (M, a) region corresponds to the Kerr metric rotation parameter, $a = 0.65 \pm 0.05$, and mass, $M = (4.2 \pm 0.2)10^6 M_\odot$, for the supermassive black hole in the Galactic center. Note also, that the additional three QPOs observed in the X-rays with periods around 1.7, 3.6 and 37.5 min.[2] These QPOs are less prominent than the used ones and seemingly related with the resonances in the accretion disk[1,2] or being the harmonics of the used QPOs with periods 11.5 and 19 min, respectively.

A supermassive black hole in the Galactic center acquires its angular momentum by accretion of tidally destructed stars and gas clouds with accidentally orientated individual angular momenta. For this reason a moderate spin value of the supermassive black hole Sgr A* is quite natural due to specific conditions in the Galactic center. At the same time, the black hole mass, derived here from the QPOs, is in a good agrement with a quite independent estimation, $M = (4.1 \pm 0.4)\, 10^6 M_\odot$, measured from observations of the fast moving $S0$ stars.[2,12,13]

References

1. R. Genzel et al., Near-infrared flares from accreting gas around the supermassive black hole at the Galactic Centre, *Nature* **425**, 934 (2003).
2. B. Aschenbach et al., X-ray flares reveal mass and angular momentum of the Galactic Center black hole, *Astron. Astrophys.* **417**, 71 (2004).
3. R. A. Syunyaev, Variability of X rays from black holes with accretion disks, *Sov. Astron.* **16**, 941 (1973).
4. B. Carter, Global structure of the Kerr family of gravitational fields, *Phys. Rev.* **174**, 1559 (1968).
5. J. M. Bardeen, W. H. Press and S. A. Teukiolsky, Rotating black hole: locally nonrotating frames, energy extraction, and scalar synchrotron radiation, *Astrophys. J.* **178** 347 (1972).
6. C. W. Misner et al., Gravitational synchrotron radiation in the Schwarzschild geometry, *Phys. Rev. Lett.* **28**, 998 (1972).
7. C. T. Cunningham and J. M. Bardeen, The opticla appearance of a star orbiting an extreme Kerr black hole, *Astrophys. J.* **173**, L137 (1972).
8. A. G. Polnarev, Radiation spectrum of a source moving in a stable circular orbit near a rotating black hole, *Astrophysics* **8**, 273 (1972).
9. D. C. Wilkins, Bound geodesics in the Kerr metric, *Phys. Rev.* **D 5**, 814 (1972).
10. A. Merloni et al., On gravitomagnetic precession around black holes, *Mon. Not. Roy. Astron. Soc.* **304**, 155 (1999).
11. V. I. Dokuchaev, Spin and mass of the nearest supermassive black hole, *Gen. Relativ. Gravit.* **46**, 1832 (2014).
12. S. Gillessen et al., Monitoring stellar orbits around the massive black hole in the Galactic center, *Astrophys. J.* **692**, 1075 (2009).
13. R. Genzel et al., The Galactic center massive black hole and nuclear star cluster, *Rev. Mod. Phys.* **82**, 3121 (2010).

Possible observational sequences for Randall–Sundrum black holes

K. A. Rannu[1] and S. O. Alexeyev[2]

Sternberg Astronomical Institute, Lomonosov Moscow State University, Moscow, 119991, Russia
[1] *E-mail: rannu@xray.sai.msu.ru*
[2] *E-mail: alexeyev@sai.msu.ru*

N. S. Erin

Department of Astrophysics and Stellar Astronomy, Physics Faculty, Lomonosov Moscow State University
Moscow 119991, Russia
E-mail: erin@sampo.ru

We consider the specific properties of the accretion disk around astrophysical black hole in Randall-Sundrum (RS) II model and show that they fully agree with the Schwarzschild case for the permissible values of the model parameters.

Keywords: Gravity; black hole solutions; accretion; Randall-Sundrum model.

1. ACPY black holes in Randall-Sundrum II

Several black hole solutions in RS model[1,2] were presented in Refs. 3–7. However for a long time it was argued that stable static black hole solutions were applicable for RSII.[8–12] Obviousely, such an argument serves an extremely strong evidence against the viability of the RSII model.

Nevertheless new solutions for large stable black holes were found recently.[13,14] In this work we consider a spherically symmetric static and asymptotically Schwarzschild black hole solution obtained by Abdolrahimi, Cattoën, Page and Yaghoobpour-Tari (ACPY-solution)[14]:

$$ds^2 = \gamma_{\mu\nu} dx^\mu dx^\nu = \left[1 - \frac{1}{-\Lambda_5 r^2} \frac{r - 2M}{r - 1.5M}\left(\mathcal{F} - r\frac{d\mathcal{F}}{dr}\right)\right]\left(1 - \frac{2M}{r}\right)^{-1} dr^2 - \left(1 - \frac{2M}{r}\right) dt^2 + \left[r^2 + \frac{\mathcal{F}}{-\Lambda_5}\right] d\Omega^2, \quad (1)$$

where Λ_5 is the 5D cosmological constant characterizing the vacuum's model qualities and $\mathcal{F}$ is a numerically found 11th-order polinom given by

$$\mathcal{F} \approx 1 - \ 1.1241 \left(\frac{2M}{r}\right) + 1.956 \left(\frac{2M}{r}\right)^2 - \ \ldots \ + 2.900 \left(\frac{2M}{r}\right)^{11}. \quad (2)$$

In our previous work we considered the weak field limit for the RSII black hole solutions by Figueras and Wiseman and by Abdolrahimi et al.[15] We showed the good agreement with GR in PPN limit for both solutions. In this work we compare the accretion properties of the ACPY solution with the case of the standard Schwarzschild metric.

2. Accretion in Randall-Sundrum II

In the present work we consider a thin disk accretion.[16–23] We adopt the following values for the two fundamental mass classes: $M = 14.8M_\odot$, $\dot{M} = 7.484 \times 10^{-7} M/M_\odot$, which are the best estimates available for the well-known galactic black hole Cygnus-X1,[24,25] and $M = 4 \times 10^6 M_\odot$, $\dot{M} = 3 \times 10^{-5} M/M_\odot$ for the Galactic centre black hole candidate Sgr A*.[26–28]

The plots in Fig. 1 show that the energy flux profiles of the disks within the RSII model deviate from the Schwarzschild black hole case for limiting values Λ_5. The inner edge of the accretion disk is located at higher radius than the inner edge of the disk around Schwarzschild black hole.

The variation of the emission spectrum for different values of the bulk cosmological constant is presented in Fig. 2. The plots show that while Λ_5 increases the cut-off frequency of the accretion disk spectrum decreases.

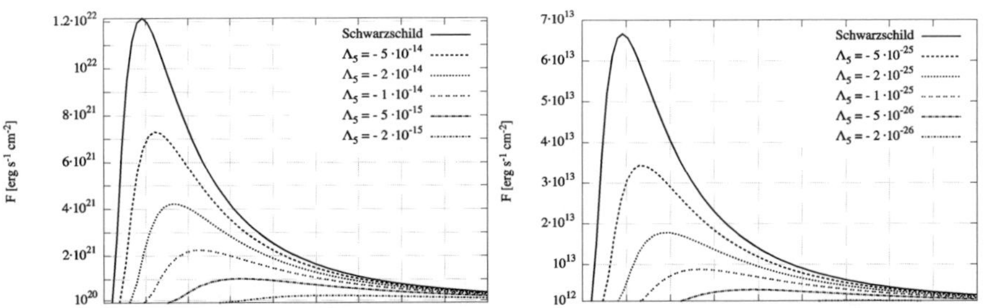

Fig. 1. The energy flux F[erg/(s cm^2)] distribution over the disks for Schwarzschild black holes and the ACPY black holes depending on cosmological constant in the bulk Λ_5 for $M = 14.8M_\odot$ (left) and $M = 4 \times 10^6 M_\odot$ (right).

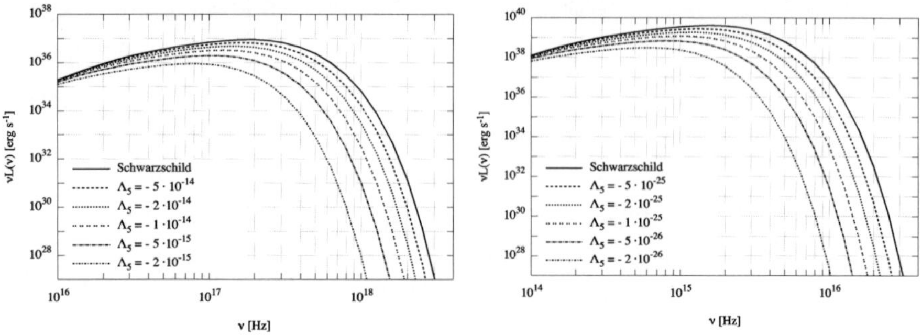

Fig. 2. The accretion disks spectra $\nu L(\nu)$[erg/s], ν[Hz] for Schwarzschild black holes and the ACPY black holes depending on cosmological constant in the bulk Λ_5 for $M = 14.8 M_\odot$ (left) and $M = 4 \times 10^6 M_\odot$ (right).

However all these differences occur at very specific values of bulk cosmological constant Λ_5. As the ACPY-solution is stable for $\Lambda_5 \sim 6$ the differences from Schwarzschild case cannot occur in reality. Therefore the radiation of the accretion disk around ACPY black holes should produce the spectra similar to the standard Schwarzschild case.

3. Conclusions

In this work the flux and spectrum of the accretion disk around the ACPY black holes in Randall-Sundrum II model are obtained. For possible bulk values of the bulk cosmological constant Λ_5 the ACPY solution shows full agreement with GR. From the other hand these solutions cannot be distinguished from it by accretion observations.

Acknowledgments

Authors would like to thank professors N. I. Shakura, K. S. Thorne and T. S. Harko for useful discussions and great help on performing of this work. This work was partially supported by individual grant from Dmitry Zimin Foundation "Dynasty" (S.A.).

References

1. L. Randall and R. Sundrum, *Phys. Rev. Lett.* **83**, 3370 (1999).
2. L. Randall and R. Sundrum, *Phys. Rev. Lett.* **83**, 4690 (1999).
3. N. Dadhich, R. Maartens, P. Papadopoulos and V. Rezania, *Phys. Lett. B* **487**, 1 (2000).

4. S. O. Alexeyev and D. A. Starodubtseva, *JETP* **111**, 576 (2010).
5. R. Maartens, *Phys. Rev. D* **62**, 084023 (2000).
6. A. Chamblin, S. W. Hawking and H. S. Reall, *Phys. Rev. D* **61**, 065007 (2000).
7. N. Dadhich, *Phys. Lett. B* **492**, 357 (2000).
8. T. Tanaka, *Phys. Suppl.* **148**, 307 (2002).
9. R. Emparan, A. Fabbri and N. Kaloper, *J. High Energy Phys.* **08**, 043 (2002).
10. R. Emparan, J. Garcia-Bellido and N. Kaloper, *J. High Energy Phys.* **01**, 079 (2003).
11. H. Yoshino, *J. High Energy Phys.* **01**, 068 (2009).
12. B. Kleihaus, J. Kunz, E. Radu and D. Senkbeil, *Phys. Rev. D* **83**, 104050 (2011).
13. P. Figuras and T. Wiseman, *Phys. Rev. Lett.* **107**, 081101 (2011).
14. S. Abdolrahimi, C. Cattoën, D. Page and S. Yaghoobpour-Tari, *Phys. Lett. B* **720**, 405 (2013).
15. S. O. Alexeyev, K. A. Rannu, P. I. Dyadina, B. N. Latosh and S. G. Turyshev, *JETP*, **147**, 1120 (2015).
16. I. D. Novikov and K. S. Thorne, *in Black Holes, ed. C. DeWitt and B. DeWitt, New York: Gordon and Breach* (1973).
17. N. I. Shakura and R. A. Sunyaev, *Astron. Astrophys.* **24**, 33 (1973).
18. N. I. Shakura and R. A. Sunyaev, *Mon. Not. R. astr. Soc.* **175**, 613 (1976).
19. R. A. Syunyaev and N. I. Shakura, *Sov. Astron. Lett.* **12**, 117 (1986).
20. D. N. Page and K. S. Thorne, *Astrophys. J.* **191**, 499 (1974).
21. D. N. Page and K. S. Thorne, *Astrophys. J.* **191**, 507 (1974).
22. T. Harko, Z. Kovacs and F. S. N. Lobo, *Phys. Rev. D* **78**, 084005 (2008).
23. T. Harko, Z. Kovacs and F. S. N. Lobo, *Phys. Rev. D* **79**, 064001 (2009).
24. J. A. Orosz, J. E. McClintock, J. P. Aufdenberg, R. A. Remillard, M. J. Reid, R. Narayan and L. Gou, *Astrophys. J.* **742**, 84 (2011).
25. M. J. Reid, J. E. McClintock, R. Narayan, J. A. Orosz, R. A. Remillard and L. Gou, *Astrophys. J.* **742**, 83 (2011).
26. A. E. Broderick and R. Narayan, *Astrophys. J.* **638**, 21 (2006).
27. J. Dexter and P. C. Fragile, *MNRAS* **432**, 2252 (2013).
28. H. Falcke, S. B. Markoff, *Class. Quant. Grav.* **30**, 244003 (2013).

Statistical properties of states in QED with unstable vacuum

S.P. Gavrilov[1,2;*], D.M. Gitman[2,3;†], and A.A. Shishmarev[2,4;‡].

[1] *Herzen State Pedagogical University of Russia, St. Petersburg, Russia;*
[2] *Department of Physics, TSU, Tomsk;*
[3] *P. N. Lebedev Physics Institute, Moscow, Russia;*
[4] *Instituto de Fisica, USP, Sao Paulo, Brazil*
[*] *E-mail: gavrilovsergeyp@yahoo.com*
[†] *E-mail: gitman@if.usp.br*
[‡] *E-mail: a.a.shishmarev@mail.ru*

Using nonperturbative expressions for the reduced density operators for electron and positron subsystems created by external uniform electric-like fields,[1] and the density operator of the system that undergo a decoherence[2] in course of the evolution due to an intermediate measurement of number of particles, we calculate the entropy and study the information loss in QED states due to partial reductions and a possible decoherence. We consider the so-called T-constant external electric field as an example.

Keywords: Entropy; particle creation, T-constant electric field.

The von Neumann information entropy for the reduced density operators of the system with vacuum or thermal initial state has the form

$$S\left(\check{\rho}_{\pm}(J)\right) = -k_B \mathrm{tr}_{\pm}\left(\check{\rho}_{\pm}(J) \ln \check{\rho}_{\pm}(J)\right). \tag{1}$$

We have shown that the entropy (1) can be presented in a following form:

$$S\left(\check{\rho}_{\pm}(J)\right) = -k_B \sum_n \left[\kappa\left(1-\kappa N^J_{n,\pm}\right)\ln\left(1-\kappa N^J_{n,\pm}\right) + N^J_{n,\pm}\ln N^J_{n,\pm}\right], \tag{2}$$

where $N^J_{n,\pm}$ are the mean numbers of electrons and positrons in the final state of the system with initial state defined by sourses J, correspondingly, n is the full set of quantum numbers. For the Dirac case $\kappa = +1$, for Klein-Gordon $\kappa = -1$. We have shown[2] that entropy of the system with initial vacuum state that undergoes a decoherence in course of the evolution due to an intermediate measurement of number of particles has exactly the same form as the entropy of reduced density operators in vacuum case (2).

To illustrate this result we consider the so-called T-constant electric field as an external background. Such a field acts only during a finite time

T and it is constant within this time interval. In $d = (D+1)$-dimensional space the T-constant electric field $\mathbf{E}(t) = E$ is constant inside the time interval $T = t_{\text{out}} - t_{\text{in}}$, and is zero outside of it. Processes of pair creation in such field were studied before.[3–6]

The summation over the quantum numbers can be reduced to an integration over momenta, $\sum_n \to \frac{\gamma_{(d)} V}{(2\pi)^{d-1}} \int d\mathbf{p}$, where V is the D-dimensional spatial volume, and $\gamma_{(d)} = 2^{[\frac{d}{2}]-1}$ is a factor that appears after summation over spin quantum number (for Klein-Gordon case $\gamma_{(d)} = 1$). The result of the integration can be written as

$$S(\check{\rho}_{\pm}(0)) = \gamma_{(d)} k_B \frac{(eE)^{\frac{d}{2}} TV}{(2\pi)^{d-1}} A\left(d, E_c/E\right), \tag{3}$$

where the factor TV can be considered as d-dimensional volume. To get finite and correct expressions, one should use the volume normalization. The factor $A\left(d, E_c/E\right)$ can be estimated for strong $E_c/E \ll 1$, critical $E_c/E = 1$, and weak $E_c/E \gg 1$ field strength limits for any given d^2.

Acknowledgement

SPG thanks FAPESP for support and University of Sao Paulo for the hospitality. DMG is grateful to the Brazilian foundations FAPESP and CNPq for permanent support. This work is also partially supported by the Tomsk State University Competitiveness Improvement Program. AAS thanks CAPES for support. The reported study of SPG, DMG, and AAS was partially supported by RFBR, research project No. 15-02-00293a.

References

1. S. P. Gavrilov, D. M. Gitman, and J. L. Tomazelli, Nucl. Phys. B **795**, 645 (2008).
2. S. P. Gavrilov, D. M. Gitman, and A. A. Shishmarev, Phys. Rev. A **91**, 052106 (2015).
3. S. P. Gavrilov and D. M. Gitman, Phys. Rev. D **53**, 7162 (1996).
4. S. P. Gavrilov and D. M. Gitman, Phys. Rev. D **78**, 045017 (2008).
5. V. G. Bagrov, D. M. Gitman, and Sh. M. Shvartsman, Zh. Eksp. Teor. Fiz. **68**, 392 (1975) [Translation: Sov. Phys. JETP **41**, 191 (1975)].
6. D. M. Gitman, V. M. Shachmatov, and Sh. M. Shvartsman, Izw. VUZov Fizika (Sov. Phys. Journ.) **18**, 23 (1975).

Type D stationary axisymmetric solutions to Einstein–Maxwell equations conformal to Carter spacetimes

Alberto A. Garcia-Diaz and Gustavo Gutierrez-Cano

Departamento de Física, Centro de Investigación y de Estudios Avanzados del IPN, A.P. 14-740, 07000 México D.F., Mexico

It is established that the conformal type D electrovacuum stationary axisymmetric Carter class of metrics splits into three families of solutions: the Plebański–Demiański family, the Carter–Plebański spacetimes, and the trigonometric–hyperbolic conformal class. This last class, via coordinate transformations, can be brought to the C–P form.

These metrics admit an Abelian group of motions G_2 with commuting stationary and angular Killing vectors, are of Petrov type D; the null eigenvectors of the general electromagnetic field coincide with the directions of the Debever null vectors of Weyl curvature tensor. These solutions exhibit, for certain ranges of the coordinate variables, a black hole behavior; a "ring" singularity similar to the one of the Kerr metric is present. The limiting transition of these solutions leads to the Kerr–Newman black hole solution as a subbranch.

Keywords: Exact solutions, black holes.

1. Introduction

Hundred years ago, Einstein reported at the Berlin Academy of Science his theory of gravity at November 15, 1915. One month latter, December, 2015, Schwarzschild derived the first exact solution for a static point mass to the Einstein equations. The same Einstein reported, on behalf of Schwarzschild, the solution at the Berlin Academy of science on January 13, 1916. The term horizon was introduced to denote the Schwarzschild singularity and Wheeler, December 1967, coined the name of black hole for the gravitational field of an imploding (collapsed) star. The deep understanding of Schwarzschild solution was achieved with its maximal extension achieved independently by Kruskal, 1959, and Finkelstein, 1958.

After practically fifty years, September 1963, Kerr published his outstanding solution for a stationary rotating mass. The full understanding of the Kerr solution was achieved only after the Carter's work,[1] 1968, on the separability of the Hamilton-Jacobi equations shearing light on the extended structure of the Kerr solution.

There were reported four families of solutions allowing for the mentioned–above separability in the presence of a Maxwell field and a cosmological constant too. Later, in 1975, Plebański, starting from the Carter[A] class of solutions, was able to reproduce via limiting transitions all the Carter's subclasses. In this contribution a conformal representation of the Carter[A] class is reported; the metric of this solution is endowed with a conformal factor function, allows for the separability of the massless Klein–Gordon and Hamilton–Jacobi equations.

2. Generalized Carter metric

The famous Carter metric contains the relevant Schwarzschild and Kerr black hole solutions, and their electromagnetic generalizations–the Reissner–Nordstrom and Kerr-Newman black holes. It also allows for their generalizations with cosmological constant.

The Carter[A] [1]–Plebański [2] metric can be given

$$ds^2 = \frac{1}{H^2}\left[\frac{\Delta}{P(x)}dx^2 + \frac{P(x)}{\Delta}(d\tau + N(y)d\sigma)^2 + \frac{\Delta}{Q(y)}dy^2 - \frac{Q(y)}{\Delta}(d\tau + M(x)d\sigma)^2\right], H = H(x,y),\ \Delta := M - N. \quad (1)$$

It is convenient to work with the null tetrad formalism, with a null tetrad such that

$$ds^2 = g_{ab}\, e^a\, e^b = 2\, e^1 e^2 + 2\, e^3 e^4, \quad (2)$$

$$\left.\begin{matrix} e^1 \\ e^2 \end{matrix}\right\} = \frac{1}{H(x,y)\sqrt{2}}\left\{\sqrt{\frac{\Delta}{P(x)}}dx \pm i\sqrt{\frac{P(x)}{\Delta}}(d\tau + N(y)d\sigma)\right\},$$

$$\left.\begin{matrix} e^3 \\ e^4 \end{matrix}\right\} = \frac{1}{H(x,y)\sqrt{2}}\left\{\sqrt{\frac{\Delta}{Q(y)}}dy \pm \sqrt{\frac{Q(y)}{\Delta}}(d\tau + M(x)d\sigma)\right\},$$

2.1. *Petrov type D family of metrics*

Requiring type D with aligned null basis tetrad vectors along the eigenvectors of the Weyl tensor is equivalent to demand additionally the vanishing of the Weyl coefficients $C^{(2)} = 0 = C^{(4)}$, which leads to the equation

$$\frac{\partial}{\partial x}\left(\frac{M_{,x}}{\Delta}\right) - \frac{\partial}{\partial y}\left(\frac{N_{,y}}{\Delta}\right) = 0 \rightarrow \Delta(M_{,xx} - N_{,yy}) - (M_{,x})^2 + (N_{,y})^2 = 0. \quad (3)$$

Differentiating this equation twice with respect to x and y one gets

$$N_{,y}M_{,xxx}+N_{,yyy}M_{,x}=0\rightarrow\frac{M_{,xxx}}{M_{,x}}=\alpha_0=-\frac{N_{,yyy}}{N_{,y}},\tag{4}$$

hence, integrating one arrives at

$$M_{,xx}-\alpha_0\,M=M_0,N_{,yy}+\alpha_0\,N=N_0,\tag{5}$$

2.1.1. *The polynomial structure functions' class; vacuum and electromagnetic solutions*

This class, containing the most relevant black hole solutions of the General Relativity Gravitational theory, arises for $\alpha_0=0$. In such case, the integrals of (5), by shifting the coordinates x and y, can be brought to the form

$$M=M_0x^2+m_0,\;N=N_0\,y^2+n_0.\tag{6}$$

The substitution of these functions into the original constraint (3) impose some conditions on the remaining constants, $M_0=-N_0=\kappa_0^2$, $m_0=n_0=:\rho_0$. Finally,

$$M=-\kappa_0^2\,x^2+\rho_0,\;N=-\kappa_0^2\,y^2+\rho_0,\,\Delta=M-N=\kappa_0^2\,(x^2+y^2).\tag{7}$$

2.1.2. *The trigonometrical structure functions' class*

This class, containing perfect fluid coupled to electromagnetic field solutions, arises for $\alpha_0=\pm\nu^2$. In these cases, the integrals of (5) are

$$\begin{aligned}&M(x)=m_0\sin\nu x+m_1 cos\nu x+\rho_0,\;\alpha_0=-\nu^2,\\&N(y)=n_0\sinh\nu y+n_1 cosh\nu x+\rho_0,\;-n_1^2+m_1^2+m_0^2+n_0^2=0,\end{aligned}\tag{8a}$$

$$\begin{aligned}&M(x)=m_0\sinh\nu x+m_1 cosh\nu x+\rho_0,\;\alpha_0=\nu^2,\\&N(y)=n_0\sin\nu y+n_1 cos\nu x+\rho_0,\;-m_1^2+n_1^2+m_0^2+n_0^2=0.\end{aligned}\tag{8b}$$

2.2. *Null tetrad basis aligned to the electromagnetic eigenvectors*

The most simple description of the linear–nonlinear electrodynamics can be accomplished in the tetrad basis coupled to the null eigenvectors of the electromagnetic field, which simplify considerably the traceless Ricci tensor; for such alignment the only non–vanishing components are $C_{12}=C_{34}$. The wider known classes of aligned type D solutions are the Carter[A] [1]–Plebański [2] metric with structural functions $M(x)=x^2$, $N(y)=$

$-y^2$, $H(x, y) = 1$, and the Plebański–Demiański metric[3] with $M(x) = x^2$, $N(y) = -y^2, H(x, y) = 1 + \mu\, x\, y$. A detailed derivation of all the subfamilies of solutions belonging to these metrics and their reductions can be found in Ref. 4.

The equations arising from $C_{13} = 0$ for the trigonometric–hyperbolic M and N functions (8) allows for variable separable function $H(x, y)$ of the form $M(x) \to \cos^2 \nu x \to x^2$, $N(y) \to -\sinh^2 \nu x \to -y^2, H\,(x, y) \to \sin\,(\nu\, x) \cosh\,(\nu\, y) \to \sqrt{1 - x^2}\,\sqrt{1 + y^2}$ The aligned electrovacumm type D stationary axisymmetric class of solutions derived on the basis of the above structural functions can be given in the last set of spatial coordinates in terms of radicals and polynomials.

3. The metric

This new representation of stationary axisymmetric solutions can be given by the metric

$$
\begin{aligned}
ds^2 = \frac{1}{H^2} \Bigg[& \frac{\Delta\, dx^2}{P(x)(1 - x^2)} + \frac{P(x)}{\Delta}(d\tau - y^2 d\sigma)^2 + \frac{\Delta\, dy^2}{Q(y)(1 + y^2)} \\
& - \frac{Q(y)}{\Delta}(d\tau + x^2 d\sigma)^2 \Bigg],
\end{aligned}
$$

$$
\begin{aligned}
P(x) = {} & 2\, L_0\, x\, \left(1 - x^2\right)^{3/2} + \left(\gamma_0 - {B_0}^2\right)\, \left(1 - x^2\right)^2 \\
& + K_0\, \left(1 - x^2\right) \left(1 - 2\, x^2\right) - \frac{\lambda}{3}\, \left(4 x^4 - 6\, x^2 + 3\right),
\end{aligned}
$$

$$
\begin{aligned}
Q(y) = {} & -2\, M_0\, \left(1 + y^2\right)^{3/2} y + \left(\gamma_0 + {E_0}^2\right)\, \left(1 + y^2\right)^2 \\
& + K_0\, \left(1 + 2\, y^2\right) \left(1 + y^2\right) - \frac{\lambda}{3}\, \left(4 y^4 + 6\, y^2 + 3\right),
\end{aligned}
$$

$$
\Delta = x^2 + y^2, \; H(x, y) = \sqrt{1 - x^2}\sqrt{1 + y^2}, \tag{9}
$$

where the parameters, roughly speaking, represent: M_0–the mass, L_0–the magnetic mass, E_0–the electric charge, B_0–the magnetic charge, λ–the cosmological constant, and K_0 is related to the angular momentum.

The real null vectors ${e^3}_\mu$ and ${e^4}_\mu$ are identified with the double null Debever eigenvectors k_μ and l_μ of the Weyl tensor of this Petrov type D gravitational field. Because of this alignment, the only nonvanishing complex curvature coefficient is $C^{(3)} = 2\Psi_2$.

3.1. *Coordinate transformations to the Carter solution*

The way of finding the transformations of the spatial coordinates is related to the invariant character of the Weyl complex coefficient $C^{(3)}$; recalling that stationary axisymmetric vacuum or Maxwell charged Carter solutions are unique, the derived generalized solution has to reduce to them for the quoted sources without cosmological constant. It is easy to establishes that the spatial coordinates transform according to

$$x = \frac{p}{\sqrt{1+x^2}},\ p = \frac{x}{\sqrt{1-x^2}},$$
$$y = \frac{q}{\sqrt{1-q^2}},\ q = \frac{y}{\sqrt{1+y^2}}, \tag{10}$$

Thus, the derived solution is not other but just the Carter solution in another coordinate representation.

This work has been partially supported by Grant CONACyT 178346. We appreciate the advises and comments of Dr. Vladimir Manko.

References

1. B. Carter, Commun. Math. Phys. **10**, 280 (1968); Phys. Lett. A **26**, 399 (1968).
2. J.F. Plebański, Ann. Phys. (USA) **90**, 196 (1975).
3. J.F. Plebański and M. Demiański, "Rotating, charged, and uniformally accelerating mass in general relativity," Ann. Phys. (N.Y.) **98**, 98 (1976).
4. A. García and A. Macías, "Black Holes as Exact Solutions of the Einstein-Maxwell Equations of Petrov Type D," in *Black Holes: Theory and observation*, proceedings of the 179th W.E. Heraeus Seminar, Bad Honnef, Germany, edited by F.W. Hehl, C. Kiefer, R.J.K. Metzler (Eds.): Lectures Notes in Physics (LNP) 514, 203. (Springer–Verlag, Berlin–Heidelberg, 1998).
5. H. Stephani, D. Kramer, M. MacCallum, C. Honselaers, and E. Herlt, *Exact solutions to Einstein's Field Equations, Second Edition* (Cambridge University Press, 2003).

The dependence of tidal forces on the polar angle in the Kerr metric

A. M. Rasulova

Theoretical Physics and Astronomy Department, The Herzen University, Moika 48
St. Petersburg, 191186, Russia,
Institute of Limnology, RAS, Sevastyanov st. 9.
St. Petersburg, 196105, Russia
E-mail: ARasulova@gmail.com

In the paper the dependence of the tidal forces for the two protons at a distance of the order of the Compton wavelength from the polar angle in the Kerr space-time is considered. It is shown that in approach to the equatorial plane the tidal forces are increase.

Keywords: Geodesic deviation; Kerr metric; tidal forces; work of the tidal forces.

We used the equation of geodesic deviation for structureless particles to obtain a more accurate expression for the tidal forces[1,2]:

$$\frac{d^2\eta^i}{ds^2} - R^i_{jkm} u^j \eta^k u^m = 0, \tag{1}$$

where η^i — infinitesimal vector deviation, R_{ljkm} — Riemann tensor, u^j— 4-speed.

A solution of equation (1) is the vector of deviations of world lines that covariantly describes the relative acceleration between them.

At substituting in the equation (1) components of the curvature tensor[3] for the Kerr metric and taking into account only the radial velocity component:

$$u^i = \Gamma(1, V, 0, 0), \tag{2}$$

where $\Gamma = \frac{1}{\sqrt{1-V^2}}$ is Lorentz factor, we obtain the explicit dependence of the vector deviation.

For numerical calculations of tidal forces of the proton in the Kerr metric, we use the following assumptions: let proton's velocity is $V = (1 - 10^{-15})c$ and deviation is proportional to the Compton wavelength $\lambda_C^p = 1.32 \cdot 10^{-15}$m; the black hole has the following parameters: $M = 10^6$, $a = 0.98$; motion occurs at a coordinate of distance of r from center of black hole $r = r_h + 10^{-5} r_h$ (r_h – the horizon).

Under the above assumptions a graph of dependence of the tidal forces is given on Fig.1.

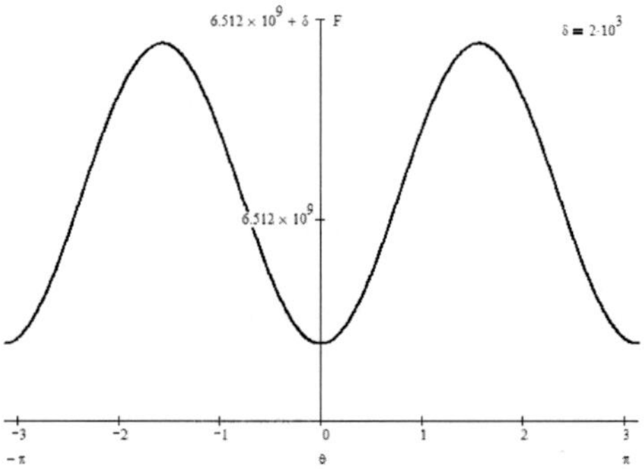

Fig. 1. The dependence of the tidal forces on the polar angle.

It is clear from the graph that the tidal forces depend on the plane of incidence, the difference between the maximal value of tidal forces (at $\theta = \pi/2$) and minimal (if $\theta = 0$ or π) ones is approximately 0.3 Newtons (or $5.762 \cdot 10^{-10}\%$). With the decrease in the mass of the black hole the difference between the tidal forces at the equator and at the pole will increase, for example, at the mass of the black hole $M = 10^2 M_\odot$ it is about $10^{-2}\%$.

References

1. J. L. Synge, On the deviation of geodesics and nullgeodesics particularly in relation to the properties of spaces of constant curvature and indefinite line-element *Element Ann. Math.* **35**, 705 (1934).
2. J. L. Synge, Relativity: The General Theory (North-Holland, Amsterdam, 1960).
3. S. Chandrasekhar, The Mathematical Theory of Black Holes (Oxford Univ. Press, Oxford, 1983).

The self-force effect of a static charge in a long throat

O. Aslan

Institute of Mathematics and Mechanics, Kazan Federal University,
Kazan, 420008, Russia
E-mail: ulsucuk@gmail.com

We compute the self-force on a scalar charge at rest in the spacetime of a long throat. We consider arbitrary values of the mass of the scalar field and the constant of nonminimal coupling of the scalar field to the curvature of spacetime. We also show the coincidence of explicit calculations of self-force in the limit of large mass of the field with known results.

Keywords: Self-force; scalar field; wormhole.

A charge in a curved spacetime may become subjected to self-interactions. The origin of this induced self-interaction resides on the non-local structure of the field caused by the spacetime curvature or non-trivial topology.

In flat spacetime this effect is produced by a local distortion of the field lines associated with the particle's acceleration. For electrically charged particles in flat spacetime, the self-force is given by the Abraham-Lorentz-Dirac formula. In the gravitational field the self-energy problem becomes more complicated. The reason is that contribution to the self-energy in this case is non-local. The self-force problem for an electric charge in a curved space background was first investigated by DeWitt and Brehme and later by Hobbs. The gravitational self-force was first calculated by Mino, Sasaki and Tanaka. Later, Quinn derived the equivalent formula for a charge, coupled to a minimally-coupled massless scalar field.

A number of simple static configurations has been analyzed, including the self-force acting on scalar or electric charges held static in the spacetime of a Schwarzschild black hole, electric or magnetic dipoles which are static outside a Schwarzschild black hole, a static electric charge outside a Kerr black hole, or a Kerr-Newman black hole, a static electric charge in a spherically-symmetric Brans-Dicke field. The self-force can be nonzero for a static particle in flat spacetimes of the topological defects.[1–5] In curved spacetimes with nontrivial topological structure the investigations of this type have additional interesting features.[6–11]

Unfortunately, the authors do not know the results of calculation of the self-force of the charge, which is the source of a massive field. In this paper we consider the problem of computing the self-force on a scalar charge at rest in the spacetimes of long throats, allowing for the arbitrary values of the mass of field and coupling constant. The considered approach gives the possibility to compute the approximate expression for the self-potential and the self-force on a scalar charge at rest in the spacetime of long throat. The used WKB approximation is valid for all the modes (including $l = 0$ mode). This implies also that the approximate expression for the self-force does not depend on the conditions at infinity and in the considered situation the effect of self-action is local even in the limit of massless field.

We also show that the asymptotic behavior of the self-potential in the limit of large mass of the field coincides with the corresponding result of paper.[12]

References

1. B. Linet, *Phys. Rev. D***33**, 1833 (1986).
2. A. G. Smith, in *The Formation and Evolution of Cosmic Strings*, G. W. Gibbons, S. W. Hawking, and T. Vachaspati (eds.) (Cambridge University Press, Cambridge, 1990).
3. N. R. Khusnutdinov, *Class. Quantum Grav.* **11**, 1807 (1994).
4. N. Khusnutdinov, *Teor. Mat. Fiz.* **103**, 339 (1995). [*Theor. Math. Phys.* **103**, 603 (1995).]
5. V. De Lorenci, Jr E. Moreira, *Phys. Rev. D* **65**, 085013 (2002).
6. N. Khusnutdinov and I. Bakhmatov, *Phys. Rev. D* **76**, 124015 (2007).
7. B. Linet, *Electrostatics in a wormhole geometry*, arXiv:0712.0539 [gr-qc]
8. S. Krasnikov, *Class. Quantum Grav.* **25**, 245018 (2008).
9. V. Bezerra and N. Khusnutdinov, *Phys. Rev. D* **79**, 064012 (2009).
10. M. Casals, S. Dolan, A. Ottewill and B. Wardell, *Self-Force Calculations with Matched Expansions and Quasinormal Mode Sums*, arXiv:0903.0395 [gr-qc].
11. A. Popov, Physics Letters B, **693**, 180 (2010).
12. A. Popov, *Phys. Rev. D* **84**, 064009 (2011).

Black-and-white holes

E. V. Mikheeva*, V. N. Lukash and V. N. Strokov

Astro Space Center, P. N. Lebedev Institute of Russian Academy of Sciences, Moscow, 117997, Russia
**E-mail: helen@asc.rssi.ru*
www.lebedev.ru

We use the phenomenological approach to study a space-time in the neighborhood of the singularity situated inside the Schwarzschild horizon. Requiring boundedness of the Schwarzschild-like metrics we come to the notion of the integrable singularity that allows the matter to pass to the white-hole region. It leads to production of a new universe that is born already inflated ('singularity-induced inflation').

Keywords: Black-and-white hole; integrable singularity.

1. Black-and-White Hole

Solutions of Einstein's equations which correspond to black holes are known many years. The simplest of them is the Schwarzschild vacuum metric describing a gravitational field generated by a massive body without electric charge and rotation. The common feature of all these solutions is a singularity situated at radial coordinate $r = 0$.

White hole is described by the same metric but with reversed initial condition. While the black hole is the finite stage of evolution of a massive body, the white hole corresponds to the cosmological expansion of the universe.

In this short paper we discuss the so-called black-and-white hole (BWH) and argue that it is not just a glueing together of black and white holes but a new configuration of gravitational fields where the singularity at $r = 0$ exists but is much softer than in classical black or white holes. The key concept behind it is an integrable singularity (IS), that caused by vacuum polarization and matter creation in strong gravitational fields.

2. Integrable Singularity

Let us consider Schwarzschild-like metric

$$ds^2 = N^2 (1 + 2\Phi) dt^2 - \frac{dr^2}{1 + 2\Phi} - r^2 d\Omega^2$$

where N and Φ are real and bounded functions of t and r.

Assuming by definition

$$\Phi = -\frac{Gm(t,r)}{r} \qquad m(t,0) = 0,$$

we derive the solution of Einstein equation as follows:

$$m(t,r) = 4\pi \int_0^r T_t^t r^2 dr.$$

As the integral is bounded we deal with integrable singularity at $r = 0$. Note that region $r > 0$ corresponds to the "black part" of BWH, whereas region $r < 0$ is an extension of the black part and similar to a white hole.

To demonstrate that geodesics in such a geometry do not terminate at $r = 0$ we need to calculate the Riemann tensor. In locally inertial reference frame

$$ds^2 = -d\hat{t}^2 + d\hat{r}^2 - d\hat{\theta}^2 - d\hat{\phi}^2$$

the non-zero Riemann tensor component are

$$R_{\hat{t}\hat{\phi}\hat{\theta}\hat{\phi}} = \Phi'', \qquad R_{\hat{t}\hat{\theta}\hat{t}\hat{\theta}} = R_{\hat{t}\hat{\phi}\hat{t}\hat{\phi}} = -R_{\hat{r}\hat{\theta}\hat{r}\hat{\theta}} = -R_{\hat{r}\hat{\phi}\hat{r}\hat{\phi}} = \frac{\Phi'}{r},$$

and the unique divergent component is:

$$R_{\hat{\theta}\hat{\phi}\hat{\theta}\hat{\phi}} = -\frac{2\Phi}{r^2}.$$

Thus, tidal forces in matter are bounded on radial geodesics:

$$\frac{D^2 \xi^{\hat{i}}}{d\hat{r}^2} = -R_{\hat{r}\hat{i}\hat{r}\hat{j}} \xi^{\hat{j}} \propto \frac{GM}{r_0^3} \xi^{\hat{i}},$$

where $0 < r_0 \ll 2GM$.

It is rather evident that the pressure of matter generated near $r = 0$ has to be anisotropic. Namely, the stress-energy tensor can be written as follows:

$$T_\mu^\nu = diag(-p, \varepsilon, -p_\perp, -p_\perp).$$

In the simplest case $p = -\varepsilon$, which results in

$$\frac{d(\varepsilon r^2)}{r\,dr} = -2p_\perp,$$

and to go on we need to model behavior of $p_\perp$.

Two examples has been analysed

$$p_\perp^{(A)} = p_0\theta\left(r_0^2 - r^2\right)$$

and

$$p_\perp^{(B)} = p_0\theta\left(rr_0 - r^2\right) - p_1\theta\left(-r\right) ,$$

where $p_{0,1} > 0$ and $\theta\left(x\right)$ is Heaviside function.

The case (A) corresponds to symmetric step for r and $-r$ and means an ideal BWH. The case of the model (B) is not symmetric and deals with a universe at $r < 0$.

A more detailed description of BWH can be found in Refs. 1–4.

3. Conclusions and Cosmological Consequences

The approach used to study the space-time in the neighborhood of the singularity situated inside the Schwarzschild horizon is phenomenological. The reasons are, first, lack of knowledge about physical processes near the inner singularity of BHs, and second, large amount of models trying to describe the physics near the singularity of black holes.

Requiring a boundedness of the Schwarzschild-like metrics we come to the notion of integrable singularity which allows the matter to pass through this horizon to the white-hole region. It leads to production of new universe which is born already inflated (the so-called "singularity-induced inflation").

In such a picture the ultra-high curvature quantum state arisen inside black hole, forms at the same time the initial state of a new universe. So, the new universe expansion is launced by collapse of a star in the parent universe. As a result the cosmological (Hubble) flow of new universe is made of the effective matter created inside black-and-white hole.

Acknowledgments

One of us (VNS) is grateful to the Rissian Foundation of Fundamental Research (research project 15-02-00554) for financial support. VNL and EVM were partially supported by the grant of President of Russian Federation for support of leading scientific schools (NSh-4235.2014.2).

References

1. V. N. Lukash, E. V. Mikheeva and V. N. Strokov, From the cosmological model to the generation of the Hubble flow,*Phys. Usp.* **55**, 204 (2012).
2. V. N. Lukash, E. V. Mikheeva and V. N. Strokov, Generation of cosmological flows in General Relativity, *Phys. Usp.* **55**, 831 (2012).
3. V. N. Lukash, E. V. Mikheeva and V. N. Strokov, The early Universe and cosmogenesis, *Phys. Part. Nuc.* **44**, 415 (2013).
4. V. N. Lukash and V. N. Strokov, Space-Times with Inregrable Singularity: Black-white holes and astrogenic universes, *Int. J. Mod. Phys. A* **28**, 1350007 (2013).

Section 4

Strings, Branes, Higher Spin Fields, Quantum Gravity

Gravitational collision of particles with domain walls: Energy-momentum balance

D. V. Gal'tsov*, E. Yu. Melkumova and P. Spirin

Moscow State University, Faculty of Physics,
119991, Leninskie gory 1, str. 2, Moscow, Russia
**E-mail: galtsov@phys.msu.ru*

We show that the relativistic momentum balance in the point particle - domain wall collision can be formulated in analogy with non-relativistic theory introducing mutual gravitational potentials.

Recently we have shown[1,2] that perforation of the domain wall by the point particle can be described in linearized gravity in terms of distributions. One novel feature of this collision is excitation of the branon wave propagating along the wall and contributing to the momentum balance of the system. Another peculiarity is the absence of asymptotically free states since the interaction potential between the colliding objects grows with the distance. To establish the momentum balance in the collision perturbatively, we expand the particle world-line and the domain wall world-volume in $\varkappa = \sqrt{16\pi G}$ obtaining the following equation for the second order metric perturbation:

$$\Box\left(\delta H^{MN} - \frac{1}{2}\eta^{MN}\delta H\right) = -\varkappa\tau^{MN} \equiv -\varkappa(\delta T^{MN} + \delta\bar{T}^{MN} + S^{MN}).$$

Here $\delta T^{MN}, \delta\bar{T}^{MN}$ are perturbations of the wall and the particle stress-tensors, and S^{MN} stands for the quadratic in the linear order metric perturbation $H^{MN} = h^{MN} + \bar{h}^{MN}$ term in the expansion of Einstein tensor (h^{MN} and $\bar{h}^{MN}$ being due to the wall and the particle) in which only the crossed contributions have to be kept.

Tensor τ^{MN} is interpreted as the total energy-momentum tensor in the lowest non-trivial order in $\varkappa$. It is conserved: $\partial_N\tau^{MN} = 0$. To convert this into the the energy-momentum balance equation one integrates over the world-tube Ω:

$$0 = \int_\Omega \partial_N\tau^{MN} = \int_{\partial\Omega} \tau^{MN} d\Sigma_N \tag{1}$$

bounded by the closed hypersurface, consisting of two space-like surfaces associated with the moments of time t_0, t_f, and lateral surface Σ_∞ at spatial infinity. Excitation of the internal degrees of freedom of the domain wall gives rise to non-zero flux of longitudinal spatial momentum through the lateral surface Σ_∞ so that the conservation equation takes the form

$$P^M_{\rm tot}(t_f) - P^M_{\rm tot}(t_0) = -\int_{\Sigma_\infty} \tau^{MN} d\Sigma_N\,, \tag{2}$$

where $P^M_{\rm tot}(t) = \int_{\Sigma_t} \tau^{M0} dz\, d^{D-2}\mathbf{r}$ as usual ($\mathbf{r}$ is the coordinate on the wall, and z is orthogonal to the wall). One can not absorb the right hand side into redefinition of the momentum, so only the instantaneous momentum balance equation in terms of the derivatives makes sense:

$$\frac{d}{dt}\left(\bar{P}^M + P^M + S^M\right) =- \lim_{R\to\infty}\left(\int_{-\infty}^{\infty} dz \int_{S_R} \delta T^{Mr} R d\varphi\right) \equiv f^M\,, \tag{3}$$

where three terms correspond to the first order contribution of the particle, the wall and gravitational stresses. The lateral flux can be split similarly. Only the contribution from the wall δT^{Mr} does not vanish.

Since in the linear order the gravitational potentials of the freely propagating particle and wall are rigidly tight to their sources, one finds that the non-local stresses can be unambiguously split between two colliding objects: $S^M = S^M_p + S^M_w$. Then combining the derivative of the wall total energy $\dot{\mathcal{P}}^0 = \dot{P}^0 + \dot{S}^0_w$ and the particle total energy $\dot{\bar{\mathcal{P}}}^0 = \dot{\bar{P}}^0 + \dot{S}^0_p$, and taking into account that $f^0 = 0$, one finds the instantaneous energy conservation law $\dot{\mathcal{P}}^0 + \dot{\bar{\mathcal{P}}}^0 = 0$. Thus, the stress contributions S^M_p, S^M_w have the meaning of the potential energy of each object in the gravitational field of the partner.

The z-component of the momentum can be treated similarly, but in this case $f^z \neq 0$. So we have a non-conservation equation for the momentum perpendicular to the wall: $\dot{\mathcal{P}}^z + \dot{\bar{\mathcal{P}}}^z = f^z$.

Thus we observe that the momentum balance equation describing collisions of extended objects may differ substantially from the form familiar for the particle collisions.

This work was supported by RFBR under the project 14-02-01092-a.

References

1. D. V. Gal'tsov, E. Yu. Melkumova, P. Spirin, *Phys. Rev. D* **89**, 085017 (2014).
2. D. V. Gal'tsov, E. Yu. Melkumova, P. Spirin, *Phys. Rev. D* **90** 125024 (2014).

Vacuum polarization of a quantized scalar field in the thermal state in a long throat

A. A. Popov

Institute of Mathematics and Mechanics, Kazan Federal University,
Kazan, 420008, Russia
E-mail: apopov@ksu.ru

Vacuum polarization of scalar fields in the background of a long throat is investigated. The field is assumed to be both massive or massless, with arbitrary coupling to the scalar curvature, and in a thermal state at an arbitrary temperature. It is shown that $\langle \phi^2 \rangle$ does not depend on the temperature of thermal state.

Keywords: Vacuum polarization; scalar field; wormhole.

The study of vacuum polarization effects in strong gravitational fields is a pertinent issue since such effects may play a role in the cosmological scenario and the construction of a self-consistent model of black hole evaporation. These effects may be taken into account by solving the semiclassical backreaction equations,

$$G_\nu^\mu = 8\pi \langle T_\nu^\mu \rangle, \tag{1}$$

where $\langle T_\nu^\mu \rangle$ is the expectation value of the stress-energy tensor operator for the quantized fields.

The main difficulty in the theory of semiclassical gravity is that the effects of the quantized gravitational field are ignored. The popular solution for this problem is to justify ignoring the gravitational contribution by working in the limit of a large number of fields, in which the gravitational contribution is negligible. Another problem is that the vacuum polarization effects are determined by the topological and geometrical properties of spacetime as a whole or by the choice of quantum state in which the expectation values are taken. It means that the calculation of the functional dependence of $\langle T_\nu^\mu \rangle_{ren}$ on the metric tensor in an arbitrary spacetime presents formidable difficulty. Only in some spacetimes with high degrees of symmetry for the conformally invariant fields $\langle T_{\mu\nu} \rangle_{ren}$ can be computed and equations (1) can be solved exactly.[1–5] Let us stress that the single parameter of length dimensionality in such a problem is the Planck length l_{Pl}. This implies that the characteristic scale l of the spacetime curvature (which correspond to the solution of equations (1)) can differ from l_{Pl} only if there is a large dimensionless parameter. As an example of such a parameter one can consider a number of fields, the polarization of which is a source of spacetime curvature.[a] For the quantized scalar field such a parameter can be the coupling constant ξ of scalar field to the curvature of spacetime. In the case of $|\xi| \gg 1, \xi < 0$ the vacuum fluctuations of a quantized scalar field can determine the curvature of spacetime so-called *a long throat.*[6]

In this work we have shown that in a long throat the effect of vacuum polarization of a quantized scalar field in the thermal state does not depend on temperature and conditions at infinity. This implies that in considered situation $\langle\varphi^2\rangle$ is a local quantity for any finite mass m of the quantized field, including $m = 0$.

Acknowledgments

This work was supported in part by grant 13-02-00757 from the Russian Foundation for Basic Research.

References

1. A. Starobinsky, *Phys. Lett. B* **91**, 99 (1980).
2. S. Mamayev and V. Mostepanenko, *Sov. Phys.–JETP* **78**, 20 (1980).
3. L. Kofman, V. Sahni and A. Starobinsky, *Sov. Phys.–JETP* **85**, 1876 (1983).
4. L. Kofman, V. Sahni, *Phys. Lett. B* **127**, 197 (1983).
5. L. Kofman, V. Sahni, *Phys. Lett. A* **117**, 275 (1986).
6. A.A. Popov, *Class. Quantum Grav.* **22**, 5223 (2005).

[a]Here and below it is assumed, of course, that the characteristic scale of change of the background gravitational field is sufficiently greater than l_{Pl} so that the very notion of a classical spacetime still has some meaning.

Hamiltonian cosmology of bigravity

V. O. Soloviev

Theory Division, Institute for High Energy Physics
NRC "Kurchatov Institute",
Protvino, Moscow region, 142281, Russian Federation
E-mail: Vladimir.Soloviev@ihep.ru
www.ihep.ru

The purpose of this talk is to give an introduction both to the Hamiltonian formalism and to the cosmological equations of bigravity. In the Hamiltonian language we provide a study of flat-space cosmology in bigravity and massive gravity constructed mostly with de Rham, Gabadadze, Tolley (dRGT) potential. It is demonstrated that the Hamiltonian methods are powerful not only in proving the absence of the Boulware-Deser ghost, but also in addressing cosmological problems.

Keywords: Hamiltonian cosmology; bigravity.

1. Friedmann Cosmology in Hamiltonian Form

The Friedmann cosmology of General Relativity is given by a Hamiltonian system with one gravitational variable (the scale factor) a, and one matter field, let be scalar ϕ. The total Hamiltonian is a first class constraint (the celebrated Friedmann equation or "the Differential Equation by God") multiplied by a Lagrange multiplier (the lapse N) due to the time reparametrization invariance

$$\mathrm{H} = Na^3\left(-\frac{6H^2}{\kappa} + \rho + \frac{2\Lambda}{\kappa}\right).$$

The scalar field may be treated as an ideal fluid with $\mathrm{H_M} = Na^3\rho$, and $\mathrm{L_M} = Na^3 p$. Then we can take the scale factor a and the Hubble constant H as a couple of almost canonical variables and the Poisson brackets are

$$\{a, H\} = -\frac{\kappa}{12a^2}, \qquad \{\phi, \pi_\phi\} = 1.$$

Two of Hamiltonian equations are kinematical ones, and other two are dynamical. In particular, the gravitational equation is as follows

$$\dot{H} = \{H, \mathrm{H}\} = -\frac{\kappa}{4}N(\rho + p),$$

and the scalar dynamical equation is equivalent to the conservation law

$$\dot{\rho} + 3NH(\rho + p) = 0.$$

The bigravity is formed as a sum of two copies of General Relativity accompanied by a potential

$$\mathcal{L} = \frac{1}{\kappa_f}\sqrt{-f}R_f + \frac{1}{\kappa}\sqrt{-g}R + \mathcal{L}_{\mathrm{M}}(g_{\mu\nu}, \phi) - U(f_{\mu\nu}, g_{\mu\nu}),$$

where the dRGT potential is formed by means of the symmetric polynomials $e_n(X)$ of matrix $X^{\mu}_{\nu} = \left(\sqrt{g^{-1}f}\right)^{\mu}_{\nu}$:

$$U = \frac{2m^2}{\kappa}\sqrt{-g}\sum_{n=0}^{4}\beta_n e_n(X) = \frac{2m^2}{\kappa}\left(\beta_0\sqrt{-g} + \ldots + \beta_4\sqrt{-f}\right).$$

Given the Friedmann ansatz for both metrics $f_{\mu\nu} = \mathrm{diag}(-N^2, r^2\chi^2\delta_{ij})$, $g_{\mu\nu} = \mathrm{diag}(u^2N^2, \chi^2\delta_{ij})$ we obtain

$$X = \sqrt{Y} = \mathrm{diag}\left(+\sqrt{u^{-2}}, +\sqrt{r^2}\delta_{ij}\right) \equiv \mathrm{diag}\left(u^{-1}, r\delta_{ij}\right),$$

where e_i are as follows

$$e_0 = 1, \quad e_1 = \frac{1}{u} + 3r, \quad e_2 = \frac{3r}{u} + 3r^2, \quad e_3 = \frac{3r^2}{u} + r^3, \quad e_4 = \frac{r^3}{u}.$$

In bigravity there are two Lagrange multipliers N, u, so we get two primary constraints, the first one is first class, the second is second class

$$H^2 = \frac{\kappa\rho}{6} + \frac{\Lambda(r)}{3}, \qquad (H_f)^2 = \frac{\Lambda_f(r)}{3},$$

where Λ and Λ_f are cubic polynomials of r and r^{-1} respectively

$$\Lambda = m^2\left(\beta_0 + 3\beta_1 r + 3\beta_2 r^2 + \beta_3 r^3\right), \; \Lambda_f = m^2\frac{\kappa_f}{\kappa}\left(\frac{\beta_1}{r^3} + 3\frac{\beta_2}{r^2} + 3\frac{\beta_3}{r} + \beta_4\right).$$

There is one secondary constraint, it is second class and it is factorized, Lagrange multiplier u is determined from the consistency. Only one factor

$$H_f = r^{-1}H,$$

provides nontrivial cosmology, allowing to get $H(r)$, $\rho(r)$, and reducing the Universe evolution to one equation for the “hidden parameter”

$$\dot{r} = Nr(1 - ur)H.$$

Reference

1. V. O. Soloviev, Hamiltonian cosmology in bigravity and massive gravity, Arxiv:1505.00840.

Massless-particle gravitational bremsstrahlung

P. Spirin

Moscow State University, Physics Department,
119991, Leninskie gory 1, str. 2, Moscow, Russia
E-mail: pspirin@physics.uoc.gr

The angular and frequency characteristics of the gravitational radiation emitted in collisions of massless particles are studied perturbatively in the context of classical General Relativity for small values of the ratio $\alpha \equiv 2r_S/b$ of the Schwarzschild radius over the impact parameter. The particles are described with their trajectories, while the contribution of the leading nonlinear terms of the gravitational action is also taken into account. The old quantum results are reproduced in the zero frequency limit $\omega \ll 1/b$. The radiation efficiency $\epsilon \equiv E_{\rm rad}/2E$ *outside* a narrow cone of angle α in the forward and backward directions with respect to the initial particle trajectories is given by $\epsilon \sim \alpha^2$ and is dominated by radiation with characteristic frequency $\omega \sim \mathcal{O}(1/r_S)$.

Keywords: gravitational bremsstrahlung.

The purpose of the work Ref. 1 is to extend the method used in Ref. 2,3, to the study of gravitational radiation in collisions of massless particles with center-of-mass energy $2E$ and impact parameter b. The formal limit $m \to 0$ (or equivalently $\gamma_{\rm cm} \to \infty$ for the Lorentz factor) of the massive case leads to nonsensical answers for the radiation efficiency $\epsilon \sim (r_S/b)^3\gamma_{\rm cm}$ of the radiated to the available energy, as well as for the characteristic radiation frequency $\omega \sim \gamma_{\rm cm}^2/b$, or the characteristic emission angle $\vartheta \sim 1/\gamma_{\rm cm}$. We shall study classically the gravitational radiation in the collision of massless particles using the same perturbative approach as in Ref. 2. We consider the two massless particles of energy $\mathcal{E}$ and their gravitational interaction in the CM frame. Thus, the particles move on null geodesics and each produces the Aichelburg – Sexl metric ($\varkappa^2 = 16\pi G$):

$$h_{\mu\nu}(x) = -\varkappa E u_\mu u_\nu \delta(t-z)\,\Phi|\mathbf{r}-\mathbf{b}/2|\,, \qquad \Phi(r) \equiv -1/2\pi \ln r\,. \qquad (1)$$

In the external field by the partner the correction to trajectory reads:

$${}^1\!z^0(\sigma) = \frac{1}{2}\, e\varkappa^2\Phi(b)\,\theta(\sigma) = -\,{}^1\!z^z(\sigma)\,, \qquad {}^1\!z^x(\sigma) = e\varkappa^2\Phi'(b)\,\sigma\,\theta(\sigma)\,, \quad (2)$$

which, indeed, vanishes for all $\sigma < 0$. Thus, we reproduce the leading order expressions of the two well-known facts about the geodesics in an

Aichelburg-Sexl metric, namely: (i) the *time delay* at the moment of shock equal $\Delta t = \varkappa^2 E\,|\Phi(b)|$; (ii) the *refraction* caused by the gravitational interaction by an angle $\alpha = 8GE/b$ in the direction of the center of gravity.

Defining in the center-of-mass frame the radiation wave-vector by $k^\mu = \omega(1, \sin\vartheta\cos\varphi, \sin\vartheta\sin\varphi, \cos\vartheta)$ and contracting with the two polarizations (defined as usual), we obtain the final (finite) expressions for the source of the gravitational radiation separately for the two polarizations ($\hat{K}_\nu(x) \equiv x^\nu K_\nu(x), \zeta \equiv \omega b \sin\vartheta\sqrt{x(1-x)}$, $C \equiv 16GE^2/\sqrt{2}$):

$$\tau_+(k) = C\int_0^1 dx\, \mathrm{e}^{-i(kb)x}\left[-K_0(\zeta) + \sin^2\varphi\,\hat{K}_1(\zeta)\right], \tag{3}$$

$$\tau_\times(k) = C\sin\varphi\int_0^1 dx\, \mathrm{e}^{-i(kb)x}\left[2i\frac{\hat{K}_2(\zeta) - \hat{K}_1(\zeta)}{\omega b\sin\vartheta} + (2x-1)\cos\varphi\,\hat{K}_1(\zeta)\right].$$

In the low-frequency regime ($\omega \to 0$) the amplitude $\tau_\times$ dominates and has the form $\tau_\times \simeq -\dfrac{16\sqrt{2}iGE^2\sin\varphi}{\omega b\sin\vartheta}$. The angular integration leads to $dE_{\mathrm{rad}}/d\omega = (2^8G^3E^4/\pi b^2)\int d\vartheta/\sin\vartheta$, which diverges and implies that our formulae are not valid for ϑ close to zero. A quick way to repair, is to impose a small-angle cut-off $\vartheta = \vartheta_{\min} \equiv \alpha$ on the ϑ–integration, and thus to obtain for $dE_{\mathrm{rad}}/d\omega|_{\omega=0}$ the value computed quantum mechanically in Ref. 4:

$$\left(\frac{dE_{\mathrm{rad}}}{d\omega}\right)_{\omega=0} = \frac{Eb}{\pi}\,\alpha^3\ln\frac{2}{\alpha}, \tag{4}$$

which agrees with our expression for $\vartheta > \alpha$. It will be convenient to treat separately the six angular and frequency regimes shown in Fig. 1.

Thus, in regimes II+IV+VI we use our formalism.

The angular distribution of total radiation reads:

$$\frac{dE_{\mathrm{rad}}}{d\vartheta} = \frac{\eta\alpha^3 E}{8\pi^2}\frac{1}{\sin^2\vartheta}, \qquad \eta \approx 89.9\,. \tag{5}$$

Integration over $\vartheta \in (\alpha, \pi - \alpha)$ gives

$$E_{\mathrm{rad}} = \frac{\eta\alpha^2 E}{4\pi^2}, \qquad \epsilon = \frac{E_{\mathrm{rad}}}{2E} \simeq 1.14\,\alpha^2\,. \tag{6}$$

Using the same approach as in Ref. 2, based on standard GR, with the leading non-linear gravity effects taken into account, we studied collisions of massless particles and computed the gravitational energy of arbitrary frequency, which is emitted outside the cone of angle $\alpha = 2r_S/b \ll 1$ in the forward and backward directions. The value $\epsilon \simeq 1.14\,\alpha^2$ was obtained for the radiation efficiency, with characteristic frequency $\omega \sim 1/r_S$. In

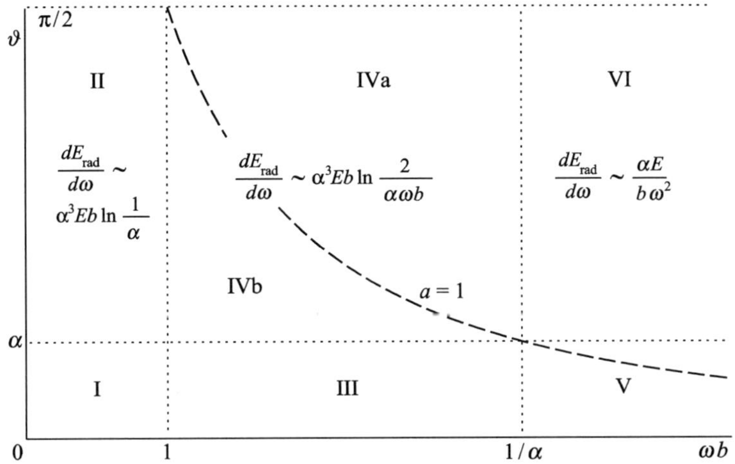

Fig. 1. The characteristic angular and frequency regimes ($a \equiv \omega b \sin\vartheta$).

fact, this value represents a lower bound of the efficiency, since it does not include the energy emitted inside that cone. The frequency distribution of radiation in the characteristic angle-frequency regimes is shown in Fig. 1. In particular, in regime IV we have good qualitative agreement with Ref. 5. The emitted radiation through the regime V is, therefore, represents the main open question for the forthcoming studies.

The Author would like to acknowledge discussions with Th. Tomaras, D. Gal'tsov, G. Veneziano and D. Colferai. This work was supported by EU programs "Thales" (MIS 375734) and "ARISTEIA II" as well as by the RFBR grant 14-02-01092. Finally, we acknowledge the non-commercial "Dynasty" Foundation (Russian Federation) for financial support.

References

1. P. Spirin and T. N. Tomaras, JHEP **06** (2015) 153.
2. D. Gal'tsov, P. Spirin and T. N. Tomaras, JHEP **01** (2013) 087.
3. D. V. Gal'tsov, G. Kofinas, P. Spirin and T. N. Tomaras, Phys. Lett. B **683** (2010) 331.
4. S. Weinberg, Phys. Rev. **140** (1965) B516.
5. A. Gruzinov and G. Veneziano, arXiv:1409.4555v1 [gr-qc].

String motion in Schwarzschild space-time surrounded by quintessence

A. V. Zhovtan

Crimean Astrophysical Observatory,
298409, Nauchnuy, Republic of Crimea, Russia
E-mail: astroalex2012@gmail.com

S. N. Roshchupkin

Crimean Federal University,
295007, Simferopol, Republic of Crimea, Russia
E-mail: rsnstring@gmail.com

In this article we consider motion of cosmic string in Schwarzschild background surrounded by quintessence. We present cosmic string equations of motion and find some particular solutions.

Keywords: Cosmic string; quintessence; nonlinear differential equations.

1. Introduction

Spontaneous symmetry breaking in gauge theories leads to phase transitions in the early Universe.[1] In this period can appear topological structures in the Universe like strings, vacuum domain walls and monopoles.[1,2] Strings of astrophysical relevance were formed during phase transitions at the grand-unified-theory (GUT) scale (10^{15} GeV). At the present stage of development of the theory of cosmic strings on the foreground the problems connected with research of dynamics of strings in curved spaces and search of effects which could form a reliable basis for detection of cosmic strings in the Universe act.

Since the motion of cosmic string is described by complicated system of the nonlinear differential equations, it is often quite difficult to obtain exact classical solutions in variety curved backgrounds. So, analytical solutions are found for special type of ansatz, in particular, for case of closed string.

2. Equations of Motion and Its Solutions

In the paper[3] was investigated the Einstein equations for static spherically symmetric quintessence surrounding a black hole and was found exact solution for a specific choice of free parameter, characterizing the energy-momentum tensor of the quintessence: $T_t^t = T_r^r = \rho_q$, $T_\theta^\theta = T_\varphi^\varphi = -\frac{1}{2}\rho_q(3\varepsilon + 1)$. This solution is given by metric

$$ds^2 = \Delta dt^2 - \Delta^{-1} dr^2 - r^2(d\theta^2 + \sin^2\theta d\varphi^2) \tag{1}$$

Here $\Delta = 1 - 2M/r - r_0/r^{3\varepsilon+1}$, M – mass of the black hole, r_0 – dimensional normalization constant, $-1 \lesssim \varepsilon \lesssim -1/3$ – quintessential state parameter.

To describe the cosmic string dynamics when ρ_s (radius of the string cross section) much smaller characteristic radius of curvature of the string can use Nambu-Goto string action $S = -\frac{2}{\alpha'}\int d\tau d\sigma \sqrt{-\det G_{ij}}$, where $1/\alpha'$ is string tension parameter, $G_{ij} = \partial_i x^\mu g_{\mu\nu} \partial_j x^\nu$ – induced metric and $g_{\mu\nu}$ – metric tensor of curved space-time. In general case the variational Euler-Lagrange equations of motion and constrains have the form

$$x^\alpha_{,\tau\tau} - \gamma^2 x^\alpha_{,\sigma\sigma} + \Gamma^\alpha_{\mu\nu}[x^\mu_{,\tau} x^\nu_{,\tau} - \gamma^2 x^\mu_{,\sigma} x^\nu_{,\sigma}] = 0, \tag{2}$$

$$x^\mu_{,\tau} g_{\mu\nu} x^\nu_{,\tau} + \gamma^2 x^\mu_{,\sigma} g_{\mu\nu} x^\nu_{,\sigma} = 0, \tag{3}$$

where $\Gamma^\alpha_{\mu\nu}$ are Christoffel symbols for metric $g_{\mu\nu}$, γ can be considered as dimensionless parameter characterizing the power of string tension.

We write the equations (2-3) for metric (1) and find some particular solutions for ansatz: $t = t(\tau), r = r(\tau), \theta = \pi/2, \varphi = \sigma$. For instance, if to use substitution $u = 1/r$, then solution can be write in the form

$$\frac{du}{dt} = \pm \frac{1 - 2Mu - r_0 u^{3\varepsilon+1}}{E} \sqrt{E^2 u^4 + 2Mu^3 - u^2 + r_0 u^{3\varepsilon+3}} \tag{4}$$

Because particular solutions does not show the whole picture of string motion, also was performed some numerical calculations. It show the presence of non-stability and the possible existence of dynamic chaos. In particular, phase space of initial conditions have complicated fractal structure.

References

1. P. J. E. Peebles, *Principles of Physical Cosmology* (Princeton University Press, 1994).
2. A. Vilenkin, E. P. S. Shellard, *Cosmic strings and Other Topological Defects* (Cambridge University Press, 1994).
3. V. V. Kiselev, Quintessence and black holes, *Class. Quantum Grav.* **20**, 1187 (2003).

Section 5

Extra Dimensions and Variation of Constants

Space-time foam and the cosmological constant problem

Sergey G. Rubin
National Research Nuclear University "MEPhI",
(Moscow Engineering Physics Institute)
Moscow, Russia
E-mail: sergeirubin@list.ru

Solution of the small value of the cosmological constant problem is presented based on multidimensional gravity.

Keywords: Cosmological constant; extra space; fine tuning; quantum foam; nonlinear gravity.

There are two challenging phenomena - the hierarchy of masses[1,2] and the fine-tuning of the physical parameters.[3–5] Most known example of fine-tuning is extremal smallness of the cosmological constant (CC).

In terms of the multidimensional gravity the problem is formulated as follows. Let we have a d-dim extra space with a metric G_{AB} and initial Lagrangian with physical parameters $\{\Lambda\}$. No fine tuning is assumed so that all parameters are of the same order of magnitude,

$$\Lambda_k \sim \Lambda_m.$$

After reduction to the 4-dim space we obtain a Lagrangian containing the observable physical parameters $\lambda_k(\{\Lambda\}, G)$. These new parameters depend on initial parameters Λ_k and an extra metric G. If the latter is different for different universes, conditions

$$\lambda_k(\Lambda, G_-) < \lambda_k(\Lambda, G) < \lambda_k(\Lambda, G_+) \qquad (1)$$

are satisfied in spite of the initial parameters $\{\Lambda\}$ are fixed.

It points the way to solve the problem of CC smallness. Indeed, if $\lambda_K(\Lambda, G_-) < 0$ and $\lambda_K(\Lambda, G_+) > 0$ for specific $K \in \{k\}$ then there exists a metric G_0 such that $\lambda_K(\Lambda, G_0) = 0$. Hence there exists a set of metrics $\{\tilde{G}\}$ such that $\lambda_K(\Lambda, \tilde{G})$ is arbitrarily close to zero provided the set $\{G\}$ of stationary metrics has continuum measure.

In this paper the idea discussed above is realized. To this end specific example of metric

$$ds^2 = \mathfrak{G}_{AB} dZ^A dZ^B = g_{\mu\nu}(x) dx^\mu dx^\mu + G_{ab}(y) dy^a dy^b \qquad (2)$$

of a space $M = M_4 \times V_n$ is considered. Here M, M' are the manifolds with metrics $g_{mn}(x)$ and $G_{ab}(x,y)$ respectively. x and y are the coordinates of the subspaces M and M'. We will refer to 4-dim space M and n-dim compact space M' as a main space and an extra space respectively. Here the metric has the signature (+ - - - ...), the Greek indices $\mu, \nu = 0, 1, 2, 3$ refer to 4-dimensional coordinates). Latin indices run over $a, b, ... = 4, 5....$

Consider a gravity with higher order derivatives and action in the form,

$$S = \frac{m_D^{D-2}}{2} \int d^D Z \sqrt{|\mathfrak{G}|} \left[f(R) + L_m \right]; \tag{3}$$

$$f(R) = \sum_k a_k R^k$$

with arbitrary parameters a_k, $k \neq 1$ and $a_1 = 1$ and $D = n + 4$. Here L_m is a Lagrangian of a scalar field. Almost uniform distribution of matter in our space is expressed as $\partial_\mu \ll \partial_a$ or as $R_4 \ll R_n$

Suppose that $R_n \equiv R$ is a solution to classical equation

$$R_{ab} f' - \frac{1}{2} f(R) G_{ab} - \nabla_a \nabla_b f_R + G_{ab} \Box f' = \frac{1}{m_D^{D-2}} T_{ab}, \tag{4}$$

+ additional conditions.

for the metric of extra space. Here $\Box$ stands by the d'Alembert operator.

Let the dimensionality of extra space equals 2, its topology is sphere and the function

$$f(R) = u_1 (R - R_0)^2 + u_2 \tag{5}$$

Evidently, there is a set of solutions to system (4) depending on additional conditions. [5]

The scalar field with Lagrangian

$$L_m = \frac{1}{2} \partial_A \varphi \mathfrak{G}^{AB} \partial_B \varphi - U_\varphi(\varphi), \quad U_\varphi(\varphi) = \sum_i g_i \varphi^i \tag{6}$$

contributes to the dark energy. The solution of the classical equation of motion in the form $\varphi(x, Y) = \phi(x) Y(y)$ gives the following equation for $Y(y)$

$$\Box_n Y(y) + U''_\varphi(\varphi_m) Y(y) = 0. \tag{7}$$

due to uniformity of our 4-dim space ($\phi(x) = const$). This equation should be solved together with Eq.4 with the trace of stress tensor

$$T = G^{ab} T_{ab} = 2 U_\varphi(\varphi) \simeq U_\varphi(\varphi_m). \tag{8}$$

Let the functions $G_{ab}(y)$ and $Y(y)$ are solutions to (4) and (7) correspondingly. Their substitution into the action leads to the following form of effective action

$$S = \frac{m_D^{D-2}}{2} \int d^4x \left[-U_\phi + d^n y \sqrt{|g(x)G(y)|} [R_4(x) f'(R_n) + f(R_n)] \right], \quad (9)$$

$$U_\phi(\phi) = \sum_i \tilde{g}_i \phi(x)^i. \quad (10)$$

Here kinetic term is neglected due to the uniformity of solution and

$$\tilde{g}_2 = \int d^n y \sqrt{|G(y)|} \left[g_1 Y(y)^2 - \frac{1}{2} \partial_a Y(y) G^{ab} \partial_b Y(y) \right] \quad (11)$$

$$\tilde{g}_{i \neq 2} = g_i \int d^n y \sqrt{|G(y)|} Y(y)^i \quad (12)$$

Comparison of expression (9) with the Einstein-Hilbert action

$$S_{EH} = \frac{M_{Pl}^2}{2} \int d^4x \sqrt{|g(x)|} (R - 2\Lambda) \quad (13)$$

gives the expression

$$\Lambda_{eff} \equiv \frac{m_D^{D-2}}{2M_{Pl}^2} \left[U_\phi(v_\phi) - \int d^n y \sqrt{|G(y)|} f(R_n) \right] \quad (14)$$

represents the cosmological Λ term. Here $v_\phi(\tilde{g}_i)$ is a vacuum state of the field ϕ.

The Λ term depends on a stationary geometry $G_{ab}(y)$ and hence on the additional conditions which are different for universes created by space time foam at the Planck scale. As an additional condition a value of 2-dim extra space metric at point $\theta = \pi$ is chosen. Dependence of the Λ term on radius of extra space $r(\pi)$ is shown in Fig.1 Various universes are created having different initial condition ($r(\pi)$ in our case). Evidently, there exist universes with the Λ term be arbitrary close to zero. Our Universe belongs to such a set.

Acknowledgments

This work was performed within the framework of the Center FRPP supported by MEPhI Academic Excellence Project (contract 02.03.21.0005, 27.08.2013) and was supported by the Ministry of Education and Science of the Russian Federation, Project No. 3.472.2014/K.This work was performed within the framework of the Center FRPP supported by MEPhI Academic Excellence Project (contract 02.03.21.0005, 27.08.2013) and was

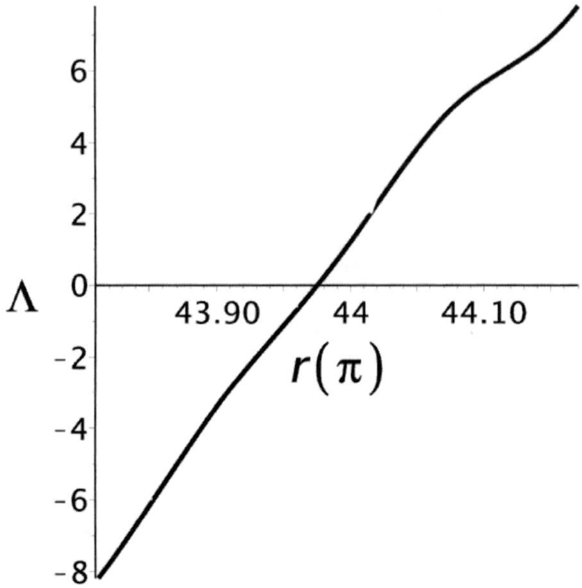

Fig. 1. Parameters: $r(\pi) = 43.7, Y(\pi) = 100, u_1 =, u_2 =, R_0 =, g_2 = 10^{-5}, g_3 = 10^{-6}, g_4 = 2.2 \cdot 10^{-7}$.

supported by the Ministry of Education and Science of the Russian Federation, Project No. 3.472.2014/K.

References

1. Gia Dvali, Black Holes and Large N Species Solution to the Hierarchy Problem, *Fortsch.Phys.* **58**, 528 (2010); arXiv:0706.2050.
2. Yuki Watanabe and Eiichiro Komatsu, Gravitational inflaton decay and the hierarchy problem, *Phys.Rev.D* **77**, 043514 (2008).
3. Leonard Susskind, The Anthropic Landscape of String Theory, arXiv:hep-th/0302219.
4. S.G. Rubin, The role of initial conditions in the universe formation, *Gravit.& Cosmol.* **21**, 143 (2015) arXiv:1403.2062.
5. Sergey G. Rubin, Scalar field localization on deformed extra space, *Eur.Phys.J. C* **75**, 333 (2015).

The effective rest mass concept and magnetic monopole problem in 5D Theory

B. G. Aliyev

Moscow State University for Design and Technology
E-mail: bgaliyev@mail.ru

In this article we have considered and analyzed the new aspects of the rest mass concept of the 5D charged test particle which interacts with the scalar gravitational field and also it was shown that the 5D Ricci identities give us the way to clear the magnetic monopole problem.

Keywords: 5D Unified Gauge Invariant Theory (5DUGIT); effective rest mass (ERM); scalar gravitational field (SGF); geodesic equation; cylindrical symmetry condition (CSC); Ricci identities; Maxwell equations; magnetic monopole.

It is easy to show that geodesic equation in the 5DUGIT of gravitation[1] being presented with help of the monad method in $(4+1)$ form gives us due to the imposed CSC along the 5-th coordinate an integral, connected with an electric charge of the test particle $Q = ne$. As a result of this $(4+1)$ splitting we can also obtain an ERM of the Lorentz type for the 5D test particle.[2] We can represent here the 5D interval as $dI^2 = ds^2 - d\lambda^2$ and then write down, introducing the velocity along the 5-th coordinate $\hat{u} = \frac{d\lambda}{ds}$, the next simple formula for the ERM: $\hat{m}_0 = \frac{m_0}{\sqrt{1-\hat{u}^2}}$. Here m_0 is an initial 5D rest mass. In these terms the 4D projection of the geodesic equation will have the compact form as follows[2] (here and then the cross marks Gauge Invariant derivative and the line over the symbols marks 4D projected values; thus the 4D momentum $\bar{p}^\alpha = \hat{m}_0 \bar{u}^\alpha$):

$$\frac{D^+ \bar{p}^\alpha}{ds} = \frac{Q}{c^2} \bar{u}^\beta F_{\beta\cdot}^{\ \alpha} + \partial^{+\alpha} \hat{m}_0. \tag{1}$$

The necessity of the ERM $\hat{m}_0$ introducing occurs especially important when then we are making (3+1+1) splitting in 5D theory, eliminating additionally to the 5-th coordinate the zero one. Let us mark that we always use here the signature $(+----)$. Then and in this case only the geodesic equation (1) is possible to give the form of the following system[2]:

$$\frac{D^+\hat{m}}{d\tau} = \hat{\bar{p}}^i(\bar{F}_i - \bar{v}^k\bar{D}_{ik}) - \frac{Q}{c^2}\bar{v}^i\bar{E}_i + \frac{Q\sqrt{1-\bar{v}^2}}{2\sqrt{k_0}\phi}\partial^+_\tau\hat{m}_0, \tag{2a}$$

$$\frac{D^+\hat{\bar{p}}^i}{d\tau} = \frac{Q}{c^2}(\bar{E}^i - \bar{v}^k\bar{H}^i_{\cdot k}) - \hat{m}\bar{F}^i + 2\hat{\bar{p}}^k\bar{D}^i_k + \frac{Q\sqrt{1-\bar{v}^2}}{2\sqrt{k_0}\phi}\partial^{+i}\hat{m}_0. \tag{2b}$$

Here $\frac{D^+}{d\tau} = \partial^+_\tau + \hat{v}\partial^+_\Lambda + \bar{v}^i\partial^+_i$ is a full covariant derivative; $\phi = \sqrt{-G_{55}}$ and the 4D effective mass $\hat{m} = \frac{\hat{m}_0}{\sqrt{1-\bar{v}^2}}$; $\bar{v}^i = \frac{d\bar{x}^i}{d\tau}$; $\hat{v} = \frac{d\lambda}{d\tau}$; $\hat{\bar{p}}^i = \hat{m}\bar{v}^i$; $\hat{p} = \hat{m}\hat{v} = \hat{m}_0\hat{u}$; $d\tau = c\,dt$. This system gives us additionally such a fact that 3D scalar force here is equal zero for the particles with the zero 4D rest mass (the same result we have in the all of the scalar-tensor theories of gravitation). Then the ERM can be written as follows below and may be expressed with help of the mass angle χ_n in a very short and suitable form:

$$\hat{m}_0 = \sqrt{m_0^2 + \frac{Q^2}{4k_0\phi^2}} = m_0\sqrt{1 + \frac{n^2\hat{m}_{pl}^2}{m_0^2\phi^2}} = m_0\cosh\chi_n. \tag{3}$$

Here in general $m_0 \neq 0$ is an above-mentioned initial 5D rest mass of the 5D test particle, $k_0 = 6.67\cdot 10^{-11}\frac{N\cdot m^2}{kg^2}$ is the Newtonian gravitational constant, the 5D “Planck mass” $\hat{m}_{pl} = \frac{e}{2\sqrt{k_0}} = 9.3\cdot 10^{-7}g$, and $\sinh\chi_n = \frac{n\hat{m}_{pl}}{m_0\phi}$. In singular case $m_0 = 0$ we have here two variants: $Q = 0$ and $\hat{m}_0 = 0$ (it should be a photon) or $Q \neq 0$ and $\hat{m}_0 = \frac{n\hat{m}_{pl}}{\phi}$ (we believe that this case corresponds for example to any exotic particle as a magnetic monopole or something like that). In case $m_0 \neq 0$, $Q \neq 0$ and $n = 1$ we would obtain an electron or a proton. In another case, when $m_0 \neq 0$ and $Q = 0$ we would obtain a neutrino. We also believe here that the SGF ϕ from the 5D viewpoint realizes the fundamental universal interaction as well as we have in case of the 4D gravitational one. In this sense we may call that the 5D equivalence principle. Thus we may rather establish here that every massive 5D particle would have an electric charge. It requires that every neutral 5D particle must be the composite one and contain of the charged parts. Thus only really neutral particles as a photon or a neutrino are in a sense the 4D ones.

Then for the further analysis it is easy to transform the 4D projected geodesic equation (1) to the following form:

$$\hat{m}_0\frac{D^+\bar{u}^\alpha}{ds} = \frac{Q}{c^2}\bar{u}^\beta F_{\beta\cdot}^{\ \alpha} - \frac{Q^2P^{\alpha\beta}\Phi_\beta}{4k_0\hat{m}_0\phi^2}, \tag{4}$$

where $P^{\alpha\beta} = g^{\alpha\beta} - \bar{u}^\alpha \bar{u}^\beta$ is a tensor of the orthogonal projection upon the 4-velocity $\bar{u}^\alpha$ direction and $\Phi_\beta = \nabla^+_\beta \ln \phi$. Here we have in the r.h.s. well known 4D Lorentz Force (LF) which is caused due to the interaction with an electromagnetic field (EMF), and also the so called 4D scalar gravitational force, which we propose to name Brans–Dicke Force (BDF). This BDF acts on every charged particle because the ERM of the 5D one and the BDF itself depends on the Q^2, thus that is quite different than the LF, which is proportional to the Q and does not act on a neutral particle. Because of this the BDF may be significant being accumulated for the bodies with the big masses (big masses effect) although $\partial^+_\alpha \phi$, as well known, has a very small value. We call here this rest mass "an effective" one in the same sense as we usually do it in the case of the quasi-particle moving in a crystal and thus we take into account that this charged particle is always in the neighborhood of the SGF ϕ, in some sense we believe that the 4D World is put into the scalar ocean. Thus we may now establish that it is just an electric charge which is most probably responsible in the Universe both for electromagnetic and scalar gravitational interactions and it seems that there is no any other the so called scalar fundamental charge, which has not been found so far. Also in this case yet we are able to generalize now the braking radiation force (BRF) concept and obtain for it the following relativistic formula[3]:

$$g^\alpha_E = \frac{2e^3 \bar{u}^\gamma}{3\hat{m}_0 c^3} \left(\bar{u}^\beta \nabla^+_\gamma F_{\beta\cdot}^{\ \alpha} + \frac{e P^{\alpha\delta}}{\hat{m}_0 c^2} F_{\beta\delta} F_{\gamma\cdot}^{\ \beta} \right). \tag{5}$$

Additionally yet we are able to obtain the hypothetical scalar and mix analogies of BRF:

$$g^\alpha_S = -\frac{e^4 P^{\alpha\beta} \bar{u}^\gamma}{6ck_0 \phi^2 \hat{m}_0^2} \left(\nabla^+_\gamma \Phi_\beta - 2\Phi_\beta \Phi_\gamma + \frac{3e^2 \Phi_\beta \Phi_\gamma}{4k_0 \phi^2 \hat{m}_0^2} \right); \tag{6}$$

$$g^\alpha_{ES} = -\frac{e^5}{6c^3 k_0 \phi^2 \hat{m}_0^3} \left(g^{\beta\delta} P^{\alpha\gamma} - 3g^{\alpha\gamma} \bar{u}^\beta \bar{u}^\delta \right) F_{\beta\gamma} \Phi_\delta. \tag{7}$$

But it should be marked here, that the above-mentioned "big masses effect" must be more significant than the SGF influence upon the BRF because of the vanishing value of the $\partial^+_\alpha \phi$.

Another interesting result we may obtain considering the 5D Ricci identities.[3,5–7] When we are making $(4+1)$ splitting, reducing $V_5 \to V_4$, we obtain firstly the 4D Ricci identities:

$$R^\alpha_{\cdot\,\beta\gamma\delta} + R^\alpha_{\cdot\,\delta\beta\gamma} + R^\alpha_{\cdot\,\gamma\delta\beta} = 0 \tag{8}$$

and then additionally two relations between electromagnetic and SGF characteristics:

$$F_{\alpha\beta;\gamma} + F_{\gamma\alpha;\beta} + F_{\beta\gamma;\alpha} = -2(F_{\alpha\beta}\Phi_\gamma + F_{\gamma\alpha}\Phi_\beta + F_{\beta\gamma}\Phi_\alpha), \tag{9}$$

$$\sqrt{\bar{\kappa}_0} F_{\alpha\beta}\partial^+_\Lambda \phi = \Phi_{\alpha;\beta} - \Phi_{\beta;\alpha}. \tag{10}$$

Here $\bar{\kappa}_0 = \frac{4k_0}{c^4}$, $\partial^+_\Lambda \phi = \partial_5 \ln \phi$. The l.h.s. in relation (9) due to 5D GI derivatives is equal identically to zero only when we impose the CSC along the 5-th coordinate, thus we obtain here in this case the first pair of the well known Maxwell equations, which we can write down as a determinant because $F_{\alpha\beta} = \nabla^+_\alpha A_\beta - \nabla^+_\beta A_\alpha$:

$$\begin{vmatrix} \nabla^+_\alpha & \nabla^+_\alpha & A_\alpha \\ \nabla^+_\beta & \nabla^+_\beta & A_\beta \\ \nabla^+_\gamma & \nabla^+_\gamma & A_\gamma \end{vmatrix} = F_{\alpha\beta;\gamma} + F_{\gamma\alpha;\beta} + F_{\beta\gamma;\alpha} = 0. \tag{11}$$

Also in this case we obtain an additional relation between EMF and SGF, which is easy to represent as a determinant, using as in (11) the connection between tensor of the EMF and 4D vector-potential A_α:

$$\begin{vmatrix} \Phi_\alpha & \nabla^+_\alpha & A_\alpha \\ \Phi_\beta & \nabla^+_\beta & A_\beta \\ \Phi_\gamma & \nabla^+_\gamma & A_\gamma \end{vmatrix} = 0. \tag{12}$$

And finally, because due to the CSC $\partial^+_\Lambda \phi = 0$, we obtain here that the scalar curls of the SGF gradient become equal to zero (it is valid only due to the CSC) and it gives us:

$$\Phi_{\alpha;\beta} = \Phi_{\beta;\alpha} \tag{13}$$

But in case when we have not the CSC along the 5-th coordinate, that is possible in the very beginning of the Universe evolution, the relation (9) is valid and we obtain here the generalized first pair of the Maxwell equation (we propose to name them Maxwell–Dirac equations) with r.h.s. connected with a hypothetical magnetic monopole, earlier proposed by Dirac. It will have, if we will take into account the equation (10) (in this case $\partial^+_\Lambda \phi \neq 0$) and also represent the r.h.s. as a determinant, the following form:

$$F_{\alpha\beta;\gamma} + F_{\gamma\alpha;\beta} + F_{\beta\gamma;\alpha} = \frac{2}{\sqrt{\bar{\kappa}_0}\partial^+_\Lambda \phi} \begin{vmatrix} \Phi_\alpha & \nabla^+_\alpha & \Phi_\alpha \\ \Phi_\beta & \nabla^+_\beta & \Phi_\beta \\ \Phi_\gamma & \nabla^+_\gamma & \Phi_\gamma \end{vmatrix}. \tag{14}$$

Equation (14) shows us that the hypothetical magnetic monopoles are probably a product of the SGF curls at the very early stage of the 5D Universe life, when the SGF had the turbulences. Thus from this viewpoint we may draw the following scenario, which helps us to establish that at the initial moments of the 5D Universe life the scalar curls initiated the EMF changing process due to the equation (10), acting here as a specific "starter," and it has led to the creation of the magnetic monopoles age. Soon, due to the some quantum effects (one of possible and preferable is the Casimir effect[8]) the above-mentioned here cylindrical symmetry along the 5-th coordinate has at last established. Then consequently an age of the electrons, because of the above-mentioned integral of the 5D geodesic equation, has come. It should be marked that another way how the 5-th coordinate may be eliminated and compactificated, based on the 5D Kasner solution, were proposed in Ref. 9. Also in Ref. 10 the scenario with fluctuating dimension of the inflationary Universe was considered. Very interesting results for the 5D Kaluza–Klein theories from the viewpoint of soliton including the magnetic monopole approach are represented in Ref. 11. Thus we can now draw a conclusion that the magnetic monopoles have most of all appeared only at the very early stage of the Universe life and now, at the modern stage of its evolution we are able probably to find only a few relict ones or may be even nothing. Thus it becomes clear why we are not able to find all these hypothetical magnetic monopoles so far and also we can understand now again that the 5D Riemannian structure of the World defines its physical properties. From this viewpoint it seems very important to consider and investigate also the 5D Bianchi identities. And even may be we should regard all of these such kinds of the identities for dimensions with more than 5D.

Thus in conclusion we should now say that may be we will be able soon yet try to answer in some sense the very old question: where has at last the magnetic monopole gone?

I want to thank Yu. S. Vladimirov and A. V. Solov'yov for the useful discussions and then also to thank A. V. Solov'yov for a great technical help in this article preparation.

References

1. Yu. S. Vladimirov, V. I. Antonov, 5D Scalar-tensor theory of gravitation, *Moscow University Physics Bulletin* **15**, 54 (1974) (in Russian).
2. B. G. Aliyev, The behavior of the charged particles in 5D theory of gravitation, in *Modern problems of the general theory of relativity* (Physical Institute of the Byelorussian Academy of Sciences, Minsk,

1979), p. 154 (in Russian).
3. B. G. Aliyev, Motion equations of the test charged particles and the 5D Ricci identities, in *Abstracts of the XV Russian Gravitational Conference* (Kazan Federal State University, Kazan, 2014), p. 89 (in Russian).
4. B. G. Aliyev, Charged particle and scalar field in 5D Theory, in *Abstracts of ICGAC-12* (PFUR, Moscow, 2015), p. 109.
5. B. G. Aliyev, Where has the magnetic monopole gone?, in *Abstracts of ICGAC-12* (PFUR, Moscow, 2015), p. 110.
6. B. G. Aliyev, 5D Ricci identities and Maxwell equations, in *Abstracts of ICGAC-5* (PFUR, Moscow, 2001), p. 23 (in Russian).
7. B. G. Aliyev, Reducing of the 5D Ricci identities and their physical consequences, in *Abstracts of the 13 Russian Gravitational Conference* (PFUR, Moscow, 2008), p. 80 (in Russian).
8. V. M. Mostepanenko, N. N. Trunov, *The Casimir effect and its applications* (Energoatomizdat, Moscow, 1990) (in Russian).
9. A. Chodos, S. Detweiler, Where has the fifth dimension gone?, *Phys. Rev. D* **21**, 2167 (1980).
10. A. D. Linde, M. I. Zelnikov, Inflationary Universe with fluctuating dimension, *Phys. Lett. B* **215**, 59 (1988).
11. D. Gross, M. Perry, Magnetic monopoles in Kaluza-Klein theories, *Nucl. Phys. B* **226**, 29 (1983).

Global structure of black hole and brane solutions in a multidimensional model with anisotropic fluid

S. V. Bolokhov* and V. D. Ivashchuk**

Peoples' Friendship University of Russia, Miklukho–Maklaya st., 6
Moscow, 117198, Russia
**E-mail: bol-rgs@yandex.ru*

Center for Gravitation and Fundamental Metrology, VNIIMS, 46 Ozyornaya ul.
Moscow 119361, Russia
***E-mail: ivas@vniims.ru*

We analyse the global causal structure of a family of multidimensional spherically-symmetric solutions with a horizon which appear in the model with 1-component anisotropic fluid. This family can be considered as a generalized analogs of the well-known black hole solutions (including the Reissner—Nordström one) and some black brane solutions. The structure of regular horizons and singular boundaries is studied, and the corresponding Carter—Penrose diagrams are constructed for various values of the parameters of the model.

Keywords: Multidimensional gravity; anisotropic fluid; black hole; brane; exact solution; global structure.

1. Introduction

In Ref. 1 the family of spherically symmetric solutions to multidimensional Einstein equations in the presence of 1-component anisotropic fluid was obtained. As special cases, it contains a family of black hole solutions (including the Reissner—Nordström one), as well as some black brane ($M2$, $M5$) solutions in the models with antisymmetric form fields, and therefore it can be regarded as a generalization of these solutions having a certain interest in the context of multidimensional cosmology and p-brane physics.[1]

Our aim is to analyse the global causal structure of the solutions obtained, i.e. to study the set of regular horizons and singularities (if any) for various values of parameters of the model, and to construct the corresponding Carter—Penrose diagrams.

2. The Model

The solutions are defined on the manifold

$$M = \mathbb{R}_{\rm rad} \times (M_0 = S^{d_0}) \times (M_1 = \mathbb{R}_{\rm time}) \times M_2 \times \cdots \times M_n \qquad (1)$$

of total dimension $D = 2 + d_0 + \sum_{i=2}^{n} d_i$ ($d_0 \geqslant 2$) containing the product of $n-1$ internal Ricci-flat spaces M_i ($i \geqslant 2$) with metrics $h^{(i)}_{m_i n_i} dy^{m_i} dy^{n_i}$.

The energy-momentum tensor is taken as

$$T^M_N = \mathrm{diag}\left(p_r,\ p_0 \delta^{m_0}_{k_0},\ -\rho,\ p_2 \delta^{m_2}_{k_2},\ ...,\ p_n \delta^{m_n}_{k_n}\right), \qquad (2)$$

and the equations of state read

$$p_r = -\rho(2q-1)^{-1},\ p_0 = -p_r,\ p_i = \rho(2q-1)^{-1}\left(1 - \frac{2U_i}{d_i}\right), \qquad (3)$$

where $d_i = \dim M_i$. The parameters q and U_i describe relations between the pressures and the density; $q > 0$, $q \neq 1/2$, $U_i \in \mathbb{R}$. In the present paper, the parameter q is taken to be a natural number to avoid the non-analytical behaviour of the metric at the horizon.

The solution has the following form:

$$ds^2 = H^{-\frac{2U^0}{(U,U)}} \left[\frac{dr^2}{1 - \frac{2\mu}{r^d}} + r^2 d\Omega^2_{d_0} - H^{-\frac{2q}{(U,U)}} \left(1 - \frac{2\mu}{r^d}\right) dt^2 \right.$$
$$\left. + \sum_{i=2}^{n} H^{\frac{-2U_i}{d_i (U,U)}} h^{(i)}_{m_i n_i} dy^{m_i} dy^{n_i} \right], \qquad (4)$$

$$\rho = \frac{(2q-1)(qd)^2 P(P+2\mu)(1-2\mu r^{-d})^{q-1}}{2(U,U) H^{2-2U^0/(U,U)} r^{2d_0}}. \qquad (5)$$

Here the following notations are used:

$$H(r) \equiv 1 + \frac{P}{2\mu}\left[1 - \left(1 - \frac{2\mu}{r^d}\right)^q\right], \quad U^0 = \frac{1}{2-D}\left(q + \sum_{i=2}^{n} U_i\right), \qquad (6)$$

$$(U,U) \equiv q^2 + \sum_{i=2}^{n} \frac{U_i^2}{d_i} + \frac{1}{2-D}\left(q + \sum_{i=2}^{n} U_i\right)^2 > 0, \qquad (7)$$

where $d = d_0 - 1 \in \mathbb{N}$; line element on the sphere S^{d_0} is denoted by $d\Omega^2_{d_0}$; parameters $P, \mu > 0$.

3. The Analysis

In order to investigate the global structure of the solutions mentioned above, it is necessary to check the behaviour of the "redshift function" (g_{00}), the "area function" (the factor at $d\Omega^2_{d_0}$), and the scale factors of internal Ricci-flat manifolds M_i at critical points corresponding to horizons or singularities.

As an additional singularity criterion, we use the method of Kasner-like asymptotics for a class of spherically-symmetric solutions which allow us to check the behaviour of the Kretschmann (i.e. Riemann tensor squared) scalar as $r \to 0$ (see Ref. 2).

As a result of this analysis, the Carter—Penrose (CP) diagrams can be constructed for various values of the parameters using the standard algorithm[3]. It turns out that the structure of diagrams depends mostly on the parity of the parameter q.

There are three classes of important critical points of the radial coordinate r for the solution from Eq. (4):

- $r = r_0 \equiv \sqrt[d]{2\mu}$. This point corresponds to a regular external horizon.
- $r = r^\star$, which is the maximal root of the equation $H(r) = 0$. This point corresponds to the singularity.
- $r = 0$. Behaviour of the metric at this point depends on the parameters of the model (it can be either a horizon or a singularity.)

We introduce the following notations. Let $\mathbf{Sch}[r_1, r_2]$ ($r_1 < r_2$) be a diagram of Schwarzschild type with a singularity at a point r_1 and a regular horizon at r_2. Similarly, we denote by $\mathbf{RN}[r_1, r_2, r_3]$ ($r_1 < r_2 < r_3$) the diagram of Reissner-Nordström type with singularity at r_1, an internal horizon at r_2, and an external horizon at r_3.

Let us consider particular cases for values of the parameters when the metric from Eq. (4) may be regarded as a generalization of the well-known solutions mentioned above. In addition, we restrict ourselves to the case $d = 1$ (or, equivalently, $d_0 = 2$) when the manifold takes the form of the product of 4-dimensional spacetime and a number of Ricci-flat spaces: $M = \mathbb{R} \times S^2 \times \mathbb{R} \times M_2 \times \cdots \times M_n$.

(i) $D = 4$, $n = 1$, $M = \mathbb{R} \times S^2 \times \mathbb{R}$.
For $q = 2m$, $m \in \mathbb{N}$, we have CP diagram of type $\mathbf{Sch}[r^\star, r_0]$.
For $q = 2m + 1$: CP diagram is of type $\mathbf{RN}[r^\star, 0, r_0]$.
We note that for $q = 1$, after some linear coordinate transformation[1], the metric from Eq. (4) yields the classic RN solution with $Q^2 =$

$P(P+2\mu)$ and $GM = P + \mu$.

(ii) $D = 5$, $n = 2$, $d_2 = 1$, $M = \mathbb{R} \times S^2 \times \mathbb{R} \times \mathbb{R}$.
For $q = 2m$: CP diagram is of type $\mathbf{Sch}[r^\star, r_0]$.
For $q = 2m+1$: CP diagram of type $\mathbf{RN}[r^\star, 0, r_0]$ when $U_2 = q/2$ and of type $\mathbf{Sch}[0, r_0]$ otherwise.

(iii) $D = 11$, $n = 3$, $U_2 = qd_2$, $U_3 = 0$. Cases $d_2 = 2$ and $d_2 = 5$ correspond to the $(M2)_q$ and $(M5)_q$ black brane solutions respectively, which contain (when $q = 1$) "electric M2 brane" and "magnetic M5 brane" solutions in the 11-dimensional supergravity.
For $q = 2m$ we have CP diagram of type $\mathbf{Sch}[r^\star, r_0]$.
For $q = 2m+1$: CP diagram is of type $\mathbf{Sch}[0, r_0]$.

4. Behaviour in T-Region

In the case $q = 2m+1$ one can obtain asymptotic behaviour of the solution from Eq. (4) as $r \to +0$ inside the T-region $0 < r < r_0$. One can express it in terms of synchronous time variable $\tau \sim r^a$, where $a \equiv 1 + d\left(hqU^0 + 1/2\right)$, $h \equiv (U,U)^{-1}$:

$$ds^2 \sim -d\tau^2 + A_0 \tau^{2\alpha_0} d\Omega^2_{d_0} + \sum_{i=1}^{n} A_i \tau^{2\alpha_i} ds_i^2, \tag{8}$$

where $\alpha_0 = (hdU^0q+1)a^{-1}$, $\alpha_1 = (hdU^0q + q^2hd - d/2)a^{-1}$, $\alpha_i = (hdU^0q + hdqU_i/d_i)a^{-1}$, $i > 1$. If $a < 0$ we would have $\tau \sim r^a \to +\infty$ as $r \to +0$, and it would mean the possibility of non-singular "cosmological evolution" in terms of synchronous time in the T-region. However, it can be directly shown that $a > 0$ for the solution from Eq. (4), so there is no such a cosmological evolution in this model.

5. Conclusion

We see that for particular choice of parameters, when the solution from Eq. (4) generalizes well-known black hole or black brane configurations, its causal structure remains quite simple. We can also conclude from the last section that this solution does not contain so-called Black Universe configurations[4,5] – a special class of regular spherically symmetric solutions combining properties of black holes, wormholes, and de Sitter universe as a result of isotropic expansion in T-region at large scales.

References

1. H. Dehnen, V. D. Ivashchuk and V. N. Melnikov, *Grav. Cosmol.* **9**, 153 (2003).
2. V. D. Ivashchuk and V. N. Melnikov, *Grav. Cosmol.* **1**, 204 (1995).
3. K. A. Bronnikov and S. G. Rubin, *Black Holes, Cosmology, and Extra Dimensions* (World Scientic, Singapore, 2008).
4. K. A. Bronnikov and J. C. Fabris, *Phys. Rev. Lett.* **96**, 251101 (2006).
5. S. V. Bolokhov, K. A. Bronnikov and M. V. Skvortsova, *Class. Quant. Grav.* **29**, 245006 (2012).

Supergravity with broken Lorentz invariance

A. O. Marakulin* and S. M. Sibiryakov**

**Institute for Nuclear Research of the Russian Academy of Sciences, Moscow, Russia, e-mail: marakulin@physics.msu.ru*

****CERN Theory division, Geneva, Switzerland; Institut de Théorie des Phénomenes Physiques, EPFL, Lausanne, Switzerland; Institute for Nuclear Research of the Russian Academy of Sciences, Moscow, Russia, e-mail: sergey.sibiryakov@cern.ch*

Keywords: Lorentz violation; Einstein-aether theory; supergravity.

Incompatibility of the principles of quantum field theory with general relativity is one of the most important problems in modern theoretical physics. A potential way out of this situation consists in restricting the domain of validity of some basic postulates of general relativity and abandoning them at high energy scales. A promising approach to quantization of gravity based on abandoning the Lorentz invariance has been proposed by Horava.[1] The low-energy limit of the Horava theory, called khrono-metric model, presents a special case of the Einstein-aether gravity.[2] In the latter model violation of the Lorentz invariance is described by the time-like vector field u^m with unit norm ($u^m u_m = -1$) called aether that minimally couples to the Einstein-Hilbert action for gravity.

$$\begin{aligned} S = S_{GR} - \frac{1}{2}\int d^4x\sqrt{-g}\Big(& c_1 \left(\nabla_n u_m\right)^2 + c_2 \left(\nabla_m u_m\right)^2 \\ & + c_3 \nabla_n u_m \nabla^m u^n - c_4 u^r u^s \nabla_r u_m \nabla_s u^m\Big) \end{aligned} \tag{1}$$

A phenomenologically acceptable theory with Lorentz violation in the ultraviolet must incorporate a mechanism that ensures emergence of Lorentz invariance at low energies. One of the possible mechanisms is based on non-relativistic supersymmetry.[3] It has been shown[4] that in the supersymmetric extension of the Einstein-aether theory based on chiral vector super-aether field interactions between the aether and the fields of the Standard Model are suppressed.

In this work we consider the coupling of supersymmetric aether to supergravity. This cannot be realized within the minimal $N=1$ supergravity which does not allow to define a chiral aether vector superfield ($\bar{\nabla}_{\dot{\alpha}} U^a = 0$): anti-commutator of two supercovariant derivatives with spinor indices acting on a vector superfield does not vanish prohibiting to impose the chirality condition. This problem is resolved in the non-minimal formulation of the $N=1$ supergravity with Ziegel parameter $n=-1$.[5] We use the latter formulation to construct a superfield Lagrangian for the aether field coupled to linearized gravity and prove its uniqueness.

Non-minimal linearized $n=-1$ supergravity is formulated[5] in terms of real superfield V^a and so-called linear compensator Γ as a gauge theory invariant under gauge transformations $\delta V_{\alpha\dot{\alpha}} = \bar{D}_{\dot{\alpha}} L_\alpha - D_\alpha \bar{L}_{\dot{\alpha}}$, $\delta\bar{\Gamma} = \frac{1}{4} D^\alpha \bar{D}^2 L_\alpha$. We construct linearized Einstein-aether supergravity theory in terms the latter superfields and the chiral vector super-aether field which is considered as consisting of the real vacuum and small perturbation: $U^a = W^a + \mathcal{U}^a$. The super-aether field gauge transformation $\delta\mathcal{U}_a = \frac{1}{4}\left(\sigma_{ab}\right)_\beta^\gamma D_\gamma \bar{D}^2 L^\beta W^b + \text{h.c.}$ is consistent with the chirality condition. A straightforward but tedious calculation yields the most general Lagrangian invariant under gauge transformations with one free parameter C:

$$\begin{aligned} L = {}& \frac{1}{4}\left(\left(\partial_k V_b\right)^2 - \left(\Delta_k V_b\right)^2\right) + \Delta_k V_k\left(\Gamma+\bar{\Gamma}\right) + i\partial_k V_k\left(\Gamma-\bar{\Gamma}\right) - \Gamma^2 - \bar{\Gamma}^2 - \Gamma\bar{\Gamma} \\ & + C\Big\{\mathcal{U}_a\bar{\mathcal{U}}^a - W^a W^c\Big[\frac{3}{16}\eta_{ac}\left(\left(\Delta_k V_b\right)^2 - \left(\partial_k V_b\right)^2\right) \\ & - \frac{1}{4}\eta_{ac}\left(\left(\Delta_b V_b\right)^2 - \left(\partial_b V_b\right)^2\right) + \Delta_a V_c\left(\Gamma+\bar{\Gamma}\right) + i\partial_a V_c\left(\Gamma-\bar{\Gamma}\right) \\ & + \frac{1}{8}\eta_{ac}\left(\Gamma^2+\bar{\Gamma}^2\right) - \frac{1}{4}\eta_{ac}\Gamma\bar{\Gamma}\Big]\Big\} \end{aligned} \tag{2}$$

where $\Delta_k V_n = \frac{1}{4}\bar{\sigma}_k^{\dot{\alpha}\alpha}[\bar{D}_{\dot{\alpha}}, D_\alpha]V_n$. This parameter is a coupling constant of the super-aether to supergravity.

Expanding in the Grassmann coordinates we obtain the bosonic part of the Lagrangian in components. Integrating out the auxiliary fields we reduce to the linearized Lagrangian of the Einstein-aether theory

$$\begin{aligned} L_{bos} = {}& \left(1+\frac{C}{4}\right)\left(\frac{1}{2}\left(\partial_k e\right)^2 - \frac{1}{2}\left(\partial_k e_{mn}\right)^2 + \partial_m e_{mp}\partial_n e_{np} - \partial_m e_{mp}\partial_p e\right) \\ & + C\Big(-\left(\partial_k v_b\right)^2 + 2w_a\partial_c v_b\left(\partial_a e_{bc} - \partial_b e_{ac}\right) \\ & - w_a w_c\left(\partial_p e_{ak} - \partial_a e_{pk}\right)\left(\partial_p e_{ck} - \partial_c e_{pk}\right)\Big), \end{aligned} \tag{3}$$

where e_{mn} is the symmetrized vielbein, w_a is the aether vacuum expectation value and v_a is the first order aether perturbation. One can find out that supersymmetry fixes all but one coupling constants in this Lagrangian to zero: $c_1 = 4C/(4+C)$, $c_{2,3,4} = 0$.

The work is supported by the RFBR grant 14-02-31429.

References

1. P. Horava, Physical Review D, **79**, 084008 (2009).
2. T. Jacobson, PoS **QG-PH**, 020 (2007).
3. S. Groot Nibbelink and M. Pospelov, Phys. Rev. Lett. **94**, 081601 (2005).
4. O. Pujolas and S. Sibiryakov, JHEP, **1201**, 062 (2012).
5. I. L. Buchbinder and S. M. Kuzenko, *Ideas and methods of supersymmetry and supergravity, or A walk through superspace* (Institute of Physics Pub., 1998).

Anisotropic exponential cosmological solutions in the 4th and 5th orders of Lovelock gravity

I. V. Kirnos

Tomsk State University, 36, Lenin prosp., Tomsk 634050, Russia
V.E. Zuev Institute of Atmospheric Optics SB RAS, 1, Academician Zuev square, Tomsk 634055, Russia
E-mail: kirnos@tspu.edu.ru, ikirnos@sibmail.com

In the 4th and 5th orders of Lovelock gravity exponential cosmological solutions of Bianchi-I type involving the lower orders are obtained.

Keywords: Modified gravity; metric of Bianchi-I type.

In Lovelock theory[1] gravitational field equations are

$$\sum_{p=0}^{\left[\frac{n-1}{2}\right]} \alpha_p G^{(p)\mu}{}_{\nu} = \frac{8\pi G}{c^4} T^{\mu}{}_{\nu},$$

where α_p are arbitrary constants, $G^{(p)\mu}{}_{\nu}$ is p-th Lovelock tensor:

$$G^{(p)\mu}{}_{\nu} = \delta^{\mu\sigma_1\sigma_2\ldots\sigma_{2p}}_{\nu\lambda_1\lambda_2\ldots\lambda_{2p}} \prod_{i=1}^{p} R^{\lambda_{2i-1}\lambda_{2i}}{}_{\sigma_{2i-1}\sigma_{2i}},$$

n is the space-time dimensionality, square brackets denote the integer part of a fraction, $\delta^{\cdots}_{\cdots}$ is the multi-index δ-symbol (equal to 1, when lower indices form an even transposition of upper ones, equal to -1, when odd, and equal to zero in other cases).

Such a theory have extra spatial dimensions as an obligate condition. But if we have extra dimensions, why are they invisible? For the explanation we should find anisotropic cosmological solutions. Power-law solutions[2,3] have shortcoming: they involve only one, p-th, order of Lovelock gravity, all the other orders, the upper and the lower, are excluded.

In opposition to this, exponential solutions involve p-th and the lower orders. For the 2nd and 3rd orders such solutions were obtained in Ref. 4.

In the 4th order equations $\alpha_1 G^{(1)}_{\mu\nu} + \alpha_2 G^{(2)}_{\mu\nu} + \alpha_3 G^{(3)}_{\mu\nu} + \alpha_4 G^{(4)}_{\mu\nu} = (8\pi G/c^4) T_{\mu\nu}$ in presence of matter with EoS $p = w\rho$ have solution

$$g_{\mu\nu} = \mathrm{diag}\{-1, e^{2H_1 t}, e^{2H_2 t}, ..., e^{2H_n t}\},$$

where Hubble parameters H_i meet the conditions (here $\sigma_s \equiv \sum_i H_i^s$, $\kappa_0 \equiv 8\pi G\varepsilon_0/c^4$)

$$\begin{aligned}
\sigma_1 &= 0,\\
\sigma_6 &= -\frac{1}{8}\sigma_2^3 + \frac{3}{4}\sigma_4\sigma_2 + \frac{1}{3}\sigma_3^2 + \frac{\alpha_2}{24\alpha_3}\left(\sigma_2^2 - 2\sigma_4\right) - \frac{\alpha_1}{32\alpha_3}\sigma_2 - \frac{1-7w}{384\alpha_3}\kappa_0,\\
\sigma_8 &= -\frac{1}{16}\sigma_2^4 + \frac{1}{4}\sigma_4\sigma_2^2 + \frac{\alpha_2}{36\alpha_3}\left(\sigma_2^3 - 2\sigma_4\sigma_2\right) - \frac{\alpha_1}{48\alpha_3}\sigma_2^2 - \frac{1-7w}{576\alpha_3}\sigma_2\kappa_0 +\\
&\quad + \frac{8}{15}\sigma_5\sigma_3 + \frac{1}{4}\sigma_4^2 - \frac{\alpha_2}{2880\alpha_4}\left(\sigma_2^2 - 2\sigma_4\right) + \frac{\alpha_1}{2880\alpha_4}\sigma_2 + \frac{1-5w}{23040\alpha_4}\kappa_0.
\end{aligned}$$

In the 5th order we have solution, where Hubble parameters meet the conditions

$$\begin{aligned}
\sigma_1 &= 0,\\
\sigma_8 &= \frac{9w-1}{23040\alpha_4}\kappa_0 - \frac{\alpha_1}{1440\alpha_4}\sigma_2 + \frac{\alpha_2}{960\alpha_4}\left(\sigma_2^2 - 2\sigma_4\right) + \frac{\alpha_3}{720\alpha_4}\left(-3\sigma_2^3 + \right.\\
&\quad \left. + 18\sigma_4\sigma_2 + 8\sigma_3^2 - 24\sigma_6\right) + \frac{1}{48}\sigma_2^4 - \frac{1}{4}\sigma_4\sigma_2 - \frac{2}{9}\sigma_3^2\sigma_2 + \frac{2}{3}\sigma_6\sigma_2 +\\
&\quad + \frac{8}{15}\sigma_5\sigma_3 + \frac{1}{4}\sigma_4^2,\\
\sigma_{10} &= -\frac{7w-1}{64\cdot 8!\alpha_5}\kappa_0 + \frac{3\alpha_1}{16\cdot 8!\alpha_5}\sigma_2 - \frac{\alpha_2}{4\cdot 8!\alpha_5}\left(\sigma_2^2 - 2\sigma_4\right) - \frac{\alpha_3}{4\cdot 8!\alpha_5}\left(-3\sigma_2^3 + \right.\\
&\quad \left. + 18\sigma_4\sigma_2 + 8\sigma_3^2 - 24\sigma_6\right) - \frac{105}{8!}\sigma_2^5 + \frac{2100}{8!}\sigma_4\sigma_2^3 + \frac{2800}{8!}\sigma_3^2\sigma_2^2 -\\
&\quad - \frac{8400}{8!}\sigma_6\sigma_2^2 - \frac{1}{3}\sigma_5\sigma_3\sigma_2 - \frac{6300}{8!}\sigma_4^2\sigma_2 + \frac{25200}{8!}\sigma_8\sigma_2 - \frac{5600}{8!}\sigma_4\sigma_3^2 +\\
&\quad + \frac{19200}{8!}\sigma_7\sigma_3 + \frac{16800}{8!}\sigma_6\sigma_4 + \frac{1}{5}\sigma_5^2.
\end{aligned}$$

Finally, I suggest similar conditions for the arbitrary order, but, unfortunately, I haven't prove this assumption yet.

References

1. D. Lovelock, *J. Math. Phys.* **12**, 498 (1971).
2. S. A. Pavluchenko, *Phys. Rev. D* **80**, 107501 (2009).
3. I. V. Kirnos, *Gravitation and Cosmology* **18**, 259 (2012).
4. I. V. Kirnos, S. A. Pavluchenko and A. V. Toporensky, *Gravitation and Cosmology* **16**, 274 (2010).

On the dependence of fundamental constants on parameters characterizing the Universe dark matter

V. M. Koryukin*

Department of Physics and Mathematics, Mari State University, Lenin sq., 1, Yoshkar–Ola, 424000, Russia
**E-mail: vmkoryukin@gmail.com*

A. V. Koryukin

Institute of Physics, Kazan Federal University, Kremlyovskaya St., 16, Kazan, 420008, Russia
E-mail: akoryukin@gmail.com

We argue that there is weighty evidence on a dependence of the gravitational constant on the parameters characterizing the physical state of dark matter in the Universe.

Keywords: Dark matter; neutrinos; topological properties of space-time; quantum gravity.

1. The Universe Neutrino Background

The assumption of the "sea" of Universe background fermions in the ground state allows to use the Landau theory of the Fermi liquid via considering charged particles as quasi-particles on the background of "sterile" neutrinos and "sterile" antineutrinos. The properties of the latter must determine the geometrical and topological properties of the space-time M_4 which must not be a simply connected one. The last property of the space-time M_4 justifies a quantization of any charges (specifically, electric charges and baryonic charges), because the wave function of the charged particle is bound to be a multivalued one (of course if $E < E_F$, where E is the particle energy and E_F is the Fermi energy of Fermi liquid). The transition to the description of the slow subsystem with the help of the space-time manifold is carried when the Fermi energy of "sterile" neutrinos tends to infinity. In this case the quotation-marks in the words "sterile" neutrinos can be discarded, because these neutrinos will not collide with the other particles even at very high energies of the latter. The all Universe matter must be characterized (as it is adopted in the statistical physics) by the statistical sum. As a result, taking into account the empirical data from the observations of type Ia supernovas, one can understand the hypothesis necessity in General Relativity for the availability "dark energy" in the Universe, connecting it with the cosmological constant. We regard, that the presence of the Universe neutrino background with the finite Fermi energy E_F is the catalytic agent of stochastic processes, but the large value of this energy causes to the determinancy of physical processes (we shall use the system of units $h/(2\pi) = c = 1$, where h is the Planck constant and c is the velocity of light). Specifically we connect the large value of the Fermi energy and the low temperature of the Universe neutrino background with the stability (or only with the metastability) of elementary particles and with conservation laws of electric and baryonic charges. Therefore cosmic rays of ultrahigh energies are not bound to interact with the cosmic radio background, so that their motion in the Universe can not be limited by the value of spatial distances.

2. The Gravitational Interaction

We assume that the energy of fundamental interactions must depend on a number of particles and quasi-particles participating in this interactions, defining its dependence to space coordinates by means of a mean number of bosons, by which two particles are exchanged. As a result in the general case (n is a number of bosons):

$$E = -2 \int \left(\sum_{\sigma_1, \sigma_2} \frac{\rho_1 \sigma_1 \rho \sigma_2 \rho_2 \sum\limits_{n=0}^{N_{\sigma_1}} n e^{-2n\rho\sigma_1 r/T}}{T \sum\limits_{n=0}^{N_{\sigma_1}} e^{-2n\rho\sigma_1 r/T}} \right) dV_1 dV_2, \quad (1)$$

where ρ_1 is the energy density of first body particles; ρ_2 is the energy density of second body particles; ρ is the energy density of quasiparticles, T is theirs temperature; σ_1 and σ_2 are cross-sections with the emission or the absorption of particles or quasi-particles. Hence (considering, that $N_{\sigma_1} \to \infty$) it can obtain the potential as

$$U(r) = -\sum_{\sigma} \left[C_\sigma / \left(e^{B_\sigma r} - 1 \right) \right] \quad (2)$$

($B_\sigma \propto \sigma_1$, $C_\sigma \propto \sigma_1 \sigma_2$) for the description of fundamental interactions. As a result we obtain the dependence of fundamental constants on parameters characterizing the Universe dark matter.

We regard that the standard gravitation interaction is not the fundamental one, but it is generated by collective oscillations in Universe's neutrino sea. [1] As a consequence of this, the weak interaction acquires the global value, because it causes the bond of neutrinos both one with the other and with charged leptons and quarks. Let us consider the gravitational interaction at large distances. In this case one can put: $\sigma_1 = \sigma_2 = \sigma_\nu$. We shall consider the scattering of the neutrino upon the charge lepton, induced by the exchange of the neutral Z° boson (taking account of the low temperature of the Universe background neutrinos) only. As a result the cross-section of the scattering for the neutrino proves to be equal to the cross-section of the scattering for the antineutrino in the low-energy approximation (the energy of the neutrino $\omega \ll m$) and they are written down as

$$\sigma_\nu = 4kG_F^2\omega^2\left[(\xi - 1/2)^2 + \xi^2 - (\xi - 1/2)\xi\right]/\pi \tag{3}$$

($G_F \approx 1.166 \cdot 10^{-5}\ GeV^{-2}$ is the Fermi constant, $\xi = \sin^2\Theta_W \approx 0.23$, and Θ_W is the weak angle), where for the charge leptons the factor $k = k_e$ depends on the fine structure constant $\alpha \approx 1/137$ ($k_e \sim 2\alpha/3$) only, while for the quarks the factor $k = k_q$ must depend on the running coupling constant α_s too, which defines the collision radiation by gluons. Note that σ_ν proves to be minimal for $\xi = 1/4$. Let us consider the scattering the background neutrino upon the electron only of the estimate of the constant σ_ν. Substituting the mean $< \omega >$ instead of ω one can receive ($T \approx 10^{-13}\ GeV$, $< \omega > \approx 3.15 \cdot T$) the crude value of the constant $\sigma_\nu \sim \alpha \cdot G_F^2 \cdot T^2 \sim 10^{-38} \cdot GeV^{-2}$, which is near to the known value of the gravitational constant G_N ($G_N \approx 6.7 \cdot 10^{-39}\ GeV^{-2}$).

Reference

1. V. Koryukin, Gravitation and Weak Interactions in the Gauge Fields Covariant Theory, Proc. of III International Symposium on Weak and Electromagnetic Interactions in Nuclei (JINR, Dubna, Moscow Region, RUSSIA, June 16-22, 1992), ed. Ts. D. Vylov (World Scientific, Singapore, 1992), pp. 456—458.

On multidimensional solutions in the Einstein–Gauss–Bonnet model with a cosmological term

A. A. Kobtsev*, V. D. Ivashchuk** and K. Ernazarov***

Moscow Mesons Factory of INR RAS, Moscow, Troitsk, 142190, Russia
* *E-mail: aak@inr.ru*

Center for Gravitation and Fundamental Metrology, VNIIMS, 46 Ozyornaya ul., Moscow, 119361, Russia
** *E-mail: ivas@vniims.ru*

Institute of Gravitation and Cosmology, Peoples' Friendship University of Russia, 6 Miklukho-Maklaya ul., Moscow, 117198, Russia
*** *E-mail: kubantai80@mail.ru*

A D-dimensional gravitational model with Gauss-Bonnet and cosmological terms Λ is considered. When ansatz with diagonal cosmological metrics is adopted, we overview recent solutions for $\Lambda = 0$ and find new examples of solutions for $\Lambda \neq 0$ and $D = 8$ with exponential dependence of scale factors which describe an expansion of "our" 3-dimensional factor-space and contraction of 4-dimensional internal space.

Keywords: Gauss-Bonnet term; multidimensional cosmology.

1. Introduction

Here we consider D-dimensional gravitational model with the Gauss-Bonnet term. The action reads

$$S = \int_M d^D z \sqrt{|g|}\{\alpha_1(R[g] - 2\Lambda) + \alpha_2 \mathcal{L}_2[g]\}, \tag{1}$$

where $g = g_{MN} dz^M \otimes dz^N$ is the metric defined on the manifold M, $\dim M = D$, $|g| = |\det(g_{MN})|$ and

$$\mathcal{L}_2 = R_{MNPQ}R^{MNPQ} - 4R_{MN}R^{MN} + R^2 \tag{2}$$

is the standard Gauss-Bonnet term. Here α_1 and α_2 are non-zero constants.

Earlier the appearance of the Gauss-Bonnet term was motivated by string theory.[1–5]

At present, the (so-called) Einstein-Gauss-Bonnet (EGB) gravitational model and its modifications are intensively used in cosmology, see Refs. 6 (for $D = 4$), 7–15 and references therein, e.g. for explanation of accelerating expansion of the Universe following from supernovae (type Ia) observational data.[16–18] Certain exact solutions in multidimesional EGB cosmology were obtained in Refs. 7- 15 and some other papers.

Here we are dealing with the cosmological type solutions with diagonal metrics (of Bianchi-I-like type) governed by n scale factors depending upon one variable, where $n > 3$. Moreover, we restrict ourselves by the solutions with exponential dependence of scale factors. We present new examples of exact solutions in dimension $D = 8$ which describe an exponential expansion of 3-dimensional factor-space and contraction of 4-dimensional internal space.

2. The Cosmological Model

Here we consider the manifold

$$M = \mathbb{R} \times \mathbb{R}^n \tag{3}$$

with the metric

$$g = -dt \otimes dt + \sum_{i=1}^{n} e^{2\beta^i(t)} dy^i \otimes dy^i, \tag{4}$$

where $\beta^i(t)$ are smooth functions, $i = 1, \ldots, n$.

We introduce "Hubble-like" variables $h^i = d\beta^i/dt$. The equations of motion for the action (1) read as follows

$$\alpha_1(G_{ij}h^i h^j + 2\Lambda) - \alpha_2 G_{ijkl} h^i h^j h^k h^l = 0, \tag{5}$$

$$\left[2\alpha_1 G_{ij}h^j - \frac{4}{3}\alpha_2 G_{ijkl} h^j h^k h^l\right] \sum_{i=1}^{n} h^i$$
$$+ \frac{d}{dt}\left[2\alpha_1 G_{ij}h^j - \frac{4}{3}\alpha_2 G_{ijkl} h^j h^k h^l\right] - L = 0, \tag{6}$$

$i = 1, \ldots, n$, where

$$L = \alpha_1(G_{ij}h^i h^j + 2\Lambda) - \frac{1}{3}\alpha_2 G_{ijkl} h^i h^j h^k h^l. \tag{7}$$

Here

$$G_{ij} = \delta_{ij} - 1, \tag{8}$$
$$G_{ijkl} = G_{ij}G_{ik}G_{il}G_{jk}G_{jl}G_{kl} \tag{9}$$

are respectively the components of two metrics on $\mathbb{R}^n$.[19,20] The first one is the well-known "minisupermetric" - 2-metric of pseudo-Euclidean signature and the second one is the Finslerian 4-metric.

Due to (5)

$$L = \frac{2}{3}\alpha_1(G_{ij}h^i h^j - 4\Lambda). \tag{10}$$

In this paper we deal with the following solutions to equations (5) and (6)

$$h^i(t) = v^i, \tag{11}$$

with constant v^i, which corresponding to the solutions

$$\beta^i = v^i t + \beta^i_0, \tag{12}$$

where β^i_0 are constants, $i = 1, \ldots, n$.

In this case we obtain the metric (4) with the exponential dependence of scale factors

$$g = -dt \otimes dt + \sum_{i=1}^{n} B_i e^{2v^i t} dy^i \otimes dy^i, \tag{13}$$

where $B_i > 0$ are arbitrary constants.

For the fixed point $v = (v^i)$ we have the set of polynomial equations

$$G_{ij}v^i v^j + 2\Lambda - \alpha G_{ijkl}v^i v^j v^k v^l = 0, \tag{14}$$

$$\left[2G_{ij}v^j - \frac{4}{3}\alpha G_{ijkl}v^j v^k v^l\right] \sum_{i=1}^{n} v^i - \frac{2}{3}G_{ij}v^i v^j + \frac{8}{3}\Lambda = 0, \tag{15}$$

$i = 1, \ldots, n$, where $\alpha = \alpha_2/\alpha_1$. For $n > 3$ this is a set of forth-order polynomial equations.

For $\Lambda = 0$ and $n > 3$ the set of equations (14) and (15) has an isotropic solution $v^1 = ... = v^n = H$, only if $\alpha < 0$[19,20] $H = \pm 1/\sqrt{|\alpha|(n-2)(n-3)}$. This solution was generalized in Ref. 15 to the case $\Lambda \neq 0$.

It was shown in Refs. 19, 20 that there are no more than three different numbers among $v^1, ..., v^n$ when $\Lambda = 0$. This is valid also for $\Lambda \neq 0$.

3. Examples of Cosmological Solutions

In this section we consider some solutions to the set of equations (14), (15) of the following form

$$v = (H, \ldots, H, h, \ldots, h). \tag{16}$$

where H the "Hubble-like" parameter corresponding to m-dimensional isotropic subspace with $m \geq 3$ and h is the "Hubble-like" parameter corresponding to l-dimensional isotropic subspace, $l > 2$.

These solutions should satisfy the following conditions: $H > 0$, $h < 0$. The first inequality $H > 0$ is necessary for a description of accelerated expansion of 3-dimensional subspace, which may describe our Universe, while the second inequality $h < 0$ is necessary for contraction of internal space volume.

3.1. *Polynomial equations*

According to our ansatz (16), we have m dimensions expanding with the Hubble parameter $H > 0$ and l dimensions contracting with the "Hubble-like" parameter $h < 0$. The set of polynomial equations (14), (15) reads

$$\begin{aligned} H^2(m-m^2)+h^2(l-l^2)-2mlHh \\ -\alpha(H^4m(m-1)(m-2)(m-3)+h^4l(l-1)(l-2)(l-3) \\ +4H^3hm(m-1)(m-2)l+4h^3Hl(l-1)(l-2)m \\ +6H^2h^2m(m-1)l(l-1))+2\Lambda=0, \end{aligned} \tag{17}$$

$$\begin{aligned} m(1-m)H^2-(1/2)lh^2(1+2l)+2lHh((3/4)-m) \\ -\alpha(H^4m(m-1)(m-2)(m-3)+H^3hl(m-1)(m-2)(4m-3) \\ +3H^2h^2l(m-1)(2lm-2l-m) \\ +Hh^3l(l-1)(4lm-3l-2m)+h^4l^2(l-1)(l-2))+2\Lambda=0, \end{aligned} \tag{18}$$

$$\begin{aligned} l(1-l)h^2-(1/2)mH^2(1+2m)+2mHh((3/4)-l) \\ -\alpha(h^4l(l-1)(l-2)(l-3)+h^3Hm(l-1)(l-2)(4l-3) \\ +3h^2H^2m(l-1)(2lm-2m-l) \\ +hH^3m(m-1)(4lm-3m-2l)+H^4m^2(m-1)(m-2))+2\Lambda=0. \end{aligned} \tag{19}$$

We put $\alpha = \pm 1$ and denote $\Lambda = \lambda$, keeping in mind the general α-dependent form of solution

$$H(\alpha) = H|\alpha|^{-1/2}, \qquad h(\alpha) = h|\alpha|^{-1/2} \qquad \Lambda = \lambda|\alpha|^{-1}. \tag{20}$$

3.2. *Solutions with* $\Lambda = 0$

Let $\Lambda = 0$ and $\alpha = 1$. It was shown in Ref. 21 that, for $m = 9$ there exists an infinite series of cosmological solutions with $l = 3000, 3001, ...,$ any of which describes an accelerated expansion of the 3-dimensional factor space with sufficiently small value of the variation of the effective gravitational constant G obeying the observational restrictions,[22] see also Ref. 23. This variation may be arbitrary small for a big enough value of l. We remind that the effective gravitational constant G is proportional to the inverse volume scale factor of the internal space, see Ref. 24–28 and references therein.

For $m = 11$ and $l = 16$ it was found in Ref. 21 a solution with

$$H = \frac{1}{\sqrt{15}}, \qquad h = -\frac{1}{2\sqrt{15}}, \tag{21}$$

which describe a zero variation of effective cosmological constant G.

Another solution of such type which was found in Rev. 21 appears for $m = 15$ and $l = 6$ with

$$H = \frac{1}{6}, \qquad h = -\frac{1}{3}. \tag{22}$$

3.3. *Solutions with* $\Lambda \neq 0$

Here we present several new cosmological solutions for $\Lambda \neq 0$, $\alpha = 1$, $m = 3$ and $l = 4$.

The first solution takes place for $\lambda = 3/16$:

$$H = \frac{1}{4}\sqrt{2}, \qquad h = -\frac{1}{4}\sqrt{2}. \tag{23}$$

The second one is valid for $\lambda = 13/48$:

$$H = \frac{1}{4}\sqrt{6}, \qquad h = -\frac{1}{12}\sqrt{6}. \tag{24}$$

The third one

$$H = \frac{2}{29}\sqrt{29}, \qquad h = -\frac{3}{58}\sqrt{29} \tag{25}$$

corresponds to $\lambda = 21/116$.

Any of these solutions describes an accelerated expansion of 3-dimensional factor space and contraction of internal space. All of these solutions take place for fixed positive values of λ (see (20)). There exist also examples of solutions with negative cosmological constant $\lambda = -21/80$:

$$H_{\pm} = \frac{1}{60320}(248 \pm 32\sqrt{30})\sqrt{68150 \mp 9280\sqrt{30}}, \tag{26}$$

$$h_{\pm} = -\frac{1}{580}\sqrt{68150 \mp 9280\sqrt{30}}. \tag{27}$$

These solutions obey inequalities $H_{\pm} > 0$ and $h_{\pm} < 0$.

4. Conclusions

We have considered the D-dimensional Einstein–Gauss–Bonnet (EGB) model with the Λ-term. By using the ansatz with diagonal cosmological type metrics, we have found new solutions with exponential dependence of scale factors with respect to synchronous time variable t in dimension $D = 1 + 3 + 4$. Any of these solutions describes an exponential expansion of "our" 3-dimensional factor-space with the Hubble parameter $H > 0$ and exponential contraction of 4-dimensional internal space. An open question arising here is to find solutions with $\Lambda \neq 0$ which obey the observational constraints on the temporal variation of the effective gravitational constant G. This question will be addressed in a separate publication.

References

1. B. Zwiebach, *Phys. Lett. B* **156**, 315 (1985).
2. D. Gross and E. Witten, *Nucl. Phys. B* **277**, 1 (1986).
3. D. J. Gross and J. H. Sloan, *Nucl. Phys. B* **291**, 41 (1987).
4. R. R. Metsaev and A. A. Tseytlin, *Phys. Lett. B* **191**, 354 (1987).
5. R. R. Metsaev and A. A. Tseytlin, *Nucl. Phys. B* **293**, 385 (1987).
6. S. Nojiri and S. D. Odintsov, *Int. J. Geom. Meth. Mod. Phys.* **4**, 115 (2007).
7. H. Ishihara, *Phys. Lett. B* **179**, 217 (1986).
8. N. Deruelle, *Nucl. Phys. B* **327**, 253 (1989).
9. E. Elizalde, A. N. Makarenko, V. V. Obukhov, K. E. Osetrin and A.E. Filippov, *Phys. Lett. B* **644**, 1 (2007).
10. K. Bamba, Z.-K. Guo and N. Ohta, *Prog. Theor. Phys.* **118**, 879-892 (2007).
11. A. Toporensky and P. Tretyakov, *Grav. Cosmol.* **13**, 207 (2007).
12. I. V. Kirnos, A. N. Makarenko, S. A. Pavluchenko and A. V. Toporensky, *Gen. Rel. Grav.* **42**, 2633 (2010).
13. S. A. Pavluchenko and A. V. Toporensky, *Mod. Phys. Lett. A* **24**, 513-521 (2009).

14. I. V. Kirnos and A. N. Makarenko, *Open Astr. J.* **3**, 37 (2010).
15. D. Chirkov, S. Pavluchenko and A. Toporensky, *Mod. Phys. Lett. A* **29**, 1450093 (2014).
16. A. G. Riess et al., *The Astronomical Journal* **116**, 1009 (1998).
17. S. Perlmutter et al., *The Astrophysical Journal* **517**, 565 (1999).
18. M. Kowalski, D. Rubin et al., Improved cosmological constraints from new, old and combined supernova datasets, arXiv: 0804.4142.
19. V. D. Ivashchuk, *Grav. Cosmol.* **16**, 118 (2010).
20. V. D. Ivashchuk, *Int. J. Geom. Meth. Mod. Phys.* **7**, 797 (2010).
21. V. D. Ivashchuk and A. A. Kobtsev, *Eur. Phys. J. C* **75**, 177 (2015).
22. E. V. Pitjeva, *Astron. Vestnik* **47**, 419 (2013).
23. P. A. R. Ade et al. [Planck Collaboration], *Astronomy and Astrophysics* **571**, A1 (2014).
24. K.A. Bronnikov , V. D. Ivashchuk and V. N. Melnikov, *Nuovo Cimento B* **102**, 209 (1998).
25. J.-M. Alimi, V. D. Ivashchuk, S. A. Kononogov and V. N. Melnikov, *Grav. Cosmol.* **12**, 173 (2006).
26. V. D. Ivashchuk, S. A. Kononogov, V. N. Melnikov and M. Novello, *Grav. Cosmol.* **12**, 273 (2006).
27. V. N. Melnikov, *Front. Phys. China.* **4**, 75 (2009).
28. V. D. Ivashchuk and V. N. Melnikov, *Grav. Cosmol.* **20**, 26 (2014).

Section 6

Gravitational Experiments

OGRAN setup in the BNO INR RAS in neutrino-gravitational correlation experiment

L. B. Bezrukov[1] and V. A. Krysanov[1,2],

[1] *Institute for Nuclear Research, Russian Academy of Sciences*
prospekt 60-letiya Oktyabrya 7a, Moscow 117312, Russia
www.inr.ru

A. M. Motylev[2], S. I. Oreshkin[2], S. M. Popov[2], V. N. Rudenko[2,*], V. V. Semenov[2], V. A. Silin[2] and I. S. Yudin[2]

[2] *Sternberg Astronomical Institute, Moscow State University*
Universitetsky pr., 13, Moscow 119991, Russia
** E-mail: rvn@sai.msu.ru*
www.sai.msu.ru

This article describes the uncooled opto-acoustical gravitational antenna constructed for a multi-channel mode operation in the deep underground laboratory of the Baksan Neutrino Observatory in parallel with Baksan underground scintillation telescope involved in the collapse searching program.

Keywords: Gravitational wave; Bar detector; multi-messaging astronomy.

A new setup of Opto-acoustical GRavitational Antenna (OGRAN) - the large scale opto-acoustical gravitational detector that structurally combines the principles of bar- and interferometric detectors was installed in underground laboratory of the Baksan Neutrino Observatory (BNO), which is located in the Caucasus Mountains inside the mine under Andyrchy mountain under 350 m of vertical rock depth (850 mwe).

The ORGAN setup[1] is based on the design type called comparator of optical standards. It is composed from two feedback loops. The first one tunes the laser generated frequency to be in the optical resonance with Fabri-Perot (FP) cavity mode at the large bar (detector). In this way any changes of the FP optical length is converted into the laser beam frequency variation. In such combination output signal has more complex structure containing separately acoustical and optical parts. It should be also noted that it is possible to achieve a sensitivity at the level of bar the Brownian noise in a relatively wide spectral frequency range due to the small back action of optical read out.

The second loop is the measuring one. It consists of the additional FP reference cavity (or frequency discriminator) pumped by the same laser beam but tuned in resonance by independent PZT driver attached to the one of discriminator's mirror. This driver is dedicated to the low frequency drift compensation. The output signal of the setup is proportional to the frequency difference of the detector and the discriminator cavity. It means that any perturbation of the detector bar cavity is reflected into the discriminator output signal. Similar scheme was tested for the AURIGA optical bar detector.[2] Meanwhile the AURIGA project used a mini gap FP resonator (a kind of optical accelerometer) placed on a displacement transducer attached to the one of the bar detectors end, OGRAN detector has a big FP cavity at the same scale as its acoustical resonance bar. This presents the OGRAN physical feature: the gravitational wave interacts with acoustical movement of the resonance bar as well as with the electromagnetic field in the FP cavity. It allows to speak about a complex structure of signal response with optical and acoustical parts. A payment for this originality is a technical problem of constructing a large scale high finesse FP cavity rigidly coupled with acoustical resonator without loosing its high mechanical resonance quality factor Q.

The potential sensitivity is defined by the thermal Brownian noise of the fundamental acoustical mode. Under optimal filtration procedure, the minimum registered perturbation of the OGRAN optical length in the bandwidth Δf can be described by the equation $h \sim 10^{-20}\sqrt{F\Delta f}\,\frac{1}{\sqrt{Hz}}$. The noise factor F describes an excess of optical readout fluctuation over the thermal noise level. So for the room temperature OGRAN detector, the integral potential sensitivity $h \sim 10^{-19}$ might be provided in the bandwidth $\Delta f \sim 10^2 Hz$ if $F \sim 1$. Calculation of the noise factor for the optical readout associated with Pound-Drever technique resulted in the formula $F = (2M/\tau)\sqrt{G_e/G_T}$ where $G_e = B\omega_0^2(2h\nu/\eta P)(\lambda/2\pi N)^2$ - spectral density of optical fluctuations with following parameters: $\eta \sim 0,7$ - the photo efficiency, N - the number of reflection of FP cavity (proportional to Finesse), $B =$ (1-1000), the phenomenological factor of exceeding the Poisson level of laser noise; the spectral density of the thermal (Brownian) bar fluctuations $G_T = 2kTM\omega_0/Q$; the parameter M which is the effective mass was roughly equal to the half of total bar weight; τ is the measurement time (the resonance signal duration for the optimal filtration) or the inverse value of filtering bandwidth Δf. Under the maximum laser power P = 2W, the incoming luminosity for each of both FP resonators cannot exceed a half of watt. But due to losses in the optical elements and the interference contrast $\sim 0,1$ the real value of effective power is P = 0,01W. Substituting all numerical data together with the measurement time $\tau =$ 0,01 s leads to the estimation of the required number of reflections (mirrors quality): $N = 10^3 - 10^4$ or better.

A perspective of the next generation is associated with a cryogenic version of the opto-acoustical bar with estimated sensitivity $h \sim 10^{-23} \frac{1}{\sqrt{Hz}}$.

It is planned to use the OGRAN setup together with the BUST setup located in the same underground tunnel BNO. BUST setup has sensitivity in the low energy neutrino channel corresponding to the OGRAN level. Both setups can continuous monitoring rare events: relativistic collapsing catastrophes in the Galaxy and close environment $\sim$ 100 kpc. The rocks above laboratories have a good filtering capability for cosmic rays. The muon flux in the underground laboratory rooms is about $10^3 - 10^4$ times lower then outside BNO. Both detectors are approximately in the same conditions and possess the low level of environmental noise that decreases false alarm level.

Effective data processing algorithm for neutrino-gravitational channels is based on the known collapsing model of astro-physical objects. According to the modern theory, gravitational waves are emitted simultaneously (after some small delay) with neutrino burst during last merging phase of collapsing astrophysical objects. The time samples when signals coincidence from the both detectors can be considered as event candidates.

During collapsing phase energy is emitted also as x-ray, γ-ray, electromagnetic wave (visual, IR), etc. According to multi-messaging astronomy principles additional independent channels can be used to improve detection possibility.

References

1. S.N. Bagaev, L.B. Bezrukov, N.L. Kvashnin et. al. A high frequency resonance gravity gradiometer. *Review of Scientific Instruments* **85**, 065114 (2014). DOI:10.1063/1.4883901
2. L. Conti, G.A. Prodi, S. Vitale et. al. *Gravitational waves and experimental gravity". Proc. 34th Ren. Moriond*, World Publish, 85 (2000).

Electromagnetic field of the Earth as a tool for astrophysical processes investigations

L.V. Grunskaya*, V.V. Isakevich, D.V. Isakevich and D.V. Rubay

Department of General and Applied Physics, Vladimir State University, Vladimir, 600005, Russia
**E-mail: grunsk@vlsu.ru*
www.vlsu.ru

Years of research have shown that all experimentally obtained time series of the electric and geomagnetic fields have components at frequencies infra low frequency gravitational-wave emission of a number of binary star systems: J 0700+6418, J 1012+5307, J 1537+1155, J 1959+2048, J 2130+1210, J 1915+160 with effective values 0,28 - 0,6 V/m. It is shown that all reviewed time series have non-coherent components on the axion frequency 5e-6 Hz with effective values of 0,71 - 1,36 V/m, and also at combination frequencies of this frequency with the frequency of the moon tides.

Interest in learning of electromagnetic fields of infra low frequency range is connected with its wide spectrum of physical mechanisms of initiation from earth to space. Purposes of current investigations are connected with learning of the signal structure, initiated by tide processes of geophysical and astrophysical nature. From theoretical and experimental problem statement about indirect detection of periodical signals of gravitational wave sources of astrophysical nature on induced by them electromagnetic fields in the resonator Earth-ionosphere is extremely relevant in modern science about the Earth. The basis of the theory of the interactions of quasi-static electromagnetic fields was inherent in the proceedings of Bocaletti D., Ginzburg V. L., Zeldovich Y. B.[1,2,5,6] These pioneering proceedings stimulated the research of many scientists both in our country and abroad, in particular the proceedings of Balakin A. B.,[3,4,7,8] which gave the theoretical basis for our nowadays investigations. The investigations in infra low frequency range of different processes require the monitoring of the electromagnetic field at spaced apart stations, because for reliable detection of the infra low frequency process in that range it should be registered in the data row of spaced apart stations (hundred and thousands kilometers). The processing of the experimental information was made using the designed and patented by our research group method of the detection of energetically non-dominant components of time series. The impact of the tidal nature of the Moon was detected on the first stage of investigations in the electromagnetic field of atmosphere boundary layer. Years of research have shown that all experimentally obtained time series of the electric and geomagnetic fields have components at frequencies infra low frequency gravitational-wave emission of a number of binary star systems: J 0700+6418, J 1012+5307, J 1537+1155, J 1959+2048, J 2130+1210, J 1915+160 with effective values 0,28 – 0,6 V/m, and also on the combination frequencies these frequencies and the frequencies of the moon tides, which cannot be found using the classical methods of spectral analysis because of non-coherent nature, but can be identified using the method of eigenvectors [9].

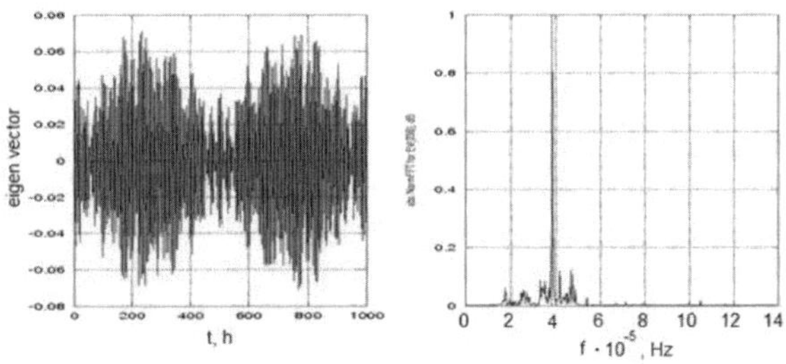

Fig. 1. Eigen vectors and their amplitude spectrums on the frequency of the gravitational wave emission of the binary star system J 1012+5307 for apart stations of the monitoring Dusheti.

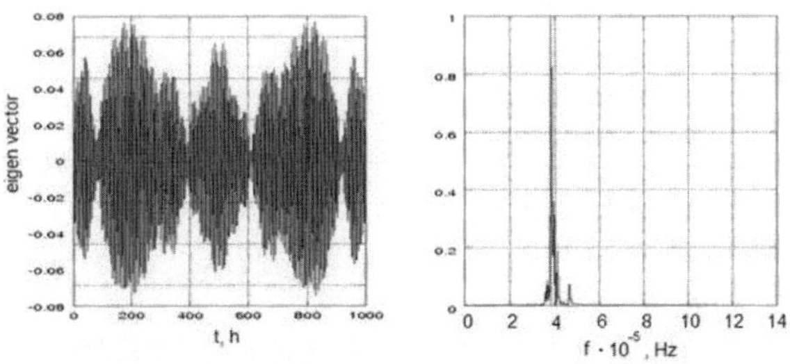

Fig. 2. Eigen vectors and their amplitude spectrums on the frequency of the gravitational wave emission of the binary star system J 1012+5307 for apart stations of the monitoring Voeikovo.

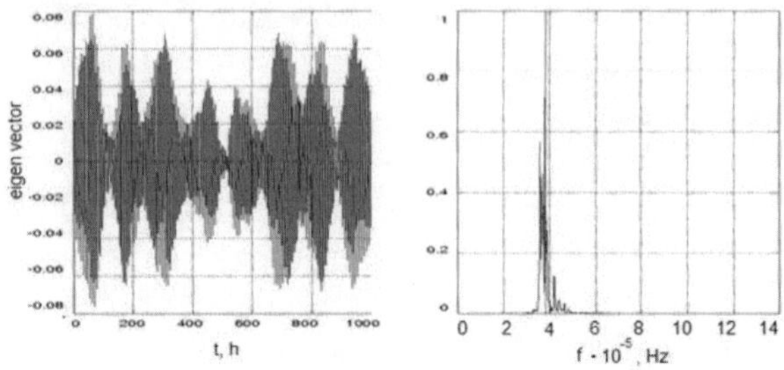

Fig. 3. Eigen vectors and their amplitude spectrums on the frequency of the gravitational wave emission of the binary star system J 1012+5307 for apart stations of the monitoring Verkhnee Dubrovo.

It is shown that all reviewed time series have non-coherent components on the axion frequency 5e-6 Hz with effective values of 0,71 – 1,36 V/m, and also at combination frequencies of this frequency with the frequency of the moon tides.

The investigations are supported by State order 2014/13,2871 and RFFI grant 14-07-97510\14.

References

1. D. Bocaletti, V. De Sabatta, P. Fortini and C. Gualdi, *Nuovo Cim.* **B70**(N2), 129 (1970).
2. V. L. Ginzburg, V. N. Cytovich, *Transition radiation and transition scattering* (Moscow, Nauka, 1984)
3. A. B. Balakin, D. V. Vahryshev, *Izvestiya VUZov (Ser. Fizika)* **9**, 28 (1993).
4. A. B. Balakin, J. P. S. Lemos, *Classical and Quantum Gravity*, **18**, 941 (2001).
5. V. K. Milukov, V. N. Rudenko, in *Results of science and technics*, Vol. 41 (Moscow, VINITI, 1991), pp. 147–193.
6. A. V. Gusev, V. N. Rudenko, in *Experimental tests of gravity* (Moscow, MSU Publ., 1989), pp. 94–101.
7. A. B. Balakin, Ni. Wei-Tou, *Classical and Quantum Gravity* **31**, 105002 (2014).
8. A. B. Balakin, L. V. Grunskaya, *Reports on Mathematical Physics*, **71**(1), 45 (2013).
9. L. V. Grunskaya, D. V. Isakevich, V. V. Isakevich, *Space, time and fundamental interactions* **2**, 54 (2014).

Viscosity noise in mirrors of gravitational wave detectors

N. M. Kondratyev* and M. L. Gorodetsky

Faculty of Physics, M. V. Lomonosov Moscow State University,
Moscow, 19991, Russia
**E-mail: kondratiev@physics.msu.ru*

The Fluctuation-Dissipation theorem is used to point out a new source of noise – viscosity of glassy state of matter. We show that fused silica should be described with Standard Linear Solid model of viscoelastisity, estimate model parameters and present a method to calculate this type of noise.

Keywords: Gravitational wave detectors; mirrors; noise; viscosity; glass.

For modern high precision measurements any source of noise can be critical. In particular, the LIGO project[1] have to account many fundamental sources of fluctuations. Thermal noise in coatings, substrates and suspensions of the interferometer's mirrors results in fluctuations of their surfaces which add a phase noise to the signal. A lot of other processes can degrade the sensitivity.[2]

Fluctuation-dissipation theorem[3] gives the spectral density of fluctuations in the mirror in the form

$$S(\omega) = \frac{4k_BT}{\omega}\Im[\alpha_s + \alpha_j^c + \alpha_j^s], \tag{1}$$

where ω is the frequency, k_B is the Boltzmann's constant and T is the temperature, α_s is dynamic permittivity of the substrate, α_j^c and α_j^s are coating and coating induced substrate dynamic permittivities.[4] The dissipation is then introduced in the form of the loss angle by substituting the Young's modulus $Y \to Y(1 - i\phi)$. Several models seek to describe this loss angle theoretically and phenomenologically.

The most obvious source of dissipation in liquids is viscosity — internal friction between layers of matter. However this property also exists in solid materials in glass state — amorphous substance that couldn't become a crystal due to high viscosity. Fused silica — the material used in substrate, coatings and suspensions of LIGO mirrors — is such a material. Having very high viscosity (ranged from 10^{17} Pa·s to 10^{40} Pa·s in literature) it can be the source of additional noise, and influence performance. Viscosity can be introduced into equations as the loss angle, providing us with a new type of noise.

We use the Standard Linear Solid Model (SLS) of viscoelastic materials[5] to describe viscosity. It consists of a first (slow or static) spring E_1 with shear and bulk moduli G and K, second (fast or dynamic) spring E_2 with parameters G' and K', and a damping component with shear and bulk viscosity η and ζ. Unfortunately, the parameters of the model G', K', G, K as viscous parameters η, ζ at room temperatures are difficult to measure and find in literature.

To estimate the SLS parameters of fused silica we used recent results, where a flow of fused silica plates during decades under their own weight was studied.[6,7] Using these data and the speed of transversal sound waves we estimate $G = 2.97$ GPa, $G' = 28.1$ GPa, $\eta = 12 \times 10^{17}$ Pa·s. Unfortunately bulk parameters still remain undetermined.

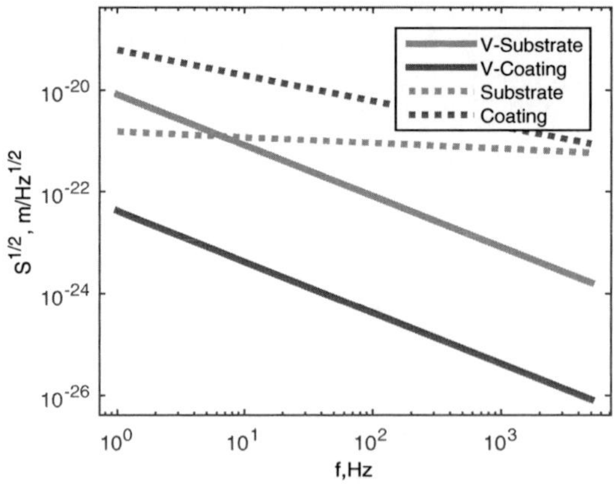

Fig. 1. Left: Brownian noises from substrate (red) and coating (blue) for LIGO parameters together with viscosity noises (solid lines).

The numerical estimates of silica part together with the standard Brownian noise are shown on fig. 1 (left). The coating part appears to be by several orders lower than substrate part due to small thickness and can be neglected until tantala has much smaller viscosity.

In conclusion the silica viscosity can be the source of additional noise exceeding the substrate Brownian noise at frequencies bellow 11 Hz. It can only exceed the coating Brownian noise if the viscosity was 10^4 times lower. The volume-viscosity part of the noise is, however, unknown due to absence of appropriate material parameters data.

The other source of uncertainty is the coating part as the viscosity value for thin films and bulk mater can differ (remember three order difference of the loss angle). Note also that viscosity is situated in the denominator, meaning that the less the viscosity, the higher is the noise. Nevertheless, the coating Brownian noise still remains prevailing as the limiting factor for LIGO interferometers.

References

1. The LIGO Scientific Collaboration, J. Aasi and B. P. A. et.al., Advanced ligo, *Classical and Quantum Gravity* **32**, p. 074001 (2015).
2. M. L. Gorodetsky, Thermal noises and noise compensation in high-reflection multilayer coating, *Physics Letters A* **372**, 6813 (2008).
3. L. Landau and E. Lifshitz, *Course of theroetical physics Volume 5. Statistical Physics* (Elsevier Science, 2013).
4. N. M. Kondratiev, A. G. Gurkovsky and M. L. Gorodetsky, Thermal noise and coating optimization in multilayer dielectric mirrors, *Phys. Rev. D* **84**, p. 022001 (Jul 2011).
5. Y. de Haan and G. Sluimer, Standard linear solid model for dynamic and time dependent behaviour of builidng materials, *HERON* **46**, 49 (2001).
6. M. Vannoni, A. Sordini and G. Molesini, Relaxation time and viscosity of fused silica glass at room temperature, *The European Physical Journal E* **34**, 1 (2011).
7. M. Vannoni, A. Sordini and G. Molesini, Long-term deformation at room temperature observed infused silica, *Opt. Express* **18**, 5114 (Mar 2010).

Single mode cavities for parametric oscillatory instability suppression in laser gravitational wave detectors

Mikhail V. Poplavskiy

Faculty of Physics, Lomonosov Moscow State University, Moscow 119991, Russia

E-mail: mv.poplavskii@physics.msu.ru

Parametric oscillatory instability will be a serious problem restricting the power circulating in laser gravitation wave detectors (aLIGO) [1]. For parametric oscillatory instability suppression we suggest to use practically "single mode" Fabry-Perot (FP) cavity with little changed mirror's specular surface.

1. State of the Problem

In laser gravitational wave detector elastic oscillations in FP cavity mirrors can interact with optical ones. High quality factor of these of two optical and one elastic modes will increase the effectiveness of the interaction between them and may give appearance of parametric oscillatory instability.[1]

This undesirable effect may create a specific upper limit for the power of circulating energy inside cavity. Level of circulating power in arms planning in aLIGO is 0.8 MWt, but recently this effect was observed in full scale aLIGO interferometer at 50 kWt.[2]

For suppression parametric oscillatory instability we suggest to use practically single mode FP cavity in which main mode has diffraction loss about 1 ppm whereas high order modes has losses about thousands ppm. Note, single mode cavities are known in literature[3] however main mode loss in them is higher ($\approx 100 \div 1000$ ppm).

We assume aLIGO parameters (arm length $L = 4$ km, wave length $\lambda = 1064$ nm, mirror radius $r_m = 17$ cm) and put mirrors with identical shapes which are described by the same functions[4]

$$A(r) = x_0 e^{-\eta(1+\alpha\eta+\beta\eta^2)}, \quad \text{where} \quad \eta = \frac{r^2}{2R_c x_0} \tag{1}$$

which turning into spherical profile with curvature radius R_c at $\alpha = \beta = 0$ and $x_0 \to \infty$.

2. Results and Conclusions

Using Hankel transform[5] we carried out numerical calculations of different modes losses under variation of the profile parameters. Several suitable parameters combinations presented in Table 1 show that:

(1) Round trip losses ($\mathcal{L}_{00}$) of main mode are similar ($\approx$ 1 ppm) as ones for spherical mirrors FP cavity with aLIGO parameters (first string in Table 1);
(2) Losses of higher order modes ($\mathcal{L}_{nm}$, $n+m>0$, where n, m are angular and radial indices) significantly increased by several orders.

Table 1. Comparison of different profile parameters and corresponding mode losses $\mathcal{L}_{nm}$, n is angular index. First string ($x_0 = \infty$) corresponds to spherical mirrors.

x_0, μm	R_c, m	α	β	$\mathcal{L}_{00}$	$\mathcal{L}_{01}$	$\mathcal{L}_{02}$	$\mathcal{L}_{10}$	$\mathcal{L}_{11}$	$\mathcal{L}_{20}$
∞	2076	0	0	0.45	180	6600	10	2100	100
3.39	2014	0.15	0.3	2.3	43 000	41 000	960	20 000	18 000
5.08	2076	0.2455	0	0.18	31 000	37 000	87	13 000	27 000
5.08	2014	0.15	0	1.1	41 000	14 000	860	19 000	11 000

We see that losses of higher order modes of single mode cavity considerably exceeds the corresponding losses for spherical mirrors cavity. It allows to suppress undesirable parametric oscillatory instability effect.

The requirements for non-spherical mirrors manufacture seem not a difficult task – parameters have to hold with accuracy about 3%.

The problem lies in fact that the radius of laser spot ($\sim$ 5 cm) is less then in aLIGO arms (6 cm) and this will enhance the density of thermoelastic noise spectrum, however, not dramatically.

We plan to investigate tolerance to tilt of mirrors in single mode cavity and to find mirror's profile corresponding to laser spot larger than 5 cm.

We acknowledge Sergey Vyatchanin, Andrey Matsko and Hiro Yamamoto for checking calculations, valuable advices and useful discussion.

References

1. V.B. Braginsky, S.E. Strigin and S.P. Vyatchanin, *Phys. Lett. A* **287**, 331 (2001); *Phys. Lett. A* **305**, 111 (2002).
2. M. Evans, S. Gras, et al, *Phys. Rev. Lett.* **114**, 161102 (2015).
3. M. Kuznetsov, M. Stern and J. Coppeta, *Optics express* **13**, 171 (2005).
4. F. Ferdous, A. Demchenko, S. Vyatchanin, A. Matsko and L. Maleki, *Phys. Rev. A* **90**, 033826 (2014).
5. J.Y. Vinet and P. Hello, *Journal of Modern Optics*, **40:10**, 1981-1993 (1993).

Fundamentals of the TianQin mission

Jianwei Mei, Chenggang Shao and Yan Wang

MoE Key Laboratory of Fundamental Quantities Measurement, School of Physics, Huazhong University of Science and Technology, 1037 Luoyu Rd., Wuhan 430074, P.R. China

This talk adds more detail to the proposed TianQin mission. Due to restriction of space, we only present one key calculation in this note.

Keywords: TianQin; spaceborne gravitational wave detector.

1. Introduction

The main purpose here is to show how the transfer function used for TianQin in Ref. 1 is calculated. To save space, we use the same notation as[2,3] to avoid explaining them here, but newly defined quantities are given explicitly.

2. Gravitational Waves from a Binary System

The gravitational wave from a binary system with circular orbit is[2]

$$\begin{aligned} h_+ &= h_0 a_+ e^{2iwt} + c.c.\,, \quad a_+ = -\Big[\frac{1}{2}(1+c_\delta^2)c_{2\gamma} + ic_\delta s_{2\gamma}\Big]e^{2i\beta}\,, \\ h_\times &= = h_0 a_\times e^{2iwt} + c.c.\,, \quad a_\times = -\Big[\frac{1}{2}(1+c_\delta^2)s_{2\gamma} - ic_\delta c_{2\gamma}\Big]e^{2i\beta}\,, \end{aligned} \tag{1}$$

where $h_0 = \frac{2G_N^2 M_1 M_2}{r_0 a}$, $c_\delta \equiv \cos\delta$, $s_\delta \equiv \sin\delta$, and so on. For the galactic white dwarf binary RX J0806.3+1527, one has

$$h_0 \approx 6.4\times 10^{-23}\Big(\frac{M_1}{0.55M_\odot}\Big)\Big(\frac{M_2}{0.27M_\odot}\Big)\Big(\frac{5\ \text{kpc}}{r_0}\Big)\Big(\frac{66000\ \text{km}}{a}\Big). \tag{2}$$

3. The Transfer Function

Instead of (21) in Ref. 3, we find considering both polarizations the gravitational wave from a binary system,

$$\langle\Delta^2\rangle = \frac{1}{T}\int_0^{+\infty}|\Delta(t)|^2 dt = \int_0^{+\infty}\frac{dtdwdw'}{16\pi^2\tau^2}\mathcal{H}(w)\mathcal{H}^*(w')\frac{e^{i(w-w')t}}{ww'}$$

$$
= \int_0^{+\infty} \frac{dw}{2\pi} \frac{h_1^2 T_1(w) + h_2^2 T_2(w) - 2h_1 h_2 T_3(w)}{(w\tau)^2},
$$

$$
\begin{aligned}
\mathcal{H} = h_1 &\left[(1-\mu_1) + 2\mu_1 e^{-iw\tau(1+\mu_1)} - (1+\mu_1)e^{-2iw\tau}\right] \\
-h_2 &\left[(1-\mu_2) + 2\mu_2 e^{-iw\tau(1+\mu_2)} - (1+\mu_2)e^{-2iw\tau}\right] \qquad (3)
\end{aligned}
$$

where $h_i = h_+(w)\cos 2\tilde{\phi}_i + h_\times(w)\sin 2\tilde{\phi}_i$.

Given (1), $\langle \Delta^2 \rangle = \frac{1}{2\pi}\int_0^{+\infty} h_0^2 R_0(w) dw$, where

$$
\begin{aligned}
R_0(w) = &\left|a_+ \cos 2\tilde{\phi}_1 + a_\times \sin 2\tilde{\phi}_1\right|^2 \frac{T_1(w)}{(w\tau)^2} + \left|a_+ \cos 2\tilde{\phi}_2 + a_\times \sin 2\tilde{\phi}_2\right|^2 \frac{T_2(w)}{(w\tau)^2} \\
&-2\Re e\left[\left(a_+ \cos 2\tilde{\phi}_1 + a_\times \sin 2\tilde{\phi}_1\right)\right. \\
&\times\left.\left(a_+ \cos 2\tilde{\phi}_2 + a_\times \sin 2\tilde{\phi}_2\right)\right] \frac{T_3(w)}{(w\tau)^2}. \qquad (4)
\end{aligned}
$$

In the low frequency limit $w\tau \to 0$,

$$
\begin{aligned}
R_0 \approx &\left|\left[\frac{1}{2}(1+c_\delta^2)c_{2(\gamma-\tilde{\phi}_1)} + ic_\delta s_{2(\gamma-\tilde{\phi}_1)}\right](1-\mu_1^2)\right. \\
&\left.-\left[\frac{1}{2}(1+c_\delta^2)c_{2(\gamma-\tilde{\phi}_2)} + ic_\delta s_{2(\gamma-\tilde{\phi}_2)}\right](1-\mu_2^2)\right|^2. \qquad (5)
\end{aligned}
$$

For TianQin, the detector plane is oriented toward the incoming radiation, so $\mu_1 = \mu_2 = 0$. With $\tilde{\phi}_2 = \tilde{\phi}_1 + \pi/3$, one has

$$
\begin{aligned}
R_0 \approx &\left|\frac{1}{2}(1+c_\delta^2)[c_{2(\gamma-\tilde{\phi}_1)} - c_{2(\gamma-\tilde{\phi}_1-\pi/3)}]\right. \\
&\left.+ic_\delta[s_{2(\gamma-\tilde{\phi}_1)} - s_{2(\gamma-\tilde{\phi}_1-\pi/3)}]\right|^2. \qquad (6)
\end{aligned}
$$

The parameter γ is difficult to measure for distant astrophysical objects, and $\tilde{\phi}_1$ varies with the orbital motion of satellites. Averaging over γ and $\tilde{\phi}_1$, we find $R_0(w) \approx 3 - 2.6\sin^2\delta$.

For the source RX J0806.3+1527, $\delta \approx 38^\circ$, and so $R_0 \approx 2.0$.

References

1. J. Luo, *et al*, "TianQin: a space-borne gravitational wave detector", Submitted to Class. Quant. Grav.
2. E. Poisson and C. M. Will, *Gravity: Newtonian, Post-Newtonian, Relativistic*, Cambridge University Press (29 May 2014).
3. S. L. Larson, W. A. Hiscock and R. W. Hellings, Phys. Rev. D **62**, 062001 (2000) [gr-qc/9909080].

Measurement of the post Newtonian parameter gamma by the cluster of spacecrafts in the gravity field of the Earth: The mission concept

V.K. Milyukov*, M.V. Sazhin, V.E. Zharov and A.P. Mironov

Sternberg Astronomical Institute, Lomonosov Moscow University,
Moscow, 119234, Russia
**E-mail: vmilyukov@yandex.ru*

Jun Luo

Sun Yat-sen University,
Guangzhou, 510275, China
E-mail: junluo@sysu.edu.cn

The paper describes the project of the space mission on measurement of PPN parameter γ in gravity field of the Earth.

Keywords: PPN parameters; space mission; cluster of spacecrafts.

1. Introduction

Recent development of space-qualified technologies opens new prospects for ultra-precise test of General Relativity. For many experiments in the area of fundamental gravitation, access to space becomes inevitable. The accuracy, with which GR is currently confirmed, is fractions of percent: 2.3×10^5, provided by microwave tracking of the Cassini spacecraft. However, in spite of the remarkable success of GR in the weak-field approximation, there are many reasons to consider alternative relativistic theories of gravity that predict the existence of effects other than GR, thus motivating new fundamental gravitational experiments. Numerically the theories of gravity can be characterized by means of the parameterized post-Newtonian formalism (PPN). In this connection, the experimental measurements of PPN of parameters play a special role.

2. Mission Concept

The project, which assume the lunch of cluster of the spacecrafts, intended to measure the Eddington PPN parameter γ with an accuracy of a part in 10^9. The mission uses four spacecrafts that are placed in high-altitude ($\sim 10^5$ km) circular orbits around the Earth. Each spacecraft is equipped with three sets of identical laser ranging transceivers that are used to measure distances between the spacecraft to high accuracy (~ 0.1 nm). The laser measurements form a trapezoid with diagonal elements such that one of the legs in the trapezoid skims close to the Earth, picking up an additional gravitational delay. In Euclidian geometry, this system is redundant; by measuring only five of the six distances one can compute the sixth one. However all six inter-spacecraft distances within the trapezoid, will be precisely measured and monitored in the project. Comparison of measured and computed values of the sixth distance gives a departure from Euclidian geometry. In the vicinity of the Earth, such a departure is primarily due to the curvature of the relativistic space-time around the Earth.[1]

3. Satellite-to-Satellite Laser Tracking

The key element of the mission is the inter-satellite laser interferometer link. The laser metrology system consists of a 1 W Nd:YAG laser, a heterodyne interferometer optical bench, a frequency stabilization bench, a phase locking optical bench, a phase-meter and three $10 - 20$ cm telescopes. Along each of the six arms, one heterodyne transponder-type laser interferometer and two inertial sensing systems will be used to measure the displacement between pairs of caged test masses, one in each spacecraft. The inertial sensing system measures the distance between the test mass and the interferometer optical bench on each spacecraft, while the displacement between the two interferometer optical benches is measured by the transponder-type interferometer. The metrology system phase-locks the outgoing laser with the incoming laser, so that the outgoing beam is synchronized with incoming beam that provides the continuous multiple-pass measurement of the inter-satellite distance.

This work is supported under Grant of RFBR No.15-52-53070.

Reference

1. S.G. Turyshev, M. Shao and A. Girerd, A search for new physics with the BEACON mission, *Int. J. Mod. Phys. D* **18**, 1025 (2009).

Section 7

Relativistic Astrophysics

Gravitational lensing properties of cosmological black holes

A. F. Zakharov*

Institute of Theoretical and Experimental Physics, Moscow, 117218, Russia
**E-mail: zakharov@itep.ru*
Bogoliubov Laboratory of Theoretical Physics, JINR,
Dubna, Moscow region, 141980, Russia

S. Capozziello and C. Stornaiolo

Universitá Federico II di Napoli, 80126 Naples, Italy
Istituto Nazionale di Fisica Nucleare, Sezione di Napoli, 80126 Naples, Italy

If really such objects like cosmological black holes exist they may be studied with a standard technique like strong and weak gravitational lensing. Cosmological voids can be explained as the result the collapse of large perturbations into black hole with masses of the order of $10^{14}M_{\odot}$ and the expansion of the universe. The resulting image of the universe is that it is more homogeneous than expected from present observations. In this paper we discuss some lensing properties related to the cosmological black holes (CBHs), namely we consider differences in gravitational lensing for point like mass and extended mass distributions. We consider the singular isothermal sphere model as a toy (illustrative) model for an extended distribution of dark matter and a slightly more complicated isothermal sphere with a core.

Keywords: Black holes; cosmology; cosmological black holes; gravitational lensing; General Relativity; black hole physics.

The Universe at scales less than 100 Mpc shows a spongy structure where together with regions with structure of galaxies there are void regions where there are underdense distribution of galaxies or where they are totally absent. The analysis of their physical properties has been done using Void finders, i.e. by using codes which defined their shapes and their actual extensions. The most recent results assign and average radius of 25 Mpc and an approximately spherical shape. There is a galaxy underdensity in the borders of the voids, but the central part is apparently without any visible matter, i.e. optically void. A simple model to explain contemporarily the presence of this dark matter and void formation is to admit that there have been large perturbations, eventually produced by a long inflationary period, that collapsed into black holes (CBH) and that voids have been created by the cosmological expansion around these CBHs. To describe the resulting model, in previous papers, the Einstein-Straus Swiss cheese model has been used. In this case dark matter can totally be identified with the CBHs, i.e. "ordinary" black holes with huge dimensions. In order to compensate the void region these black holes must have a mass of the order of $10^{14}M_{\odot}$ (the mass value is consistent with density fluctuations).[1,3]

In a perfect Swiss Cheese model, a CBH will not have any direct interaction with the other structures except for contributing to the energy density of the universe and participating to the collective cosmological expansion. To explain some deviations from this scenario which are in the actual observations, one has to consider the presence of peculiar velocities for galaxies and some perturbation of this scenario. Otherwise the CBHs can be detected only through their lensing properties.[2] If a cosmological black hole exists in a void, there are two stages in studying it as a strong gravitational lens. The first stage is to find gravitational lens systems in the void region. The second one is to confirm that it is a point-like gravitational lens (i.e. a Schwarzschild lens), and but not an extended distribution of dark matter as it has been analyzed for extended microlenses.[4] Gravitational lensing properties are different for extended and point mass distributions are different since gravitational potentials are different for photons propagating inside a region with an extended mass distribution and in empty region near a black hole with the same impact parameter.

If we follow parameters suggested earlier[2] we adopt CBH mass $M = 10^{14}M_{\odot}$ and the distance to the void $D_{void} = 50Mpc$ (or we assume $\Omega_{CBH} = 0.1$ and $R_{void} \sim 20Mpc$, then Einstein – Chwolson ring is around is a few angular minutes, while shadow diameter is around a few 10^{-2} angular seconds and we do not discuss an opportunity to find the small dark shadow inside a huge void.

References

1. S. Capozziello, M. Funaro, C. Stornaiolo, *A & A* **420**, 847 (2004).
2. M. Serpico, R. D'Abrusco, G. Longo, C. Stornaiolo, *Gen. Rel. Grav.* **39**, 1551 (2007).
3. C. Stornaiolo, *Gen. Rel. Grav.* **34**, 2089 (2002).
4. A.F. Zakharov, *Phys. Lett. A* **250**, 67 (1998).

Experiments on invariance of Planck's law and a new view of anisotropy of cosmic microwave background

Leonardo Hsu

Department of Postsecondary Teaching and Learning, University of Minnesota, Minneapolis, Minnesota 55414, USA.
E-mail: lhsu@umn.edu

Jong-Ping Hsu

Department of Physics, University of Massachusetts Dartmouth, North Dartmouth, MA 02747, USA. E-mail: jhsu@umassd.edu

We discuss an invariant Planck's law for inertial and rotating (Earth) frame. This invariant law suggests a new source for the dipole anisotropy of CMB.

The 'non-Lorentz invariant' Planck distribution is used by researchers analyzing the cosmic microwave background (CMB) using satellite data from COBE, WMAP and Planck mission.[1] One assumes a 'fundamental frame' F_f at rest relative to the CMB, in which CMB is isotropic[2] and described by the Planck law involving the usual distribution $P(\omega, T) = 1/[exp(\omega/k_B T_{CMB}) - 1], (\hbar = c = 1), T_{CMB} = 2.725K$. One also assumes that the Planck distribution remains 'the same' in the solar frame F', which moves with V in the x-direction. That is, $P(\omega', T') = 1/[exp(\omega'/k_B T') - 1]$, $\omega' = k'_0$. This Planck distribution $P(\omega', T')$ is not explicitly Lorentz invariant and the temperature $T' = T'(\theta)$ in F' has to be angle-velocity-dependent, $T'(\theta) = T\sqrt{1 - V^2}/(1 + V cos\theta')$, so that there is also a Planck spectrum in F' frame, where θ is the angle between V and the line of sight.[2,3] Such a non-invariant $P(\omega', T')$ and non-covariant T' lead to the interpretation that the dipole anisotropy arises from the motion of the solar frame[2] relative to the CMB with the velocity $V \approx 369.0 \pm 0.9 km/s$.

Let us consider a generalized invariant Planck Law for inertial and non-inertial frames to avoid shortcomings of the lack of Lorentz invariance. In general, the invariant Planck law is expressed in terms of the differential[2] of the energy momentum tensor of photons. For simplicity, let us consider first the invariant distribution for CMB radiation in inertial frames,

$$P(k_\mu U^\mu, T_{inv}) = \frac{1}{[exp(k_\mu U^\mu)/(k_B T_{inv}) - 1]}, \quad k_\mu U^\mu = \eta_{\mu\nu} k^\mu U^\nu, \tag{1}$$

where $\eta_{\mu\nu} = (1,-1,-1,-1)$ and $U^\mu = dx^\mu/ds$ is the 4-velocity of the blackbody, as measured in an inertial frame F. Clearly, when the blackbody is at rest in F, we have $U^\mu = (1,0,0,0), k_\mu U^\mu = k_0 = \omega$, and the invariant distribution (1) reduces to the usual form $P(\omega, T) = 1/[exp(\omega/k_B T) - 1]$.

The invariant blackbody distribution (1) implies that if the CMB radiation is isotropic in F, then it will be isotropic in every other inertial frame as well. Therefore, the observed dipole anisotropy of CMB cannot be due to the motion of the solar system, contrary to the usual interpretation.

The invariant distribution (1) can be generalized to non-inertial frames $F'(w', x', y', z')$ such as a rotating frame with more complicated metric tensors $P'_{\mu\nu}$, and $\eta_{\mu\nu} k^\mu U^\nu$ is replaced by $P'_{\mu\nu} k'^\mu U'^\nu$. This can be done using a limiting Lorentz-Poincare invariance of physical laws.[4] We just give the results here. To the lowest order, the metric tensors $P'_{\mu\nu}(x')$ can be calculated,[3,4] $P'_{\mu\nu} k'^\mu U'^\nu \approx k'_0[1 + \beta'_e\ sin(\Omega' w')] = \eta_{\mu\nu} k^\mu U^\nu = k_0$ for the rotating F' frame with an orbital velocity β'_e.

Let us consider a solar frame F_s, in addition to the CMB rest frame F and the rotating frame F' around the sun. The results are: (A) We can have a 'fixed blackbody' with the temperature $T_{fb} = 2,725K$ in an earth lab and measure its radiation energy k_0 and spectrum, which can be identified with CMB measured in F. (B) There is a departure of the measured energy k_{s0} in F_s from the isotropic k_0 due to the Doppler shift, $\Delta k'_0 = k'_0 - k_0 \approx k_0 \beta_s cos\theta_s$, which enables us to estimate the velocity of F_s relative to F by using satellite data. (C) The prediction $P(k_\mu U^\mu, T_{inv}) = P(k_{s\mu} U^\mu_s, T_{inv}) = P(k'_\mu U'^\mu, T_{inv})$ should be tested by satellite data.

If (C) is not confirmed, then the Planck distribution violates the Lorentz invariance, at least in the cosmic scale. If (C) is supported by experiments, then the dipole anisotropy of CMB could indicate the existence of the other half of the universe made of anti-matter, which manifests itself as a fuzzy anti-galaxies blackbody located far beyond our observable matter-galaxies of the universe.[4] (Cf. discussions by J.P. Hsu in this proceedings.)

References

1. D. Scott and G. F. Smoot, "27. COSMIC MICROWAVE BACKGROUND", section 27.2.2, in Chinese Phys. C 38 (2014) 09000.
2. S. Weinberg, *Gravitation and Cosmology* (John Wiley, 1972) p. 522.
3. L. Hsu and J. P. Hsu, Eur. Phys. J. Plus, 129 (2014) 16.
4. J. P. Hsu and L. Hsu, *Space-Time Sym and Quantum Yang-Mills Gravity* pp. 23-27, 61-78. Hsu and Cottrell, Chinese Phys. C39 105101.

Electromagnetic waves propagation nearby rotating gravitating astrophysical object with atmosphere

V. O. Gladyshev*, A. A. Tereshin,
I. V. Fomin, M. B. Chelnokov, V. L. Kauts,
T. M. Gladysheva and D. D. Bazleva

Department of Physics,
Bauman Moscow State Technical University,
Moscow, 105005, Russian Federation
**E-mail: dekan-fn@mail.ru*
http://www.bmstu.ru

The aim of the article to explore the effects of gravitational lensing and attraction of electromagnetic radiation in the description of the propagation of radiation nearby the atmospheres of rotating astrophysical objects.

Keywords: Deflection of light; gravitational lensing; rotating atmosphere.

1. The Gravitational Deflection of Light

The geometry of space–time nearby rotating gravitating object described by the Kerr metric, which we write in the following form[1]:

$$ds^2 = \left(1 - \frac{R_g r}{\rho^2}\right) c^2 dt^2 - \frac{\rho^2}{r^2 - R_g r + r_g^2} dr^2 - \rho^2 d\theta^2$$
$$- \left[r^2 + r_g^2 + \frac{R_g r}{\rho^2} r_g^2\right] \sin^2\theta d\varphi^2 + \frac{RGr}{\rho^2} r_g \sin^2\theta d\varphi dt, \qquad (1)$$

where $R_G = 2GM/c$, $r_g = Jv/MRc$, $\rho^2 = r^2 + r_g^2 \cos^2\theta$, J is the moment of inertia, G is the gravitational constant, c is the speed of light, R – radius, v – equatorial velocity, M is the mass of gravitating body.

Lagrangian $\mathcal{L}$ can be written as $\mathcal{L} = g_{\mu\nu}\dot{x}^\mu \dot{x}^\nu$ where the dot indicates derivative with respect to some affine parameter.[1] The Lagrangian of photon moving in the equatorial plane has the form

$$\mathcal{L} = (1-\alpha)\dot{\tau}^2 - \frac{\dot{r}^2}{1-\alpha+\beta^2} - r^2\left(1-\alpha+\alpha\beta^2\right)\dot{\varphi}^2 + 2\alpha\beta r\dot{\tau}^2\dot{\varphi}, \qquad (2)$$

where $\alpha = R_G/r$, $\beta = r_G/r$, $\tau = ct$.

Solving the Euler-Lagrange equation and neglecting terms α^2, $\alpha\beta$ and $\alpha\beta^2$ we get the angle of light deflection

$$\vartheta = \frac{4GM}{Rc^2} + \frac{\pi}{2}\left[\pm 1 \mp \frac{\sqrt{1-\frac{24}{25}\left(\frac{v}{c}\right)^2}}{\left(1-\frac{12}{25}\left(\frac{v}{c}\right)^2\right)^{3/2}}\right] \tag{3}$$

For the Sun the angle of deflection of light has the value $\vartheta = 4.8 \times 10^{-4\circ} \mp 2.8 \times 10^{-5\circ}$.

2. The Angular Deflection of the Electromagnetic Wave in a Rotating Disk Stellar Atmosphere

We consider the model of the light propagation from a distant source of radiation through the atmosphere of the Sun, following Ref. 2. The angle of deflection is calculated as:

$$d\vartheta = arctg\left(\frac{\sin\vartheta_0}{\sqrt{\cos^2\vartheta_0 + \kappa_2} - \kappa_2\beta_{2z}}\right) - arctg\left(\frac{\sin\vartheta_0}{\sqrt{\cos^2\vartheta_0 + \kappa_2}}\right), \tag{4}$$

where $\kappa_2 = n_2^2 - 1$, $\beta_{2z} = v_{2z}/c$, n_2^2 is the refractive index of the atmosphere and v_{2z} is the normal projection of speed.

Calculations according to the increment of the angle of escape of radiation $d\vartheta$ from the solar atmosphere from the angle of falling on the outer layer of the atmosphere ϑ_0 show that for large ϑ_0 light deflection tends to $d\vartheta = 10^{-5\circ}$.

The angular deviation of light associated with the gravitational field of a rotating object and with the rotation of the atmosphere have different signs.

By comparing the values $\vartheta = 4.8 \times 10^{-4\circ} \mp 2.8 \times 10^{-5\circ}$ and $d\vartheta = 10^{-5\circ}$ obtained for the Sun, we conclude that the rotation of the atmosphere contributes to the deflection of the light at the level 10^{-1} of the value of the light deflection in the gravitational field of the Sun.

References

1. S. Chandrasekhar, *The Mathematical Theory of Black holes* (Oxford University Press, 1983).
2. V. O. Gladyshev, V. L. Kauts, P. S. Tiunov, M. B. Chelmokov, The effect of rotation of the Earth's atmosphere and the Sun on the propagation of electromagnetic radiation, *Vestnik MSTU Ser. "Natural Sciences", Special issue*, p. 152 (2012).

On the necessity of laboratory estimation of the parameters characterizing the Universe dark matter

V. M. Koryukin

Department of Physics and Mathematics, Mari State University, Lenin sq., 1, Yoshkar–Ola, 424000, Russia
E-mail: vmkoryukin@gmail.com

A. V. Koryukin

Institute of Physics, Kazan Federal University, Kremlyovskaya St., 16, Kazan, 420008, Russia
E-mail: akoryukin@gmail.com

We propose to carry out experiments on observation of temperature waves, considering a weighty evidences on the presence of a Bose condensate in the Universe.

Keywords: The dark matter; neutrinos; temperature waves.

1. The Universe Neutrino Background

We shall divide the Universe matter into two subsystems (a slow one and a rapid one), so that the baryon matter (a rapid subsystem) behavior must be defined by the non-baryon matter (a slow subsystem) behavior. We consider that the Universe baryon matter must play role of Brownian particles whose behavior allows to study "black matter" properties. Just rapid subsystem's particles allow us to do predictions and conclusions on states and properties of particles of Universe matter's slow subsystem (a dark matter), using methods of inverse problems. The presence of dark matter was detected for the first time in the research of the galaxies motion which must be caused the gravitation interaction and it is just that effect which allows to detect the presence of the Bose condensate. Since in our approach it is laid stress on known particles, then we have the weighable evidence of the presence of Cooper pairs of Universe background neutrinos. We can use the similar mechanism to Bardeen-Cooper-Schrieffer theory of superconductivity for the explanation of formation of the Cooper neutrino pairs the more so that a dark matter is detected only in the vicinity of the Universe baryon matter. At the same time the reverse process of the Cooper neutrino pairs decay must be observed. It must cause increasing of the background neutrino density in the Fermi liquid and it can be interpreted as the expansion of the given volume of the Universe (the Hubble flux of "Local volume" ($< 10Mpc$)). We note that the presence of the Universe neutrino background with the finite Fermi energy is the catalytic agent of stochastic processes, but the large value of this energy causes determinancy of physical processes.

In first, it is necessary to give a definition of fast subsystem's particles as particles participating in strong and electroweak interactions, at the time slow subsystem's particles do not have such opportunity. When particles go from the ground state to the excited one they acquire all properties of fast subsystem fermion – the color charge and (or) the electric charge. We will emphasize the contrast of the particles in the ground state of the particle in the excited state, developing the Dirac hypothesis about the existence the universal "sea" of electrons in the ground state. Generalizing this hypothesis on fermions of arbitrary flavors in the ground state at a temperature close to absolute zero, it will be possible to use the Landau theory of the Fermi liquid, considering the remaining particles as quasi-particles, attributing them to a condition called 'excited'. In the experiments, the particles are observed directly in the excited state, so we can determine the properties of the particles in the ground state only indirectly. In the excited state a particle should emit real particles or virtual particles, so that they can not move linearly. They are easy to read in conjunction with the "coat" of virtual particles as a complex dynamic system, characterized by a drift velocity, which can be arbitrary (in particular to be zero) with respect to a macroscopic observer. Frequency emission of virtual particles will be characterized by a charge, which depends oń the type of the virtual particles which will be of a specific title. For example, if the virtual particles are photons, the charge will be called 'electric'. We assume that the emission of elementary particles, real particles or virtual particles is generated by their encounter with the background particles of the Universe. Since between the particles in the ground and excited states it is possible only weak interaction, which is characterized by chiral asymmetry, which is why it makes sense to assume that the fermions in the ground state of the universe are "sterile" neutrinos and "sterile" antineutrinos.

2. The Interaction between Baryons

We regard, that the standard gravitation interaction is not the fundamental one, but it is generated by collective oscillations in the Universe neutrino sea. In consequence of this the weak interaction acquires the global value (which is characterized by the group $SU(2)$), because it causes the bond of neutrinos both one with the other and with charged leptons and quarks. The process of the symmetry recovery (the reverse to the spontaneous breaking of symmetry) to the group $SU(3)$ must take place at nucleus distances, as a result of this the gluons will bear the main responsibility for an interaction between baryons.[1] Further we shell rely on the cold plasma theory developed for the first-kind superconductor and for the second-kind superconductor. By this a nuclear matter is an analog of second-kind superconductors with respect to gluons, which as vortices penetrate in a Bose condensate of Cooper pairs compounded from neutrinos, at the same time the vacuum with respect to gluons is an analog of first-kind superconductors. Thereby, we can consider the interaction between baryons using a potential consisting of two summands

$$U(r) = -\, C_s / \left(e^{B_s r} - 1\right) - C_w / \left(e^{B_w r} - 1\right), \tag{1}$$

where the first summand corresponds for the strong gravitation (induced by the interchange of gluons) and the second summand corresponds for the weak (standard) gravitation. It is naturally that at very small distances processes must be described by the quantum chromodynamics, but not the strong gravitation theory. As a result it is become a high-priority test of the hypothesis on the presence of the Universe neutrino background under laboratory conditions. Therefore it might be worthwhile to make experiments for the observation of the neutrino background temperature, it is possible with the employment of the Josephson effect. The result will have the fundamental importance. Particularly if the temperature waves will be detected by monitoring.

Reference

1. V. M. Koryukin, On construction methods of the strong gravitation theory, PoS(Baldin ISHEPP XXII) 024.

Long-term observations of the blazar BL Lacertae

K. S. Strigunov* and A. V. Zhovtan

Crimean Astrophysical Observatory,
Nauchny, Republic Crimea, Russia
**E-mail: sks6891@gmail.com*
http://www.crao.crimea.ua/ru/

In 1998-2011, the BL Lacertae blazar was active in various bands, including TeV energies. During the growth of activity in the gamma band, the spectrum of the object became flat, with the spectral index decreasing. This effect is associated with the conversion mechanism of particle acceleration in the BL Lacertae jet.

Keywords: Blazar; spectral index; conversion mechanism; γ-ray.

1. Observations of BL Lacertae in Optics, X-ray and TeV

At the Crimean Astrophysical Observatory (CrAO), observed in 1998-2011 on the 2nd generation GT-48 Cherenkov telescope was the active galactic nucleus (AGN) of BL Lacertae (belongs to blazar type) in >1 TeV energies with a total accuracy of around 11.8σ. A fourfold change was registered in average annual flows, together with more than 20 bursts of activity. The comparison with the optical data (B-band) also obtained at CrAO showed the presence of consistency with TeV energies data within individual time intervals. In addition, optical radiation was compared with the RXTE/ASM satellite data (2-10 keV), and the comparison showed a coincidental sharp rise, in 2011, in the flows in both bands.

2. Comparison with Fermi LAT Data and Discussion

The data of GT-48 were also compared with the results of observation by the Fermi LAT satellite (0.1-300 GeV). Earlier, it was noted in paper[1] that as the blazar activity increases, the γ-ray spectrum becomes flatter, while during perturbations in the AGN jet, the particles responsible for γ-ray generation which possess a higher energy demonstrate a more effective acceleration. In order to check this hypothesis, we compared the energy flows in the Fermi LAT telescope within the intervals of a 6-day flare observed on

GT-48 from 18 to 23 October 2009. In addition, a comparison was carried out with a 24-hour period (28 June 2011) in which a powerful 35-minute flare occurred which was registered by the VERITAS collaboration in over-200 GeV energies[2]. For that purpose, the Fermi LAT data were adjusted by a power law spectrum: dN/dE=N_0 (E/E_0)$^{-\alpha}$. As a result, the spectral index obtained (α=1.4$\pm$0.2) proved to be lower, at the moment of the 2009 flash observed on GT-48, than the spectral index over the entire interval of observations of BL Lacertae by the Fermi LAT telescope (α=1.9$\pm$0.1). The spectral index in the Fermi LAT energies over the 24-hour period during which the 2011 flare occurred and was registered by VERITAS made up α=1.8 $\pm$ 0.2. This indicates an increase in the share of high-energy γ-rays at the time of elevated activity of the object. A similar effect was detected in the 1ES 1426+428 blazar whose spectral index in the Fermi LAT energies made up, during the flare observed on GT-48 from 14 to 17 May 2009, 1.3 $\pm$ 0.2, i.e. it proved to be lower than the index over the entire five-year interval of observations on the Fermi LAT telescope (α=1.5 $\pm$ 0.1).

The study[3] put forward an assumption that an increase in the share of high-energy γ-rays at the time of activity growth is connected with the conversion mechanism of particle acceleration[4] in the blazar jet shock waves. In accordance with the said mechanism, the total energy content of the particles demonstrates a Γ^2-fold increase with each new cycle, where Γ is the Lorentz factor of the shock wave. The main contribution in this case is done by the particles with the energy closest to the maximum achievable energy. For more precise evaluations, we need to analyze the activity of an additional number of different γ-ray sources.

References

1. Yu. I. Neshpor, Particular Features of γ Radiation from Active Galactic Nuclei, *Bull. of the Crimean Astrophys. Obs.* **107**, 30 (2011).
2. M. Errando, Observations of low and intermediate-frequency-peaked BL Lacs above 100 GeV with VERITAS, *Proc. EPJ Web of Conferences*, **61**, 04012 (2013).
3. K. S. Strigunov, A. V. Zhovtan, Long-term observations of the blazar 1ES 1426+428 with the GT-48 Cerenkov telescope, *Astron. Rep.*, **59**, 175 (2015).
4. E. V. Derishev, F. A. Aharonian, V. V. Kocharovsky et al., Particle acceleration through multiple conversions from a charged into a neutral state and back, *Phys. Rev. D*, **68**, 043003 (2003).

New perspectives on an old problem: The bending of light in Yang–Mills gravity

Kazuo Ota Cottrell and Jong-Ping Hsu*
Department of Physics,
University of Massachusetts Dartmouth,
North Dartmouth, MA 02747
*E-mail address: jhsu@umassd.edu

Yang-Mills gravity with electromagnetism predicts, in the geometric optics limit, a value for the deflection of light by the sun which agrees closely with the reanalysis of Eddington's 1919 optical measurements done in 1979. Einstein's General Theory of Relativity, on the other hand, agrees very closely with measurements of the deflection of electromagnetic waves made in the range of radio frequencies. Since both General Relativity and Yang-Mills gravity with electromagnetism in the geometric optics limit make predictions for the optical region which fall within experimental uncertainty, it becomes important to consider the possibility of the existence of a frequency dependence in the measurement results for the deflection of light, in order to determine which theory more closely describes nature.

Since its birth as an observational science following the discovery of the first doubly imaged quasar in 1979,[1] the study of the bending of light by astronomical objects, more formally known as gravitational lensing,[2] has grown to become a rapidly growing area of astrophysics.[3]

The possibility that light could be attracted by matter was recognized by Newton[4] as early as the 18th century, as evidenced by his Opticks, published in 1704, and a prediction for the magnitude of this effect was published by Soldner one hundred years later.[5] Soldner's value (obtained using Newtonian physics) was reproduced by Albert Einstein in 1911, using an incomplete version of the General Theory of Relativity. After completing the General Theory of Relativity in 1916, Einstein derived the following value for the deflection angle of light grazing the solar limb[6]

$$\delta = \frac{4GM_\odot}{c^2 R_\odot} = 1.75\,\text{arcsec} \tag{1}$$

It was the well-known verification of this prediction within a margin of error of 20% by the eclipse expedition[7] organized by Sir Arthur Eddington in 1919 off the west coast of Africa on the island of Principe, and in the South American city Sobral, Brazil that made Einstein a celebrity.[7,9]

However, it may be less well-known that the Eddington expedition's original result of 1.77 ± 0.3 arcsec for the deflection of light grazing the solar limb was reexamined in 1979. Reanalysis of the Sobral plates using modern equipment found a measured deflection of 1.55 ± 0.34 arcsec.[10]

Hsu has developed a total-unified model with generalized gauge symmetry.[10–12] Within the limits of the experimental uncertainty for optical measurements of the deflection of light by the sun stated in section I above, it is clear that the prediction for the deflection of light grazing the limb of the sun made by using the relevant sector of Hsu's total-unified model, i.e., Yang-Mills gravity with electromagnetism, is also verified by experiment. Yang-Mills gravity with electromagnetism predicts, in the geometric optics limit, that an optical measurement of the deflection of light by the sun will yield a value of 1.53 arcsec.

The value of 1.75 arcsec predicted by Einstein's General Theory of Relativity agrees very well with measurements made at radio frequencies which have an uncertainty of only about 0.1%[13] This suggests an interesting possibility: is it possible that the large experimental uncertainty which thwarts making accurate measurements in the optical frequency range is also responsible for hiding the existence of a frequency dependence in measurements of the deflection of electromagnetic radiation which grazes the solar limb? Experimentally, one cannot yet answer this question.

Continuing the work begun by Hsu and Hsu[14] on the predictions of Yang-Mills gravity, further work by Cottrell and Hsu[15] suggests that a general eikonal equation which does not employ the geometric optics limit and may be adequate to describe a possible frequency dependence in the deflection measurements of the bending of light is given by

$$G_L^{\mu\nu}\left[\partial_\mu\psi\partial_\nu\psi + (\partial_\mu\partial_\nu\psi)\tan\psi\right] = 0 \qquad (2)$$

where $G_L^{\mu\nu}$ is the effective Riemann metric tensor obtained for the bending of light in the usual geometric optics approximation, and ψ is the eikonal. Thus, according to the sector of Hsu's total-unified model which describes Yang-Mills gravity with electromagnetism, measurements of the deflection of electromagnetic radiation passing in grazing incidence to the sun may show a frequency dependence, which could explain why both the 1979 revision of the optical measurements made by Eddington's team in Sobral and the predictions of Yang-Mills gravity with electromagnetism made in the geometric optics approximation are somewhat lower than the measurements made in the radio frequency range and the predictions of Einstein's General Theory of Relativity. A final explanation requires further analysis, and awaits the completion of definitive measurements of the deflection of electromagnetic radiation by the sun in the optical frequency region.

References

1. D. Walsh, R. F. Carswell, and R. J. Weymann, "0957+561 A, B: twin quasistellar objects or gravitational lens?", Nature **279**, 381–384, (1979).
2. J. Wambsganns, "Gravity's Kaleidoscope", Sci. Amer. **Nov 2001**, 64–71.
3. J. Wambsganss, "Gravitational Lensing in Astronomy", Living. Rev. Relativity **1**, 12, (1998). [Online Article]: cited on September 14, 2015, http://www.livingreviews.org/Articles/Volume1/1998-12wamb/.
4. I. Newton, Opticks, or A Treatise of the Reflections, Refractions, Inflections & Colours of Light, (New York: Dover Publications, 1952).
5. J. Soldner, "Ueber die Ablenkung eines Lichtstrals von seiner geradlinigen Bewegung, durch die Attraktion eines Weltkrpers, an welchem er nahe vorbei geht", Berliner Astronomisches Jahrbuch fr das Jahr 1804, 161–172, (1805).
6. A. Einstein, "Die Grundlage der allgemeinen Relativittstheorie", Annalen der Physik **49**, 769, (1916).
7. R. Ellis, P. Ferreira, R. Massey and G. Weszkalnys, "90 years on - the 1919 eclipse expedition at Prncipe", Astronomy and Geophysics **50**, 4.12-4.15, (August 2009).
8. F. W. Dyson, A. S. Eddington, and C. R. Davidson, "A determination of the deflection of light by the sun's gravitational field from observations made at the total eclipse of May 29, 1919", Mem. R. Astron. Soc. **62**, 291, (1920).
   9R. Ellis, P. Ferreira, R. Massey and G. Weszkalnys, "90 years on - the 1919 eclipse expedition at Prncipe", Astronomy and Geophysics **50**, 4.12-4.15, (August 2009).
9. M. Longair, "Bending spacetime: a commentary on Dyson, Eddington and Davidson (1920) 'A determination of the deflection of light by the Sun's gravitational field' ", Phil. Trans. R. Soc. A. **373**, 20140287, (2015). http://dx.doi.org/10.1098/rsta.2014.0287.

10. J. P. Hsu, "Exact space-time gauge symmetry of gravity, its couplings and approximate internal symmetries in a total-unified model", Chinese J. Phys. **52** (2), 691, (2014).
11. J. P. Hsu, "A model of unified quantum chromodynamics and Yang-Mills gravity", Chinese J. Physics C **36** (5), 403, (2012).
12. J. P. Hsu, "A unified gravity-electroweak model based on a generalized Yang-Mills framework", Mod. Phys. Lett. A **26**, 1707, (2011).
13. C. M. Will, Was Einstein Right? Putting General Relativity to the Test, 2nd ed., (New York: Basic Books, 1993).
14. J. P. Hsu and L. Hsu, Space-Time Symmetry and Quantum Yang-Mills Gravity", (Singapore: World Scientific Publishing Co. Pte. Ltd., 2013).
15. K. O. Cottrell and J.-P. Hsu, "Gauge independence of the eikonal equation in Yang-Mills gravity", Eur. Phys. J. Plus **130**, 147–153, (2015). DOI 10.1140/epjp/i2015-15147-8.

Epilogue

After 100 years, gravitation remains an interesting field for research, and there are still opportunities for intrepid young researchers.

In retrospect, Einstein's theory based on the profound principle of general coordinate invariance opened physicists' eyes to the universe as a whole. Its impact has not being limited to physicists as one can see from mathematician Hilbert's enthusiasm toward Einstein's 'powerful posing of problems', as well as his penetrating method devised for their solution. (See Preface.) In the past century, the list of new and interesting ideas and discoveries is endless. They are well described in most textbooks of gravity. But physics would not progress simply by talking about successful stories. There are still longstanding and challenging problems that have existed since the creation of the theory of gravity or general relativity.

For enthusiastic researchers, it is worthwhile to pay attention to some of these basic problems, as noted and discussed by mathematicians and physicists Hilbert, Noether, Wigner, Dyson and so on. Some of them may be summarized as follows:

In her famous 1918 paper on symmetry and conservation law, Noether proved her well-known theorem. However, she also proved an important and not-so-well-known "second theorem": *The principle of general coordinate invariance in general theory of relativity leads to the failure of the law of conservation of energy.* From the group-theoretical viewpoint in mathematics, Einstein's theory is based on the group formed by general coordinate transformations in curved space-time. It is a Lie group with a continuously infinite number of generators, in contrast to the Poincaré group. She noted that her result was consistent with Hilbert's claim (1918) that the failure of the law of conservation for energy was characteristic of Einstein's theory of gravity.

Wigner discussed a difficulty associated with the general coordinate invariance. The basic premise of general relativity is that coordinates are only auxiliary quantities and can be given arbitrary values for any events. He wrote: "Most of us have struggled with the problem of how, under these premise, the general theory of relativity can make meaningful statements and predictions at all. Evidently, the usual statements about future positions of particles, as specified by their coordinates, are not meaningful statements in general relativity. This is a point which cannot be emphasized strongly enough and is the basis of a much deeper dilemma than the more technical question of Lorentz invariance of the quantum field equations. It pervades all the general theory, and to some degree we mislead both our students and ourselves when we calculate, for instance, the Mercury perihelion motion without explaining how our coordinates system is fixed in space.... Expressing our results in terms of the values of coordinates became a habit with us to such a degree that we adhere to this habit also in general relativity, where values of coordinates are not per se meaningful."

Concerning quantum gravity, Dyson stressed another problem with the principle of general coordinate invariance: "The most glaring incompatibility of concepts in contemporary physics is that between Einstein's principle of general coordinates invariance and all the modern schemes for a quantum-mechanical description of nature." All the established or viable formalisms for describing nature quantum-mechanically use the Poincare group, i.e., "a class of preferred inertial coordinate-frames is postulated *a priori,* in flat contradiction to Einstein's principle."

In the past 100 years, starting with Hilbert's fascinating attempt to give an axiomatic and geometric foundation to unify electrodynamics and gravity, we have seen many failures in this approach for a unified theory. However, recent success of the Weinberg-Salam unified electroweak theory and quantum chromodynamics reveals a new direction: local quantum fields with gauge symmetry appears to be the most effective key to unlocking the secret of quantum gravity, including a unified gauge field theory with all interactions, based on the seminal ideas of Weyl, Yang, Mills, Utiyama, Glashow et al. The Yang-Mills approach is based on the geometry of fiber bundle. The apparent curvature of space-time appears to be a manifestation of the flat space-time translational gauge symmetry for the motion of quantum particle in the classical limit. The advantage of a Yang-Mills gravity is that since gauge fields are based on space-time with the Poincaré group, previous difficulties associated with the general coordinate invariance will automatically disappear.

Gravitation is a fundamental interaction that governs all phenomena at large and very small scales, but it is still not well understood at the quantum level. It is a missing cardinal link to unification of all physical interactions. The discovery of the present acceleration of the Universe issues with dark matter and dark energy and some results on possible time and space variations of fundamental physical constants are also a great challenge to modern physics. Resolving these challenges may lead to a new revolution in it.

During the recent decades there has been significant progress in unifying the weak and electromagnetic interactions, some more modest achievements in GUT, supersymmetric, string and superstring theories. Now, theories with membranes, p-branes and M-theory are being created and studied. Having no definite successful theory of unification, it is desirable to study the common features of these theories and their applications to solving basic problems of modern gravity and cosmology. Theories with scalar fields, F(R) and extra dimensions may be important in their low-energy limits.

Another very important feature which may be envisaged is an increasing role of fundamental physics studies, gravitation, cosmology and astrophysics in particular, in space experiments. Unique microgravity environments combined with modern technology give a nearly perfect place for gravitational experiments and observations, which are difficult to perform on Earth due to its relatively strong gravitational field and the gravitational fields of nearby objects since there is no way to screen gravity.

There are also three problems related to G, whose origin lies mainly in unified model predictions: 1) low accuracy of absolute G measurements, 2) possible time variations of G and 3) possible range variations of G or new interactions.

The knowledge of constants' values has not only a fundamental significance in physics but also in metrology. The modern system of standards is based on stable physical phenomena. Thus, the stability of constants plays a crucial role. Since all physical laws were established and tested during the last 2-3 centuries in experiments on the Earth and in the near space, i.e. at rather short space and time intervals as compared to the radius and age of the Universe, the possibility of slow variations of constants (i.e., with the rate of the Universe evolution or slower) cannot be excluded a priori. This problem becomes especially important due to the SI reform based on fixing Planck, Avogadro, Boltzmann and elementary charge constants planned in 2018.

References

1. E. Noether, Goett. Nachr., 235 (1918). Hilbert's claim was in this paper.
2. E. P. Wigner, Symmetries and Reflections, MIT press,1970, pp. 52-53.
3. F. Dyson, in 100 Years of Gravity and Accelerated Frames, The Deepest

Insights of Einstein and Yang-Mills, Eds. J. P. Hsu and D. Fine, World Scientific, 2005, pp. 348-350. For gauge invariant quantum Yang-Mills gravity, see p. xix and DOI 10.1140/epjp/i2011-11024-x.

4. P.A.M. Dirac, The Cosmological Constants, Nature, 1937, v. 139, N 35, p. 323.
5. Gravitational Measurements, Fundamental Metrology and Constants, Eds. V.N. Melnikov and V. De Sabbata, Kluwer Acad. Publ., 1988, 552p.
6. Albert Einstein Century International Conference, Eds. J.-M. Alimi and A. Fuzfa, AIP Conf. Proc., Melville, NY, 2006, v. 861, 260 p.

J.P. Hsu and V.N. Melnikov

International Organizing Committee

Vitaly Melnikov (Russia) – Chair, **Alexander P. Yefremov** (Russia) – Co-Chair, **Sang-Pyo Kim** (Korea) – Co-Chair, **Alexei Starobinsky** (Russia) – Co-Chair, **Medeu Abishev** (Kazakhstan), **Andrei Arbuzov** (Russia), **Gennady Bisnovaty-Kogan** (Russia), **Rong-Gen Cai** (China), **Yongmin Cho** (Korea), **Dmitry Fursaev** (Russia), **Jong-Ping Hsu** (USA), **Jun Luo** (China), **Masakatsu Kenmoku** (Japan), **Sung-Won Kim** (Korea), **Kei-ichi Maeda** (Japan), **Remo Ruffini** (Italy), **Misao Sasaki** (Japan), **Roland Triay** (France).

International Advisory Committee

- **Armenia:** Eduard Chubaryan
- **Australia:** David Blair, Victor Flaumbaum, Susan Scott, John Webb
- **Brazil:** Julio Cesar Fabris, Mario Novello, Nelson Pinto-Neto, Carlos Romero
- **Belarus:** Albert Minkevich
- **Canada:** Valery Frolov, Don Page
- **China:** Rong-Gen Cai, Sijie Gao, Jun Luo, Xin-zhou Li
- **France:** Jean-Michel Alimi, Thibault Damour, Roland Triay
- **Germany:** Claus Laemmerzahl, Claus Kiefer
- **Greece:** Spiros Cotsakis
- **India:** Varun Sahni
- **Italy:** Vladimir Belinski, Salvatore Capozziello, Alexander Dolgov, Alexander Kamenshchik, Remo Ruffini
- **Japan:** Yasunori Fujii, Masakatsu Kenmoku, Misao Sasaki, Katsuhiko Sato
- **Kazakhstan:** Medeu Abishev, Leonid Chechin, Ratbay Myrzakulov
- **Kuwait:** Elias Vagenas
- **Kyrgyzstan:** Vladimir Dzhunushaliev
- **Uzbekistan:** Bobomurat Ahmedov
- **Korea:** Yongmin Cho, Sang-Pyo Kim and Sung-Won Kim
- **Mexico:** Alberto Garcia, Hernando Quevedo and Vladimir Man'ko
- **RSA:** Roy Maartens, Sunil Maharaj

• **Russia:** Andrei Barvinsky, Gennady Bisnovaty- Kogan, Joseph Buchbinder, Aleksander Yefremov, Aleksander Filippov, Dmitry Fursaev, Dmitry Gal'tsov, Andrei Grib, Vladimir Ivashchuk, Maxim Khlopov, Vitaly Melnikov, Vladimir Mostepanenko, Slava Mukhanov, Valery Rubakov, Vladimir Rudenko, Alexei Starobinsky

• **Spain:** Emilio Elizalde, Sergei Odintsov

• **Taiwan:** Chiang-Mei Chen, James Nester

• **USA:** Diter Brill, Jong-Ping Hsu, Ken Nordtvedt, H.J. Paik, Douglas Singleton, Alexander Vilenkin

• **Ukraine:** Yury Shtanov, Oleg Zaslavskii, Aleksander Zhuk

• **Vietnam:** Nguyen van Hieu, Pierre Darriulat

Sections and moderators

• **Classical Gravity, GR Extensions** – *A.B. Balakin, Yu.G. Ignatyev, Yu.S. Vladimirov*

• **Classical and Quantum Cosmology** – *A.Y. Kamenshchik, V.N. Lukash, M. Sasaki, A.A. Starobinsky*

• **Dark Matter and Dark Energy** – *S.V. Chervon, K.-I. Maeda, S.D. Odintsov*

• **Black Holes, Wormholes** – *K.A. Bronnikov, Sung-Won Kim, S.V. Sushkov*

• **Strings, Branes, Higher Spin Fields and Quantum Gravity** – *I.L. Buchbinder, D.V. Gal'tsov, Rong-Gen Cai, Sang-Pyo Kim*

• **Extra Dimensions and Variation of Constants** – *V.D. Ivashchuk, V.N. Melnikov*

• **Experimental Studies of Gravity, GW and Fundamental Physics Space Projects** – *C. Laemmerzahl, Jun Luo, V.M. Mostepanenko, V.N. Rudenko*

• **Relativistic Astrophysics** – *G.S. Bisnovaty-Kogan, L.M. Chechin, A.D. Dolgov, H. Quevedo, R. Ruffini*

List of participants of the Conference ICGAC-12.
(This is not a full list of registered participants but only those who sent contributions to the Proceedings.)

No	Name	E-mail	Address
1.	Abishev M. E.	abishevme@mail.ru	Al-Farabi Kazakh National Univ, 6-line 3/1, ap 28, 050008 Almaty, Kazakhstan
2.	Aliyev B. G.	bgaliyev@mail.ru	Russia, 117437 Moscow, Profsoyuznaya ul., 114-6-897
3.	Arbuzov A.	arbuzov@theor.jinr.ru	Bogoliubov Laboratory of Theoretical Physics, JINR, Joliot-Curie str. 6, Dubna, 141980, Russia
4.	Aristov V. V.	aristovvl@yandex.ru	Vavilova str., 40, Moscow, 119333, Russia
5.	Aslan O.	alsucuk@gmail.com	Kazan Federal Univ, 18 Kremlyovskaya St., Kazan 420008, Russia
6.	Babenko I.	Inna.babenko@list.ru	Lomonosovskiy pr. 4/2/126, 119296 Moscow, Russia
7.	Berezin V. A.	berezin@inr.ac.ru	Institute for Nuclear Research of the Russian Academy of Sciences, 60th October Anniversary prospect 7a, 117312 Moscow, Russia
8.	Bolokhov S.V.	boloh@rambler.ru	Smol'naya ul. 29, kv. 137, Moscow 125195, Russia
9.	Boshkayev K.	kuantay@mail.ru	Al-Farabi avenue, 71, Almaty 050040, Kazakhstan
10.	Bronnikov K. A.	kb20@yandex.ru	VNIIMS, Ozyornaya 46, Moscow 119361, Russia
11.	Burinskii A.	bur@ibrae.ac.ru	B. Tulskaya 52, 115191 Moscow, Russia

12.	Chechin L. M.	chechin-lm@mail.ru leonid.chechin@gmail.com	V.G. Fessenkov Astrophysical Institute, Kamenskoye plato, Observatory, 050020, Almaty city, Kazakhstan
13.	Chen Chiang-Mei	cmchen@phy.ncu.edu.tw	No. 300, Zhongda Rd., Zhongli District, Taoyuan City 32001, Taiwan (R.O.C.)
14.	Cho Yongmin	ymcho7@konkuk.ac.kr	Administration Building 310-4, Konkuk University, Seoul 143-701, Korea
15.	Dokuchaev V. I.	dokuchaev@inr.ac.ru	60th October Anniversary prospect 7a, 117312 Moscow, Russia
16.	Dolgov A.D.	dolgov@fe.infn.it	University of Ferrara, Dept. Phys., Via Saragat, 1 Ferrara, Italy
17.	Fomin I.V.	ingvor@inbox.ru	Moscow,2-nd Baumanskaya street, 5, 105005
18.	Frolov B. N.	frolovbn@mail.ru	119992, Moscow, ul. Malaya Pirogovaskaya, 29
19.	Galiakhmetov A. M.	agal17@mail.ru	Kirova str. 51, Gorlovka 84646, Ukraine
20.	Gal'tsov D.V.		Faculty of Physics, Lomonosov Moscow State University, Leninskie Gory, Moscow 119991, Russia
21.	Gao Sijie	sijie@bnu.edu.cn	Beijing Normal University, Physics Department, Beijing 100875, China
22.	Garcia-Diaz Alberto A.	aagarcia@fis.cinvestav.mx	Departamento de Fisica, Centro de Investigacion y de Estudios Avanzados del IPN, A.P. 14-740, 07000 Mexico D.F., Mexico
23.	Gladyshev V.	vgladyshev@mail.ru	105005, Moscow, 2-nd Baumanskaya st., 5

24.	Gorbunov D.	gorby@ms2.inr.ac.ru	Institute for Nuclear Research of the Russian Academy of Sciences, 60th October anniversary prospect, 7a, 117312, Moscow
25.	Gorodetsky M. L.		Faculty of Physics, Lomonosov Moscow State University, Leninskie Gory, Moscow 119991, Russia
26.	Grunskaya L. V.	grunsk@vlsu.ru	Vladimir State University, Gorkogo str., 87, 600000, Vladimir, Russia
27.	Gubarev E.	e.gubarev.21@gmail.com	Intersectoral Science-Technical Centre of Venture Non-Conventional Technologies, Of. 7, 8, Shenkyrsky pr., 127349 Moscow, Russia
28.	Hohmann M.	manuel.hohmann@ut.ee	Ravila 14c, 50411 Tartu, Estonia
29.	Hsu J.-P.	jhsu@umassd.edu	Phys. Dept., UMass Dartmouth, 285 Old Wesport Rd, N. Dartmouth, MA, 02747, USA
30.	Ignat'ev Yu. G.	ignatev_yu@rambler.ru	Kazan Federal Univ, Lobachevsy Institute of Matematics and Mecanics, Kremleovskaya str. 18, 420008 Kazan, Russia
31.	Ivashchuk V. D.	ivashchuk@mail.ru	VNIIMS, 46 Ozyornaya ul., Moscow, 119361, Russia
32.	Kamenshchik A. Yu.	alexander.kamenshchik@bo.infn.it	Dipartimento di Fisica e Astronomia and INFN, Via Irnerio 46, 40126, Bologna, Italy
33.	Kassandrov V. V.	vkassan@rambler.ru	Peoples' Friendship University of Russia, Miklukho-Maklay street 6, 117198 Moscow, Russia

34.	Kim Sang Pyo	sangkim@kunsan.ac.kr	Dept. of Phy., Kunsan National Univ. 558 Daehak-ro, Kunsan 573-701, Korea
35.	Kim Sung-Won	sungwon@ewha.ac.kr	Dept. of Science Education, Ewha Womans University, Seoul 12075, Korea
36.	Kirnos I. V.	ikirnos@sibmail.com kirnos@tspu.edu.ru	Tomsk State University, 36 Lenin Prospekt, Tomsk, 634050, Russia
37.	Klimchitskaya G.	g_klimchitskaya@mail.ru	GAO RAN, Pulkovskoye Shosse 65, 196140, St. Petersburg, Russia
38.	Konstan-tinov M.	konst@bmstu.ru	N.E. Bauman Moscow State Technical University, 2nd Bauman Str. 5, Moscow, 105005, Russia
39.	Koryukin V.	vmkoryukin@gmail.com	Lenin square - 1, MarSU, Department of Physics and Mathematics, 424000, Yoshkar-Ola, Russia
40.	Krugly A. L.	akrugly@mail.ru	Nahimovskiy pr., 36, k. 1, Moscow, Russia, 117218
41.	Luo Jun	junluo@sysu.edu.cn tlc@hust.edu.cn	No. 135, Xingang Xi Road, Guangzhou, 510275, P. R. China
42.	Manko V.	vsmanko@fis.cinvestav.mx	Departamento de Fisica, CINVESTAV-IPN, A.P. 14-740, 07000 Mexico D.F., Mexico
43.	Marakulin A.	marakulin@physics.msu.ru	60th October Anniversary Prospect, 7a, 117312 Moscow, Russia
44.	Mei Jianwei	jwmei@hust.edu.cn	CGE, school of physics, Huazhong Univ. of Science and Technology, 1037 Luoyu Rd., Wuhan 430074, P.R. China
45.	Meierovich B.	meierovich@mail.ru	Kapitza Institute, 2 ul. Kosygina, Moscow, 119334, Russia
46.	Melnikov V. N.	melnikov@phys.msu.ru	VNIIMS, 46 Ozyornaya ul., Moscow, 119361, Russia

47.	Milyukov V. K.	vmilyukov@yandex.ru	Sternberg Astronomical Institute of Lomonosov Moscow State University, Universitetskii prospect 13, Moscow 119991, Russia
48.	Obukhov Yu.	obukhov@ibrae.ac.ru	IBRAE RAS, B. Tulskaya str. 52, 115191 Moscow, Russia
49.	Pavlov A. E.	alexpavlov60@mail.ru	127550, Moscow, Russia, Timiryazevskaya street 49
50.	Pervushin V. N.	pervush@theor.jinr.ru	Bogoliubov Laboratory of Theoretical Physics, JINR, Joliot-Curie street 6, 141980 Dubna, Russia
51.	Poplavskiy M.V.	mv.poplavskii@physics.msu.ru	Faculty of Physics, Lomonosov Moscow State University, Leninskie Gory, Moscow 119991, Russia
52.	Popov A.A.	arkady_popov@mail.ru	Institute of Mathematics and Mechanics, KFU, 18 Kremlyovskaya St., Kazan 420008, Russia
53.	Puetzfeld Dirk	dirk.puetzfeld@zarm.uni-bremen.de	ZARM, University of Bremen, Am Fallturm, 28359 Bremen, Germany
54.	Rannu K.A.	rannu@xray.sai.msu.ru	Sternberg Astronomical Institute of Lomonosov Moscow State University, Universitetskii prospect 13, 119991 Moscow, Russia
55.	Rasulova A. M.	ARasulova@gmail.com	Pr. Veteranov 1-1-291, St.-Peterburg 198255, Russia
56.	Rubin S. G.	sergeirubin@list.ru	Krasnogo Mayaka str, 4-3, kv. 156, Moscow 117519, Russia
57.	Rudenko V. N.	valentin.rudenko@gmail.com	Sternberg Astronomical Institute of Lomonosov Moscow State University, Universitetskii prospect 13, 119991 Moscow, Russia

58.	Rybakov Yu. P.	soliton4@mail.ru	Peoples' Friendship University of Russia, Miklukho-Maklay street 6, 117198 Moscow, Russia
59.	Shishanin A.	soliton4@mail.ru	N.E. Bauman Moscow State Technical University, 2nd Bauman Str. 5, Moscow, 105005, Russia
60.	Solenov P.	solenovps@gmail.com	105425, Russia, Moscow, Konstantina Fedaina st., 4, 67
61.	Solov'yov A. V.	a.v.solovyov@gmail.com	Faculty of Physics, Lomonosov Moscow State University, Leninskie Gory, Moscow 119991, Russia
62.	Soloviev V.	vosoloviev@gmail.com	Centralnyj Proezd 3 app.95, 142281, Protvino, Moscow region, Russia
63.	Starobinsky A.	alstar@landau.ac.ru	L. D. Landau Institute for Theoretical Physics of the Russian Academy of Sciences, Chernogolovka, Academica Semenova prospect 1a, Chernogolovka 142432, Russia
64.	Strigunov K.	sks6891@gmail.com	Crimean Astrophysical Observatory, Nauchniy, Bakhchisarai district, Crimean Republic, 298409, Russia
65.	Strokov V. N.	vnstrokov@gmail.com	ul. Profsozyuznaya, 84/32, 117997 Moscow, Russia
66.	Tegai S.	tegai_s_f@inbox.ru	Siberian Federal University, 79 Svobodny pr., 660041 Krasnoyarsk, Russia
67.	Tomaru Takayuki	tomaru@post.kek.jp	1-1 Oho, Tsukuba, Ibaraki, 305-0801, Japan
68.	Unnikrishnan C. S.	unni@tifr.res.in	Gravitation Lab, TIFR, Homi Bhabha Road, Mumbai 400005, INDIA

69.	Urazalina A.	Y.A.A.707@mail.ru	Dept. Teor. and Nucl. Phys., KazNU, Almaty, 050040, Kazakhstan.
70.	Vernov S.	svernov@theory.sinp.msu.ru	Skobeltsyn Institute of Nuclear Physics of Moscow State University, Leninskie gory, GSP-1, Moscow, 119991, Russia
71.	Vladimirov Yu. S.	yusvlad@rambler.ru	Faculty of Physics, Lomonosov Moscow State University, Leninskie Gory, Moscow 119991, Russia
72.	Yefremov A. P.	a.yefremov@rudn.ru	Peoples' Friendship University of Russia, Miklukho-Maklay street 6, 117198 Moscow, Russia
73.	Zakharov A.	alex_f_Zakharov5@mail.ru	Institute of Theoretical and Experimental Physics, Bolshaya Cheremushkinskaya 25, 117218 Moscow, Russia
74.	Zhovtan A.	astroalex2012@gmail.com	Crimean Astrophysical Observatory, Nauchniy, Bakhchisarai district, Crimean Republic, 298409, Russia

Delightful Moments at The 2015 ICGAC-12

CAPTAIN
AMERICA

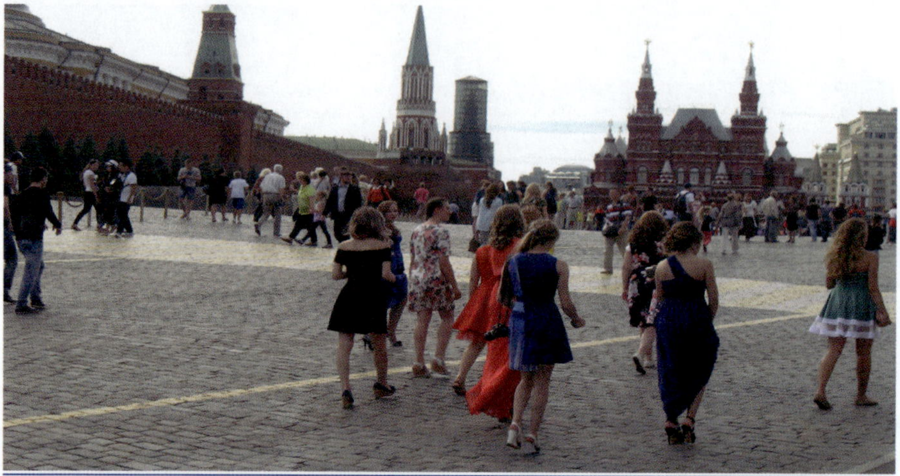

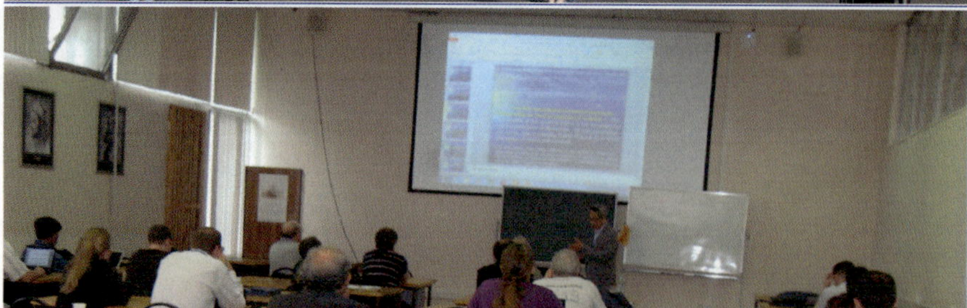